Chemistry & Ecology

GOD'S DESIGN®

- Properties of **Matter**
- Properties of **Ecosystems**
- Properties of **Atoms & Molecules**

MasterBooks Curriculum

Debbie & Richard Lawrence

Fourth Edition: January 2016
Master Books Edition second printing: August 2020

Copyright © 2003, 2008, 2016 by Debbie & Richard Lawrence and Master Books®. All rights reserved. No part of this book may be reproduced, copied, broadcast, stored, or shared in any form whatsoever without written permission from the publisher, except in the case of brief quotations in articles and reviews. For information write:

Master Books®, P.O. Box 726, Green Forest, AR 72638

Master Books® is a division of the New Leaf Publishing Group, Inc.

ISBN: 978-1-68344-125-0
ISBN: 978-1-61458-649-4 (digital)

Cover by Diana Bogardus
Book design: Diane King
Editor: Gary Vaterlaus

Unless otherwise noted, Scripture quotations are from the New King James Version of the Bible. Copyright 1982 by Thomas Nelson, Inc. Used by permission. All rights reserved.

God's Design® for Chemistry & Ecology is a complete curriculum for grades 3–8. The books in this series are designed for use in the Christian school and homeschool, and provide easy-to-use lessons that will encourage children to see God's hand in everything around them.

The publisher and authors have made every reasonable effort to ensure that the activities recommended in this book are safe when performed as instructed but assume no responsibility for any damage caused or sustained while conducting the experiments and activities. It is the parents', guardians', and/or teachers' responsibility to supervise all recommended activities.

Please consider requesting that a copy of this volume be purchased by your local library system.

Printed in the United States of America

Please visit our website for other great titles:
www.masterbooks.com

For information regarding author interviews,
please contact the publicity department at (870) 438-5288.

Properties of Matter

Unit 1: Experimental Science — 13

Lesson 1	Introduction to Experimental Science	14
Lesson 2	The Scientific Method	17
Lesson 3	Tools of Science	21
Special Feature	Lord Kelvin	24
Lesson 4	The Metric System	26

Unit 2: Measuring Matter — 30

Lesson 5	Mass vs. Weight	31
Lesson 6	Conservation of Mass	34
Lesson 7	Volume	37
Lesson 8	Density	40
Lesson 9	Buoyancy	42
Special Feature	James Clerk Maxwell	45

Unit 3: States of Matter — 46

Lesson 10	Physical & Chemical Properties	47
Lesson 11	States of matter	49
Lesson 12	Solids	53
Lesson 13	Liquids	56
Lesson 14	Gases	59
Lesson 15	Gas Laws	62
Special Feature	Robert Boyle	65

Unit 4: Classifying Matter — 66

Lesson 16	Elements	67
Special Feature	William Prout	71
Lesson 17	Compounds	73
Lesson 18	Water	76
Lesson 19	Mixtures	79
Lesson 20	Air	82
Lesson 21	Milk & Cream	85

Unit 5: Solutions — 89

Lesson 22	Solutions	90
Lesson 23	Suspensions	92
Lesson 24	Solubility	95
Lesson 25	Soft Drinks	98
Lesson 26	Concentration	101
Lesson 27	Seawater	104
Special Feature	Desalination of Water	107
Lesson 28	Water Treatment	108

Unit 6: Food Chemistry — 111

Lesson 29	Food Chemistry	112
Special Feature	Genetically Modified Foods	114
Lesson 30	Chemical Analysis of Food	116
Lesson 31	Flavors	119
Special Feature	Chocolate & Vanilla	122
Lesson 32	Additives	124
Lesson 33	Bread	127
Special Feature	Bread through the Centuries	130
Lesson 34	Identification of Unknown Substances: Final Project	132
Lesson 35	Conclusion	134
Glossary		136
Challenge Glossary		137

Properties of Ecosystems

Unit 1: Introduction to Ecosystems	**141**
Lesson 1 What Is an Ecosystem?	142
Special Feature Garden of Eden	145
Lesson 2 Niches	147
Lesson 3 Food Chains	150
Lesson 4 Scavengers & Decomposers	153
Lesson 5 Relationships among Living Things	156
Lesson 6 Oxygen & Water Cycles	159

Unit 2: Grasslands & Forests	**162**
Lesson 7 Biomes around the World	163
Special Feature Alexander von Humboldt	167
Lesson 8 Grasslands	169
Lesson 9 Forests	173
Lesson 10 Temperate Forests	176
Lesson 11 Tropical Rainforests	179

Unit 3: Aquatic Ecosystems	**182**
Lesson 12 The Ocean	183
Lesson 13 Coral Reefs	187
Lesson 14 Beaches	190
Lesson 15 Estuaries	193

Lesson 16	Lakes & Ponds	196
Lesson 17	Rivers & Streams	199
Special Feature	The Amazon River	201

Unit 4: Extreme Ecosystems — 203

Lesson 18	Tundra	204
Special Feature	Robert Peary	208
Lesson 19	Deserts	210
Lesson 20	Oases	214
Lesson 21	Mountains	217
Lesson 22	Chaparral	221
Lesson 23	Caves	224

Unit 5: Animal Behaviors — 228

Lesson 24	Seasonal Behaviors	229
Lesson 25	Animal Defenses	233
Lesson 26	Adaptation	236
Lesson 27	Balance of Nature	239
Special Feature	Eugene P. Odum	243

Unit 6: Ecology & Conservation — 244

Lesson 28	Man's Impact on the Environment	245
Lesson 29	Endangered Species	248
Special Feature	Theodore Roosevelt	252
Lesson 30	Pollution	254
Lesson 31	Acid Rain	258
Lesson 32	Global Warming	261
Lesson 33	What can you do?	265
Lesson 34	Reviewing Ecosystems: Final Project	268
Lesson 35	Conclusion	269
Glossary		270
Challenge Glossary		272

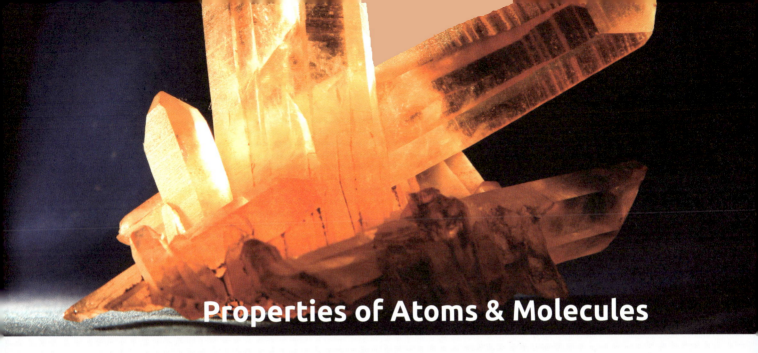

Properties of Atoms & Molecules

Unit 1: Atoms & Molecules	277
Lesson 1 Introduction to Chemistry	278
Lesson 2 Atoms	280
Lesson 3 Atomic Mass	283
Special Feature Madame Curie	285
Lesson 4 Molecules	287

Unit 2: Elements	290
Lesson 5 Periodic Table of the Elements	291
Special Feature Development of the Periodic Table	295
Lesson 6 Metals	296
Lesson 7 Nonmetals	299
Lesson 8 Hydrogen	302
Lesson 9 Carbon	305
Lesson 10 Oxygen	308

Unit 3: Bonding	311
Lesson 11 Ionic Bonding	312
Lesson 12 Covalent Bonding	316
Lesson 13 Metallic Bonding	319
Lesson 14 Mining & Metal Alloys	321
Special Feature Charles Martin Hall	324
Lesson 15 Crystals	326
Lesson 16 Ceramics	330

Unit 4: Chemical Reactions — 333

- Lesson 17   Chemical Reactions … 334
- Lesson 18   Chemical Equations … 338
- Lesson 19   Catalysts … 341
- Lesson 20   Endothermic & Exothermic Reactions … 344

Unit 5: Acids & Bases — 347

- Lesson 21   Chemical Analysis … 348
- Lesson 22   Acids … 351
- Lesson 23   Bases … 354
- Lesson 24   Salts … 357
- Special Feature   Batteries … 360

Unit 6: Biochemistry — 362

- Lesson 25   Biochemistry … 363
- Lesson 26   Decomposers … 367
- Lesson 27   Chemicals in Farming … 370
- Lesson 28   Medicines … 373
- Special Feature   Alexander Fleming … 376

Unit 7: Applications of Chemistry — 378

- Lesson 29   Perfumes … 379
- Lesson 30   Rubber … 382
- Special Feature   Charles Goodyear … 386
- Lesson 31   Plastics … 388
- Lesson 32   Fireworks … 391
- Lesson 33   Rocket Fuel … 394
- Lesson 34   Fun with Chemistry: Final Project … 397
- Lesson 35   Conclusion … 400
- Glossary … 401
- Challenge Glossary … 403
- Index … 405
- Photo Credits … 410

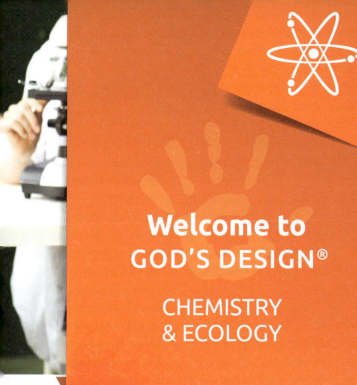

Welcome to GOD'S DESIGN®

CHEMISTRY & ECOLOGY

You are about to start an exciting series of lessons on chemistry and ecology. *God's Design® for Chemistry & Ecology* consists of: *Properties of Matter, Properties of Ecosystems,* and *Properties of Atoms & Molecules.* It will give you insight into how God designed and created our world and the universe in which we live.

No matter what grade you are in, third through eighth grade, you can use this book.

3rd–5th grade

Read the lesson.

 Do the activity in the light blue box (worksheets will be provided by your teacher).

 Test your knowledge by answering the **What did we learn?** questions.

 Assess your understanding by answering the **Taking it further** questions.

Be sure to read the special features and do the final project.

There are also unit quizzes and a final test to take.

6th–8th grade

Read the lesson.

 Do the activity in the light blue box (worksheets will be provided by your teacher).

 Test your knowledge by answering the **What did we learn?** questions.

 Assess your understanding by answering the **Taking it further** questions.

 Do the Challenge section in the light green box. This part of the lesson will challenge you to do more advanced activities and learn additional interesting information.

Be sure to read the special features and do the final project.

There are also unit quizzes and a final test to take.

When you truly understand how God has designed everything in our universe to work together, then you will enjoy the world around you even more. So let's get started!

Properties of Matter

UNIT 1

Experimental Science

1 Introduction to Experimental Science
2 The Scientific Method
3 Tools of Science
4 The Metric System

◊ **Describe** how the scientific method is used to study the world.

◊ **Distinguish** between qualitative and quantitative observations.

◊ **Identify** the proper tools and units used for measuring different properties of matter.

Properties of Matter • 13

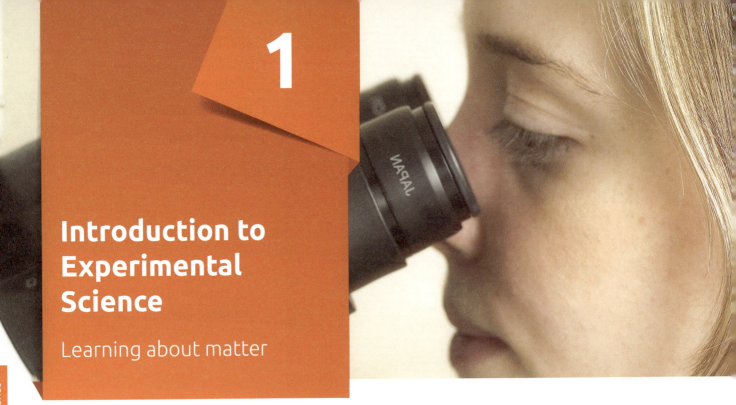

1

Introduction to Experimental Science

Learning about matter

How do scientists conduct experiments?

Words to know:

matter   experiment

Challenge words:

operational science   naturalism

origins science

Do you want to learn about science? I hope so. Science can be very exciting. Why do you want to learn about science? Maybe you really like animals and want to learn more about them, or maybe you want to be an astronaut or an engineer and you know you need to learn a lot about science to do those jobs. Biology, astronomy, and physics are important parts of science. But this is a chemistry book, so why should you want to learn about chemistry?

Chemistry is the study of the material around you and how it reacts with other material. Scientists call this material matter. **Matter** is anything that has mass and takes up space. Understanding what matter is and how it acts in different conditions is important to every other area of science. You need to understand chemistry to understand how plants and animals grow. You need to understand chemistry to understand what stars are and how they produce light. You need to understand chemistry to know how to build a computer, too. You even need to understand how one material reacts with another material to make a new recipe for dinner. So you can see that chemistry affects every area of your life.

So how do we learn about matter? One of the best ways to learn about matter is by conducting experiments. An **experiment** is a controlled test to see what happens in a certain situation. You will get to do lots of experiments in this book. Experiments will teach you what works, what doesn't work, and give you ideas of what to try next time.

Before you begin an experiment, you need to know the purpose: what are you trying to learn? You also need to know what you expect to happen. You can then design an experiment to test what you think will happen and see if you are right. Does this sound like fun? It is! So let's get started with your first experiment. ✳

What did we learn?

- What is matter?
- What do chemists study?
- What is an experiment?

Taking it further

- Why is it important to study chemistry?
- What are two things you need to know before conducting an experiment?

Chemistry is fun

Different kinds of matter act differently even under the same conditions because they have different properties. Today we are going to look at how well different types of matter conduct heat. If something is a good conductor of heat, it gets hot very quickly when it is near something hot. You might need to use pot holders to carry something that is a good conductor of heat. Something that is not a good heat conductor stays cooler for a longer period of time when it is near something hot. Most pot holders are made out of a thick cloth because cloth is not a good heat conductor. In the following experiment, you will test various substances such as wood, plastic, and metal to see which are the best conductors of heat.

Purpose: To understand how to conduct an experiment in a scientific manner

Materials: metal spoon, wooden spoon, plastic ruler, pencil, butter knife, butter, large cup, hot water, stopwatch, "Conducting Heat Experiment" worksheet

Procedure:

1. Think about what you know about different materials such as metal, glass, and wood. Which ones do you think conduct heat the best? Which ones do not conduct heat well? Place a metal spoon, a wooden spoon, a plastic ruler, a pencil, and a butter knife in order of which ones you think would conduct heat the fastest to the slowest. Write these items on your worksheet.
2. Smear a small amount of cold butter on the end of each item.
3. Place the items, buttered end up, in a large cup.
4. Fill the cup with hot water; be sure that the butter is not in the water.
5. Use a stopwatch to time how long it takes for the butter on each object to begin to melt. Record this time on your worksheet.
6. Compare your results with your predictions.

Questions:

- Did the butter melt fastest on the item you expected to conduct the heat the best?
- Which items actually conducted heat the best? Which ones conducted heat the slowest?

Conclusion: By using your knowledge to make good predictions, following a careful plan, and checking your results, you learned about the heat conductive abilities of several materials. You will learn more about how to conduct experiments when you learn about the scientific method in the next lesson.

Operational science vs. origins science

You just learned that chemistry is the study of how different materials react with each other. This is a form of observational science. You can observe what happens when you do an experiment. Observational sciences is what you normally think of as science. A scientist can measure how far away a star is. A scientist can cross-pollinate two plants and observe what kind of flowers or fruit they produce. Scientists test the strength of new metals or measure how much energy is in a particular sample of coal. Observational or *operational science* can be repeated and reproduced by other scientists. This is an important part of discovering how things work.

However, there is an area of science that is

LESSON 1 **Properties of Matter** • 15

not observable. This area of science tries to answer questions about the past, such as where all of the matter and energy in the universe came from. This area of science is often called **origins science** or historical science. Origins science does not deal with repeatable experiments like observational science does. Origins science looks at what we see today and tries to explain how it got here.

There are basically two views of origins. One view says there is a powerful God who created everything we see. The other view says there is no God, but only what we are able to see. This view is called **naturalism** and only accepts explanations of origins that exclude any supernatural being.

Experiments cannot be used to prove the ideas of origins. We cannot go back in time and observe the universe before there were planets. We cannot observe matter being created. In fact, all of the experiments that have been conducted in the past have shown that matter cannot be created or destroyed. This has been shown by so many experiments that it is called a scientific law. Because questions of origin cannot be tested or repeated, these areas must be accepted on faith. Either God created everything, or nature is all there is and everything developed by random, natural processes. God is the only eyewitness to creation and He has revealed in His Word that He created everything in six days (Genesis 1; Exodus 20:11).

It is important to recognize when someone is talking about operational science or origins science. When you read about scientific discoveries, you need to ask questions that will help you decide if it is observational science or origins science. Here are a few questions to help you get started in evaluating scientific claims.

1. What is this scientist claiming? What is the purpose of his investigation?
2. Is this claim based on experimentation that can be repeated, or is it trying to explain something that may have happened in the past?
3. What assumptions is the scientist making?
4. Does this claim contradict what the Bible says?

If a scientist claims to be able to show how something happened in the past, he is not dealing with observational science, but is dealing with origins science. This is the area where creationists and evolutionists often disagree. If a scientist claims to be able to show how something works today, then she is dealing with observational science, and others can test this claim to prove whether it is true.

Below are several scientific claims. Ask yourself the previous questions and try to determine if each claim represents observational or origins science.

1. The bones of an ancient ape-like creature show the evolution of man.
2. Carbon nanotubes have great potential in electronics because of their semiconducting properties.
3. The rocks in this area are millions of years old.

The first claim is origins science. The scientist observes some bones. He sees that the bones are similar to apes that can be observed today. But the scientist has not done, nor can he do, any experiments to show how that creature evolved into another creature. The scientist is making assumptions about what the bones show without supporting experimentation. The claim contradicts the Bible's claim that God created man.

The second claim is observational science. The semiconducting properties of carbon nanotubes can be tested. Whether this potential is ever fulfilled remains to be seen, but the properties can be tested today, so this is observational science.

The age of the rocks may appear to be observational science as well. Scientists perform many tests to try and determine the age of a rock. However, these tests only give them relative amounts of elements in the rocks. The ages they assign to rocks is their interpretation of the evidence. These tests rely on assumptions that are not always true; therefore, the scientists get unreliable answers. Because this claim contradicts the Bible, it is worth investigating further. You can find more information about rock dating by going to the website for Answers in Genesis.

The theory of evolution deals with origins and is not operational science. Be a wise scientist and ask questions to help you understand the claims of other scientists.

2

The Scientific Method

How do scientists do it?

What is the scientific method?

Words to know:

scientist   hypothesis

scientific method controlling variables

What is a scientist? A scientist is some-one who uses observations and a systematic method to study the physical world and how everything in it works. **Scientists** study things that can be observed, measured, and tested.

Over the centuries, scientists have developed a way of approaching problems that is called the **scientific method**. This is not a set of rules, but a way of thinking. It is a logical, systematic approach that involves observing and testing to gain knowledge. In general, the scientific method has five steps.

1. Learn about something—make observations
2. Ask a question—identify a problem
3. Propose a solution—make a **hypothesis** (a good or educated guess)
4. Design a way to test your hypothesis—test your idea, record your observations and results
5. Check if your results support your hypothesis—draw a conclusion

Different books will list different steps or ways to perform the scientific method, but the ideas will still be the same.

You probably use this method to solve simple problems every day without even realizing it. For example, if you own a houseplant, you probably learned that plants need sunshine and water to grow. You identify a problem when you notice that your plant looks droopy. You make a hypothesis and guess that the plant needs to be watered. You test your hypothesis by watering the plant. Then you check to see if your hypothesis was right by seeing if the plant looks better after you watered it. If it does, you conclude that your hypothesis was right. This is a very simple example of how you use the scientific method frequently.

Now let's look at each step of the scientific process in a little more detail. First, you need to learn about a topic. This can be done by direct observation using your five senses. You can also do research by reading books or reports. We can and should learn from those who have gone before us. There is no sense in "reinventing the wheel" when we can find out a great deal from others' work.

Second, we can identify a problem by asking a question. This requires curiosity. Ask, "How does this work, why did this happen, or what if we tried this?" Almost all great inventions have come about

LESSON 2   **Properties of Matter** • 17

🧪 Using the scientific method

Purpose: To test which sweetener makes bread dough rise best

Materials: three identical empty bottles, masking tape, warm water, yeast, sugar, molasses, three balloons, cloth tape measure, "Scientific Method" worksheet

Procedure: Let's apply the scientific method to a fun problem:

A. Learn: Have you ever watched bread dough rising? The dough starts out as a relatively small lump, but in a few minutes it is tall and fluffy. Bread dough rises because tiny organisms called *yeast* combine with the sweetener in the dough and give off a gas that lifts up the dough.

B. Ask a question: What is the best sweetener to use to make the fluffiest bread?

C. Make a hypothesis: Sugar and molasses are common sweeteners. Guess which one you think would work best. Write your guess on your "Scientific Method" worksheet.

D. Design and perform a test: Remember, you must control your variables.

1. Get three identical empty bottles.
2. Put a piece of masking tape on each bottle and number them 1, 2, and 3.
3. Pour 1 cup of water that is 100°F into each bottle.
4. Add 1 teaspoon of yeast to each bottle. Do not add anything else to bottle 1.
5. Add 2 tablespoons of sugar to bottle 2, and add 2 tablespoons of molasses to bottle 3.
6. Gently mix the contents of each bottle by swirling the contents for 30 seconds.
7. After mixing, place a balloon over the top of each bottle to catch any gas produced.
8. After 15 minutes, use a cloth tape measure, or string and ruler, to measure the circumference of each balloon and record your measurements on your worksheet.
9. Repeat these measurements every 15 minutes for one hour, recording your results each time.

E. Questions: Did the bottle with your chosen sweetener produce the most gas? Which sweetener would you use to make bread? Answer the questions on your worksheet.

Conclusion: Even if you guessed the sweetener that did not produce the most gas, you still learned from your experiment. Share your results with someone who did not do the experiment with you, and look for ways to apply the scientific method to other problems.

---

because the inventor asked, "How can we do this a better way?" God gave humans a great amount of curiosity, and He wants us to use it to improve our lives and to glorify Him.

Once we identify a problem that needs to be solved or have a question we want to answer, we need to make a guess as to what the answer will be. This answer should be based on what we have observed and learned. For example, if you have a plant that is not growing well, you may ask how you can make it grow better. You have learned that plants not only need water and sunshine, but they also need nutrients. So you guess that your plant is doing poorly because the soil does not have enough nutrients in it. This is a reasonable guess based on what you have learned. It would not be a reasonable guess to say that the plant needs bubble gum to grow better.

The next step is to design a test that will help you decide if your guess is correct. This test or experiment should be set up to test only one thing at a time. If you change two or more things at once, you will not know which change gave the observed results. If you move the plant from a dark room into a sunny room and feed it more plant food, and then the plant starts growing better, you will not know if the added sunlight, the plant food, or a combination of both caused the change. You need to change only one thing at a time. This is called **controlling variables**. So first, keep the plant in its current location and give it fertilizer. If that helps, you know it needed nutrients. If the plant does not improve, try moving it to a sunny location. Changing only one thing at a

Scientists follow careful procedures when conducting experiments.

time will help you determine why you got the results you did.

Finally, after you complete your test, you need to check and see if the results show that your hypothesis is correct or not. It is okay to have a wrong hypothesis! When Thomas Edison was trying to find a material that would work for the light bulb, he tried hundreds of different materials before he found one that worked. The important thing is to try to understand why your hypothesis was wrong, and to try to come up with a better idea next time. This is how scientists learn. Also, it is important to share your results so that others do not make the same mistakes you did, and so they can learn from your successes as well.

There are limitations to the scientific method. We cannot answer all questions nor can we solve all problems by experimentation. The scientific method can only be used on physical materials. So we cannot use science to establish truth, to make moral judgments, or to determine what is right and wrong. We must use God's Word to help us decide these kinds of issues.

Also, the conclusions drawn from experiments can be affected by the scientist's beliefs. For example, many scientists do not believe there is any power outside of the physical world. Therefore, they interpret the results of their tests in light of that idea. When they look at the world today, they think that it got the way it is by natural forces only. However, when a scientist who believes the Bible sees the world, he understands that God created the universe, and that our world has been affected by the curse God put on the earth when Adam disobeyed, and by the Genesis Flood when God judged man's wickedness. Our beliefs affect how we interpret our test results. As Christians, we need to take God's Word first, and then interpret our world in light of what He says, because we know that God cannot lie.

What did we learn?

- What is the overall job of a scientist?
- What are some areas that cannot be studied by science?
- What are the five steps of the scientific method?

Taking it further

- Why was it necessary to have bottle number 1 in the experiment?
- What other sweeteners could you try in your experiment?
- What sweeteners were used in the bread at your house?
- Why do you think that sweetener was used?

LESSON 2   Properties of Matter • 19

🎖 Design your own experiment

Show that you understand the scientific method by designing your own experiment. You can test any hypothesis that your parent or teacher approves of. Follow the steps you have learned in the lesson. Make your own data sheet, record your data, show how you are controlling your variables, and be sure to write up a conclusion to share with others. If you need ideas for where to start here are some suggestions of things you can test:

- Which brand of battery lasts the longest?
- Which paper towel is the strongest?
- Does music affect plant growth?
- Are childproof caps really childproof?
- Which laundry detergent gets out stains the best?

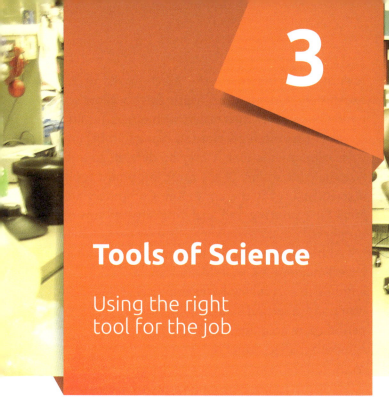

3

Tools of Science

Using the right tool for the job

What's the difference between qualitative and quantitative observations?

Words to know:

qualitative observations

quantitative observations

Challenge words:

microscope

eyepiece

objective lens

reflecting telescope

Every person who performs a job requires special tools. A cook needs measuring cups, a mixer, an oven, and pots and pans. A carpenter needs hammers, saws, sanders, and many other tools. An athlete needs weight equipment, running shoes, special clothes, ice skates, etc. So also, a scientist needs special tools to do his or her job.

A scientist's main job is to make observations. There are two kinds of observations that a scientist makes: qualitative observations and quantitative observations. **Qualitative observations** are ones that do not involve numbers. For example, a scientist may observe the color or texture of a material before, during, and after an experiment. The main tools that a scientist uses for qualitative observations are his five senses. What qualities might you observe with your five senses? You might see color, bubbles, smoke, size, etc. You could smell odors, hear popping, or taste flavors. And you could feel the texture and temperature of the object. Please note that you should **never** taste any unknown substance!

Qualitative observations are very useful; however, they are dependent on the observer. What one person considers red, another may describe as purple or pink. One person may describe a light as very bright, while another does not. Therefore, qualitative observations are limited in their usefulness. Whenever possible, scientists choose to make quantitative observations. **Quantitative observations** involve numerical data.

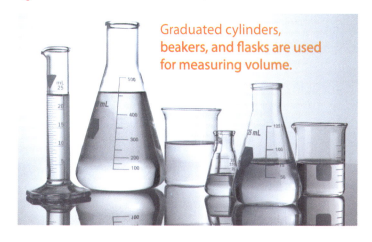

Graduated cylinders, beakers, and flasks are used for measuring volume.

LESSON 3   **Properties of Matter • 21**

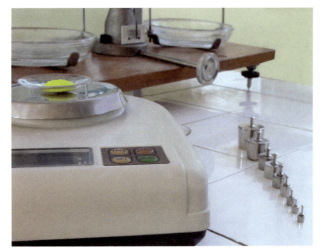

Modern electronic balances have replaced traditional beam balances.

Scientists have a number of tools to help them make quantitative observations. These include a balance for measuring mass, graduated cylinders for measuring liquid volumes, thermometers for measuring temperature, and spectrometers for measuring the wavelength of light. The list goes on and on. These tools provide unbiased data with which scientists can determine the exact results of an experiment. Quantitative measurements allow scientists to compare their results with others' results.

Scientists often generate so many numbers in their experiments that it may be difficult to analyze them all, so one of the most important tools a scientist uses is the computer. Computers are ideal for compiling and analyzing large groups of numbers.

Although quantitative data are preferred, they have limitations as well. Measurement tools have limits on their accuracy. For example, an analog wristwatch with a second hand cannot measure any time period more accurately than to the nearest second. A more accurate digital timer would be needed to measure a reaction that happens in microseconds. A scientist must know the limits of her tools and use the best tool for the job. As a scientist, you must learn to make good quantitative observations and good qualitative observations as well.

Fun Fact

Qualitative observations can sometimes be unreliable. Our eyes are easily deceived. In each picture below, tell whether you think line A or line B is longer.

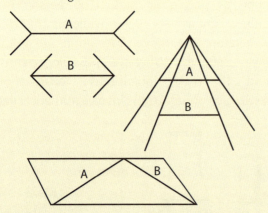

It appears that line A is longer in each picture; however, lines A and B are actually the same length in each picture. Measure each line with a ruler to convince yourself. Quantitative observations can help prevent mistakes that we might otherwise make.

What did we learn?

- What is the main thing a scientist does as she studies the physical world?
- What are the two types of observations that a scientist can make?
- What is the main problem with qualitative observations?
- What are some scientific tools used for quantitative observations?

Taking it further

- What qualitative observations might you make when observing the experiment in lesson 1?
- What quantitative observations might you make when observing the experiment in lesson 1?

Learning to use your tools

Complete each task on the "Scientific Tools" worksheet.

Magnifying objects

Two of the most important scientific instruments are the microscope and its cousin the telescope. Both of these instruments use lenses to magnify an image. **Microscopes** are used to make very small objects appear larger so you can see more detail. A microscope uses two types of lenses.

The first lens is called the **objective lens**. The objective is close to the object that is being viewed. At the top of the microscope is another lens called the **eyepiece**. Some microscopes have two eyepieces so you can use both eyes to look at the object. The objective magnifies the image a certain amount. It could be 10 times or 100 times. Then the eyepiece magnifies that larger image another 10 times. So what you see may appear to be 100 or even 1,000 times bigger than it actually is.

To properly use a microscope, place a slide containing what you wish to view on the stage. Watch the objective as you lower it all the way down until it is nearly touching the slide. Do not look through the eyepiece while lowering the objective or you may lower the lens into the slide causing damage to the slide and/or the lens. After you have lowered the objective, you can look through the eyepiece as you slowly raise the objective until the object comes into focus.

Most microscopes have a coarse and a fine adjustment for moving the objective. Use the coarse adjustment until the object is nearly in focus, then use the fine adjustment to bring the object into sharp focus. Some microscopes have knobs that can be used to move the stage to allow you to view different areas of the slide while looking through the eyepiece. Many microscopes have clips to hold the slide in place. If you have access to a microscope and slides, use what you have learned here to safely view the objects on your slides. If you do not have any prepared slides, you can view some salt and sugar crystals to see the differences in their shapes.

A telescope also uses lenses to enlarge an image, but telescopes are used to view objects that are very far away rather than to view objects that are very small. A **reflecting telescope** (pictured at right) actually uses a series of lenses and mirrors to enlarge the image without distortion. Light from a distant object enters the front of the telescope and reflects off the concave mirror near the back of the telescope. The reflected light then reflects off of a convex mirror

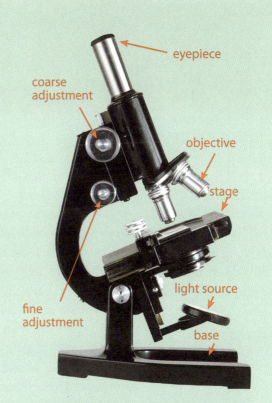

near the eyepiece where it passes through a lens.

If you have a telescope available, you can use it during the day to view objects that are far away such as a building or a tree. At night you can use it to view the moon and the stars.

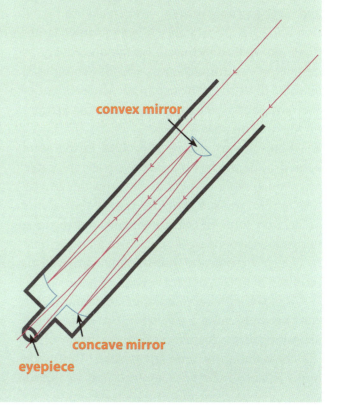

LESSON 3 **Properties of Matter • 23**

Lord Kelvin
1824–1907

SPECIAL FEATURE

William Thomson, or Lord Kelvin, was the fourth child of seven, born on June 26, 1824, in Belfast, Ireland. His mother died when he was six years old. His father wrote textbooks and taught math at a university in Belfast and he took on the task of teaching the newest math to William. William had a very close relationship with his father. Over the next several years he learned much about math and wrote several papers. By the time he was 15 he had won a gold medal for his exceptional mathematical ability.

When William Thomson was 16 he entered Cambridge, and four years later earned a bachelor's degree with high honors. At 22 years of age William was unanimously elected the Chair of Natural Philosophy (physics) at the University of Glasgow, where he stayed until he was 75 years old. Much of William's work involved a theory that electricity and magnetism are related, and that electromagnetism and light are related. His work enabled James Clerk Maxwell to formulate his famous equations describing electromagnetism, which is considered by some to be the most significant achievement of the nineteenth century. (See the Special Feature on Maxwell on page 45.)

Thomson was also very involved in the work behind laying the first transatlantic telegraph cable. After much work and controversy, his design was selected and was successful. Because of the patents he held on some of the equipment used to transmit and receive the signal, he was knighted in 1866 by Queen Victoria and thus became Lord William Thomson of Kelvin, or Lord Kelvin.

Because of the success of the transatlantic cable, Lord Kelvin became a partner in two engineering consulting firms. These firms played a major role in the construction of submarine cables. This made him a wealthy man. With some of that wealth he bought a huge yacht and a magnificent estate.

Lord Kelvin's interest in science was very broad; he did research in electricity, magnetism, thermodynamics, hydrodynamics, geophysics, tides, the shape of the earth, atmospheric electricity, thermal studies of the ground, the earth's rotation, and geomagnetism.

In 1884 Lord Kelvin developed an analog computer for measuring and determining the tides in a harbor for any hour, past or future. And he started a company to manufacture these devices. He also went on to publish a textbook on natural philosophy or physics, as it would be called today. He received honorary degrees from all over the world. He was said to be entitled to more letters after his name than any other man in the United Kingdom.

Out of all his scientific discoveries, Lord Kelvin is most remembered for his work in accurately measuring temperatures. He developed the temperature scale used by almost all scientists around the world. It is named after him, and is called the Kelvin scale. The Kelvin scale starts at absolute zero, the temperature at which all movement of molecules ceases. This means that zero degrees kelvin is the coldest anything can get; it's the lowest temperature possible.

In spite of all his great discoveries, Lord Kelvin still made mistakes. In 1900, at an assembly of physicists, he stated, "There is nothing new to be discovered in physics now. All that remains is more and more precise measurement." He also stated that heavier-than-air flying machines were impossible. This just goes to show that even a highly respected physicist can make mistakes in his field, even though most of his work has been shown to be correct.

Lord Kelvin was also a very insightful man. In his time, many new theories were being developed. Some had merit and others did not. In 1847 he heard about Joule's theory on heat and the motion of heat, which went against the accepted knowledge of the day. Kelvin studied this new theory and later gave his cautious endorsement of it. He also worked to advance the theories of Faraday and Fourier, as well as Joule.

However, when Charles Darwin's theory of evolution was first published, Lord Kelvin opposed it. He believed that all science must be subjected to the same rigors, and he applied what he knew about science to the theory of evolution. Darwin's theory was based on the assumption that life had evolved over a very long time (at that time Darwin believed the earth was millions of years old) during which the forces of nature remained fairly constant, and that nature operated millions of years ago just as it does today. Lord Kelvin based his opposition to evolution on the theory of thermodynamics, showing that the earth would have been considerably hotter only one million years ago and any life that lived at that time would have been very different from what can live today. He also showed that these conditions would have produced violent storms and floods over the earth. Also, the second law of thermodynamics, which was developed by four of Kelvin's contemporaries, shows that in a closed system the natural order of things is to become more disorganized, which clearly contradicts Darwin's theory that living things, over time, become more organized and more complex.

William Thomson (Lord Kelvin) lived an exemplary Christian life. He spent much of his time showing that science supports the idea of an intelligent creator. He wrote, "Mathematics and dynamics fail us when we contemplate the earth, fitted for life but lifeless, and try to imagine the commencement of life upon it. This certainly did not take place by any action of chemistry, or electricity, or crystalline grouping of molecules under the influence of force, or by any possible kind of fortuitous concourse of atoms. We must pause, face to face with the mystery and miracle of the Creation of living creatures." Lord Kelvin believed the Bible when it says that God spoke all life into existence.

Lord Kelvin died on December 17, 1907, at the age of 83. He is buried in Westminster Abbey in London, and a stained glass window has been installed in his honor.

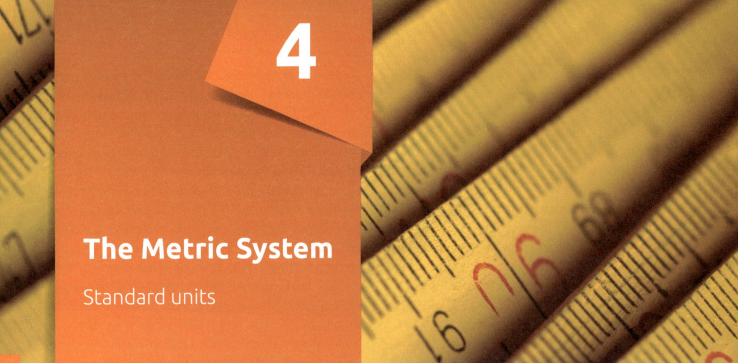

4

The Metric System

Standard units

What units do scientists use to measure things?

Words to know:

meter

gram

liter

Challenge words:

Mohs scale

Fujita scale

Beaufort scale

Saffir-Simpson scale

Scientists from around the world often work together on projects. In order to do this, they must overcome language barriers and other obstacles. One way scientists overcome the language problem is by using Latin terms whenever possible so there is less confusion. They also use numbers to reduce confusion. Because numbers and quantitative data are so important to science, a standard system for measuring has also been adopted by the scientific community.

The traditional system of measuring used in America is based on the Old English measures established during the Middle Ages. These measurements include the inch, foot, yard, mile, rod, hand, and span for length; the ounce, pound, and ton for weight/mass; and the fluid ounce, cup, pint, quart, and gallon for liquid volume. Because the conversion between units is difficult to use and remember in the Old English/American system, the scientific community has adopted the metric system. The metric system is often referred to as SI units, from the French term *Système International* (International System).

The metric system is very easy to use and to remember. All conversions from one unit to another are multiples of 10, so the math is easy. Each type of measurement is based on a standard unit. These units include the **meter** for length (hence the name metric system), the **liter** for liquid volume, and the **gram** for mass. The metric unit for time is the second; for temperature it is degrees kelvin; and for electrical current it is the ampere. If you wish to measure something that is significantly smaller than the standard unit, you use a unit that

Fun Fact

A meter was originally defined as 1/1,000,000 of the distance from the North Pole to the equator. But it has been redefined to be 1,650,763.73 times the wavelength of light given off by the element krypton-86.

is the standard unit divided by 10 or 100 or some other multiple of 10. For example, to measure the length of a paper clip you would use a unit that is a meter divided by 100—a centimeter. To measure something that is significantly larger than the standard you would use a unit that is that standard unit times some multiple of 10. For example, to measure the mass of a paper clip you can use a gram. But to measure the mass of a person you would want to use a unit that is 1,000 times bigger than a gram, called a kilogram.

Prefixes & conversions for metric units

X1,000,000
mega (M)

X1000
kilo (k)

X 100
hecto (h)

X 10
deka (da)

Basic Unit
meter (m)
gram (g)
liter (l)

1/10
deci (d)

1/100
centi (c)

1/1000
milli (m)

1/1,000,000
micro (µ)

This chart demonstrates how the metric system works. The appropriate prefix is applied to the name of the basic unit. For example, the prefix for 1,000 times is kilo, so 1,000 meters is a kilometer and 1,000 grams is a kilogram. The symbol in parentheses shows the abbreviation used for that unit. For example, a centimeter is abbreviated cm and a kilometer is shown as km. At first, using metric units may seem strange because they are different from what you are used to, but after you use them for a while, you will find that they are much easier to use than the Old English/American units. Besides, it will help you to understand what other scientists are talking about. ✷

Fun Fact

The origin of the Old English units is very interesting. In the Middle Ages, an inch was equal to the length of three barley seeds placed end to end. And a yard was equal to the distance from the tip of the king's nose to the end of his outstretched hand. As you can imagine, these lengths varied from time to time so they were never quite accurate. However, the units we use in America today no longer vary with the size of a barley seed or the size of a king. Instead, we have standard measurements. The information necessary for verifying these standard measurements is provided by the National Institute of Standards and Technology (NIST) in Boulder, Colorado. NIST develops and supplies references that companies and other organizations use to check the accuracy of their equipment.

LESSON 4 **Properties of Matter**

🧠 What did we learn?

- What are some units used to measure length in the Old English/American measuring system?
- What is the unit used to measure length in the metric system?
- What metric unit is used for measuring mass?
- What metric unit is used for measuring liquid volume?
- Why do scientists use the metric system instead of another measuring system?

🚀 Taking it further

- What metric unit would be best to use to measure the distance across a room?
- What metric unit would you use to measure the distance from one town to another?
- What metric unit would you use to measure the width of a hair?

🧪 Using metric units

Purpose: To become more familiar with the terminology of the metric system

Materials: 1-liter container, metric measuring cup, meter stick, paper clip, pencil

Activity 1—Procedure:

1. Take a giant step—that is about equal to a meter.
2. Now start at one side of the room and measure how many giant steps it takes to cross the room. That is close to how many meters it is from one wall to another.
3. Use a meter stick to get a more accurate measurement after you have walked the distance.

Activity 2—Procedure:

1. Use your hands to show how big a soft drink bottle is. Most bottles of soft drink are 2 liters.
2. Now show how big half a bottle of soft drink would be. This is about 1 liter.
3. Find a container that holds approximately 1 liter of liquid. Verify the container's volume with a measuring cup marked in milliliters or liters.

Activity 3—Procedure:

1. Hold a paper clip in your hand. A small paper clip is about ½ gram.
2. Now hold a pencil. How does a pencil compare to a paper clip? (It is heavier.) How many grams do you think a pencil is? (It depends on the size of the pencil, but could be about 5–10 grams.)
3. If you have a gram scale, measure the mass of the pencil to see how close your guess is.

Activity 4—Procedure:

1. For older children with sufficient math skills: use the conversion chart to answer the following questions. Each step on the chart represents a multiple of 10.

    - If you pour 1,000 ml of water into a bottle, how many liters of water do you have?
    - If you weigh 20 kg, how many grams do you weigh?
    - If it is 40 hectometers from your house to your best friend's house, how many meters must you walk to get from your house to his?
    - If your pet hamster is 60 mm long, how long is she in cm?
    - If you have 5 dekagrams of chocolate to share, how many decigrams do you have?

Measuring scales

Although using the metric system is ideal for many applications in science, not everything can be easily measured and described using meters, liters, or grams. For example, how do you measure the hardness of a rock or mineral? There are SI units for area, volume, length, density, force, and voltage, but there are no units for hardness. So scientists have developed a special scale to describe relative hardness of a rock or mineral. This scale is called the Mohs hardness scale. The **Mohs scale** assigns a number from 1 to 10 with 1 being the softest and 10 being the hardest. Talc is the softest mineral and diamond is the hardest. Other rocks and minerals have been assigned numbers based on how their hardness compares to these and other minerals.

Storms are also difficult to measure with the metric system. It is possible to measure wind speed in kilometers per hour and the amount of rainfall in centimeters, but those measurements do not fully describe a storm. It is helpful to have some idea of what effects a storm will have if it passes through an area, so scientists have developed several different scales to describe storms. The **Beaufort scale** describes wind, assigning numbers from 0 to 12 where 0 is calm—no wind—and 12 is hurricane force winds. Each number has a wind speed range associated with it as well as visible effects. For example, if the wind is a level 4 the wind would be described as a moderate breeze, the wind speed would be 21–29 km/h and the effects would be dust raised and small branches moving.

The **Fujita scale** measures tornado intensity on a scale of 0 to 5. An F0 tornado is called a gale tornado and does some damage such as breaking down chimneys, breaking tree branches, and damaging billboards. The Fujita scale describes the damage done by the storm rather than the size or shape of the funnel. There is a similar scale for describing hurricanes called the **Saffir-Simpson scale**, which rates the intensity of the hurricane from 1 to 5.

Earthquakes are another natural phenomenon that is not easily measured. Do some research and see if you can find out two different ways that earthquakes are measured.

UNIT 2
Measuring Matter

5 Mass vs. Weight
6 Conservation of Mass
7 Volume
8 Density
9 Buoyancy

◊ **Distinguish** between mass and weight.
◊ **Describe** the relationship between mass, volume, and density.
◊ **Describe** the relationship between density and buoyancy

5

Mass vs. Weight

What's the difference?

What is the difference between mass and weight?

Words to know:

mass
weight
newton

Challenge words:

slug
pound
ounce

When studying matter, one of the first questions scientists ask is, "How much matter do we have?" The amount of matter in a sample is called its mass. The mass of an object does not depend on its shape. If you start with a ball of Silly Putty and you flatten it into a disk you still have the same amount of Silly Putty. You have not changed its mass. You have not changed how much you have.

People often confuse weight with mass. In everyday, nonscientific applications you may see weight and mass treated as the same quantity, but in scientific applications there is a distinct difference. **Mass** is how much of something there is. **Weight** is a measure of how strongly one thing is attracted to another by gravity. On Earth, weight is a measure of how much the earth's gravity pulls on an object such as your body. But in space, where there is no gravity, a person becomes weightless; however, his mass remains the same.

Mass is measured by using a balance to compare an object's mass with a known mass. You place the object to be measured on one side of the balance and known masses on the other side of the balance until both sides are even. For example, to measure the mass of a pencil, you would place it on one side

Many bathroom scales have springs inside to measure your weight.

LESSON 5   **Properties of Matter • 31**

of a balance, and then add 1-gram mass pieces to the other side until both sides are balanced. This would show you the mass of the pencil in grams.

Because weight is a measurement of gravitational pull, it is measured using a spring scale. The object to be measured is attached to one end of a spring and the other end is held up. The amount that the spring stretches indicates the weight of the object. With many spring scales the object can be placed on top of a spring and the amount the spring compresses shows the object's weight. This is how most bathroom scales work.

In SI units, mass is measured using grams for small objects such as an eraser or a cherry, and kilograms (1,000 grams) for larger objects such as people or cars. The metric system unit for weight is the **newton**—named after Sir Isaac Newton.

So even though we often use the terms weight and mass interchangeably, there is a difference. Mass is how much material you have, and weight is how much gravity pulls down on that material.

What did we learn?

- What is the difference between mass and weight?
- How do you measure mass?
- How do you measure weight?

🚀 Taking it further

- What would your weight be in outer space?
- What would your mass be in outer space?
- Name a place in the universe where you might go to increase your weight without changing your mass.

🧪 Measuring mass

Mass is measured with a balance. You will build a balance and measure the mass of various object by comparing their masses to the mass of a penny. The standard unit of mass in the metric system is the gram.

Purpose: To build a balance in order to measure mass

Materials: ruler with holes punched in it, pencil, two paper cups, paper clips, hole punch, string, tape, pennies

Procedure:

1. Punch a hole near the top of a paper cup. Punch a second hole directly across from the first hole.
2. Tie the ends of a 50 cm piece of string to the cup through the holes.
3. Tape the center of the string to the ruler approximately 2 cm from one end.
4. Repeat steps 1–3 and tape the string of the second cup to the other end of the ruler.
5. Insert a pencil through the center hole of the ruler.
6. Set the pencil on a table with the balance hanging off the edge and, see if the empty cups balance so that the ruler is level. If not, adjust one cup by moving it

closer to or farther from the center until the balance is level.

7. Now you can use your balance to find the mass of small objects. Place a small object in one cup.
8. Now place pennies one at a time in the other cup until the ruler is level. A penny has the mass of approximately 3 grams, so multiply the number of pennies by 3 to determine the approximate mass of the object. If the object has a mass that is less than that of a penny, you can use paper clips. Small paper clips are about ½ gram each. Save this balance to use in lessons 6 and 8.

Measuring weight

Weight is measured with a spring scale that measures how much the weight of an object stretches a spring. You will build a spring scale (or a rubber band scale if you don't have a good spring) and will measure the weight of various objects by seeing how much they make the spring stretch. The units used to measure weight are newtons, pounds, or ounces. We will use ounces for the units on our spring scale.

Purpose: To build a spring scale in order to measure weight

Materials: rubber band or spring, paper cup, hole punch, string, tape, pen or pencil, 25 pennies

Procedure:

1. Punch a hole near the top of a paper cup. Punch a second hole directly across from the first hole. Tie one end of a 50 cm string to one of the holes. Thread the string through your spring or rubber band, then tie it to the second hole
2. Place a pencil through the rubber band to hold up the top of your scale.
3. Set the pencil on a table near the edge so the scale hangs in front of the table.
4. Tape a piece of paper to the edge of the table so that it is behind the rubber band and cup.
5. Make a mark on the paper showing the bottom of the rubber band when there is nothing in the cup.
6. Place five pennies in the cup and mark the bottom of the rubber band on the paper.
7. Repeat this for 10 pennies, 15 pennies, 20 pennies, and 25 pennies, marking where the bottom of the rubber band is at each measurement.
8. Remove the pennies. Now your spring scale is ready to use.
9. Place a small object in the cup and see how much it stretches the rubber band. You can compare this to the weight of the pennies. A penny weighs about 0.1 ounces, so if your object stretches the rubber band the same amount that 10 pennies did, it weighs about 1 ounce.

Note: If you have a spring scale you can use it to directly measure the weight of an object instead of using the rubber band and paper cup scale. Just be sure the scale reads zero before putting anything in the cup.

Mass & weight units

As you have learned, the units used to describe something are very important. This is especially true when you are measuring mass and weight. The units used will help you determine whether a measurement is mass or weight. Since it is easy to confuse weight and mass, the units help to make things clear.

If you are using the metric system to measure mass the units will be in grams. If the object has a large mass the units may be given in kilograms, but the basic unit is the gram. The unit for weight in the metric system is the newton. If a measurement is given in newtons, you know that the weight of the object is being measured, not its mass.

In the English system mass is measured with the unusual unit called a **slug**. Most of the people who work with mass measurements use the metric system so you may never have heard of the slug, but it is the proper unit for measuring mass in the English system. Weight, on the other hand, is something you hear about all the time so you probably already know that the unit for weight in the English system is the **pound** for larger weights and the **ounce** for smaller weights. You probably know how many pounds you weigh, and have a pretty good idea of how much five pounds weighs based on how heavy a bag of sugar or flour feels.

To get some practice using these units, complete the "Mass and Weight Units" worksheet.

LESSON 5   **Properties of Matter • 33**

6

Conservation of Mass

Where does it go?

How is mass conserved?

Words to know:

law of conservation of mass

law of conservation of energy

first law of thermodynamics

In the previous lesson you learned that mass is how much of a substance is in a sample or object. You may think that mass changes when something changes form because it looks different or it takes up a different amount of space, but this is not true. For example, let's say you have 10 grams of water in a cup. That water can change form—it can be frozen and become ice, or it can be boiled and become steam—but you will still have 10 grams of water. Similarly, if you have a 20 gram rock and you hit it with a hammer until it is broken into tiny pieces, you will still have 20 grams of rock. The pieces will be very small but if you put them all together the mass remains the same.

Sometimes matter doesn't just change form; sometimes it combines with other matter to make a new substance. Hydrogen and oxygen molecules combine to form water. But if you have 50 grams of hydrogen and 50 grams of oxygen before they combine, you will have 100 grams of water after. In general, the amount of mass does not change even in a chemical reaction. Occasionally, a nuclear reaction does produce a slightly smaller amount of mass because some mass has been converted into energy.

The fact that matter does not go away is a law called the **law of conservation of mass**. Energy also does not go away and this fact is called the **law of conservation of energy**. These two laws taken together are called the **first law of thermodynamics**. This law states, "Matter and energy cannot be created or destroyed, they can only change form." The law of conservation of mass is very important and is apparent all around us. The water cycle is one example and is God's way of providing for our needs. Water evaporates from the oceans, lakes, and rivers. It then forms into clouds and eventually falls as rain or other precipitation to provide water for crops and people. If water was used up in this process the world would eventually run out of water.

Another example of the conservation of mass occurs in plants. Plants absorb nutrients such as nitrogen from the ground. Then when they die, they decay and return those nutrients to the soil. An animal may eat the plant and absorb the nutrients into its body, but when the animal dies, it decays and returns the nutrients to the soil to be used by

Changing form without losing mass

Purpose: To help you understand that changing the form of something does not change its mass

Materials: paper cup, sugar cubes, balance, spoon

Procedure 1:

1. In a paper cup, dissolve a sugar cube in a small amount of water. Where did the sugar go? (It is still there, in the water. It just looks different because it has been broken into very small pieces by the water molecules.)
2. Set the cup in a place where it will not be disturbed. Check the cup every day until all of the water is evaporated. What do you see in the bottom of the cup? (You should see sugar crystals.)
3. Set up the balance you made in lesson 5.
4. Place the cup with the sugar crystals in the cup on one side of the balance and place a second identical cup with one sugar cube in it in the other cup of the balance. Do both cups balance? (They should.)

Conclusion: This demonstrates the conservation of mass. The sugar cube did not go away when it dissolved in the water. It just changed form into crystals instead of a cube, but it still has the same mass.

Procedure 2:

1. Put a sugar cube in each cup of your balance. Be sure that the ruler is level. If necessary adjust the position of one cup to make it balance.
2. Remove one sugar cube and set is on a piece of paper. Using the back of a spoon, crush the sugar cube. Pour the crushed sugar back into the cup. Be sure to get all of the sugar off of the paper.
3. Which cup appears to have the most sugar in it? Observe the balance. The ruler should still be balanced.

Conclusion: Changing the form or appearance of a substance does not change its mass. The crushed sugar has the same mass even though it is more spread out than when it was in the form of a cube. This is another example of the conservation of mass.

other plants. This reusing of nutrients is part of the conservation of mass. It is God's way of recycling so that we do not run out of essential materials.

The law of conservation of mass raises an important question. If matter cannot be created or destroyed, then where did it come from in the first place? This is a question that cannot be answered by operational science. The Bible tells us in Genesis 1 that God spoke the universe into existence. This is the only viable explanation for the matter that exists today. God created it and it continues to exist because man cannot destroy it. God set up a wonderful system to maintain life on Earth through the reusing of all matter on the planet.

What did we learn?

- What is the law of conservation of mass?
- How is the mass of water changed when it turns to ice?

Taking it further

- If you start with 10 grams of water and you boil it until there is no water left in the pan, what happened to the water?
- Why is the law of conservation of mass important to understanding the beginning of the world?

🏅 Mass of gases

Scientists use the law of conservation of mass to help them calculate the mass of the substances in a chemical reaction. Since the mass of the beginning substances must be equal to the mass of the ending substances, you can do some simple calculations to determine the mass of certain substances in the reaction. Gases are often difficult to measure without special equipment, but if you do an experiment that produces a gas, you can use the masses of the substances before the reaction and the masses of the substances after the reaction to calculate the mass of the gas that was released during the reaction.

Purpose: To get more practice using the law of conservation of mass

Materials: bottle with narrow neck, gram scale, vinegar, baking soda, paper, balloon, "Conservation of Mass" worksheet

Procedure:

1. Place a bottle with a narrow neck on a gram scale.
2. Pour ¼ cup of vinegar into the bottle.
3. Measure ½ teaspoon of baking soda and pour it onto a small piece of paper.
4. Place the piece of paper onto the scale with the bottle containing the vinegar. What is the total mass of the bottle, vinegar, paper, and baking soda? Record your measurements on a copy of the "Conservation of Mass" worksheet.
5. Now, pour the baking soda into the bottle and place the piece of paper back on the scale. What do you observe happening? You should see lots of bubbles being formed. The bubbles are carbon dioxide that is being produced by the chemical reaction.
6. Once there are no more bubbles being formed, read the mass of the bottle, the resulting liquid, and the paper. Record this number on your worksheet. How does this number compare with the original number? Why is it less? Where did the mass go? If you correctly decided that the "missing" mass was the mass of the carbon dioxide gas that escaped into the air you are correct. To prove this is true we will repeat the experiment but will not let the gas escape.
7. Empty the bottle and then place ¼ cup of vinegar in it and place it on the scale.
8. Measure ½ teaspoon of baking soda into an empty balloon and place the balloon on the scale as well. Write down the total mass of the bottle, vinegar, balloon, and baking soda.
9. Carefully place the end of the balloon around the mouth of the bottle without allowing any of the soda to go into the bottle. Make sure that the balloon is tight around the bottle.
10. Gently lift the balloon and allow the soda to go into the bottle. What do you observe happening? You should see the balloon begin to fill with gas.
11. Once the reaction has stopped, measure the total mass of the bottle, balloon, and substances inside. How does this measurement compare with your original measurement? They should be very close. If they are different, what might be the cause? Complete your worksheet.

7

Volume

How much space does it take up?

How do we measure the volume of something?

Words to know:

volume
meniscus
displacement method
overflow method

When scientists study matter, they make many measurements to help them understand more about the material they are studying. They measure mass so they know how much matter they have. They also measure volume.

Volume is how much space the matter takes up. For example, if you want to know how much room you need to stack boxes in a warehouse, you would need to know the volume of each box or how much room it takes up. If a box is 10 cm wide, 15 cm long, and 2 cm high it has a volume of 300 cubic centimeters (10 x 15 x 2 = 300). You would need to allow 300 cubic centimeters of room for each box you want to store.

Measuring the volume of something solid that has a regular shape like a rectangle, cube, or sphere is easy to do. You just need to measure the height, length, and width, or circumference, and then use a little math to calculate the volume. However, scientists work with many materials that are not solids, and many materials that are not regular shapes.

To measure the volume of a liquid, you need a container that has marks on its side. For example, you can measure the volume of water in a liquid measuring cup. Scientists usually use beakers or graduated cylinders instead of kitchen measuring cups to make accurate measurements of liquids. The units that scientists use for liquid volume are milliliters (1 ml = 1 cubic centimeter). To get

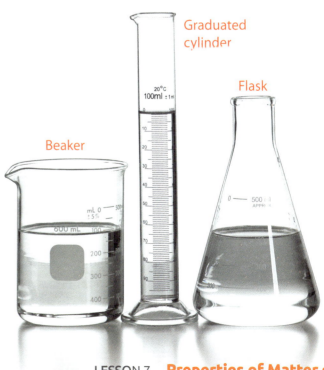

Beaker
Graduated cylinder
Flask

LESSON 7 **Properties of Matter • 37**

a correct measurement of any liquid, you need to set the measuring cup or cylinder on a level surface. Then you need to get down so your eye is level with the liquid. This will allow you to correctly read the amount of liquid in the container. If you are using a graduated cylinder or other device that is relatively narrow, the liquid may be slightly higher on the sides than in the center like the picture shown here. This curve is called a **meniscus**, and you should use the lowest point of the meniscus as your measurement.

To measure the volume of a gas, you must measure the volume of its container because gas expands to fill its container. For example, the air in a room fills up the room. In a balloon, the volume of air is equal to the volume of the balloon because the air spreads out to fill the whole balloon.

All of these volume measurements are relatively simple to make. But how do scientists measure the

🧪 Measuring volume

Purpose: To practice finding the volume of several objects

Materials: metric ruler, small box, drinking glass, liquid measuring cup, meter stick, irregularly shaped object

Activity 1—Procedure:

Measure the volume of a small box.

1. Use a metric ruler to measure the length, width, and height of the box in centimeters.
2. Multiply these three numbers together to obtain the volume of the box in cubic centimeters.

Activity 2—Procedure:

Measure the volume of water that a glass can hold.

1. Fill a glass with water.
2. Carefully pour the water into a liquid measuring cup.
3. Read the volume measurement. Be sure to place the measuring cup on a level surface and bend down until your eye is even with the water level to determine the volume of the water.

Activity 3—Procedure:

Measure the volume of air in your classroom or a room in your house.

1. Use a meter stick to measure the length, width, and height of your room.
2. Multiply these numbers together to determine the volume of the room in cubic meters. This is not quite equal to the volume of air because

   you are in the room and you probably have furniture and other items in the room that are taking up space. But it tells you the volume of air that would be in the room if the room were empty.

Activity 4—Procedure:

Determine the volume of an irregularly shaped object.

1. If you have an object that is small enough to fit in the measuring cup, use the displacement method to determine its volume.
2. If you want to measure something that does not fit in the cup, use the overflow method described in the lesson.

volume of an unusually shaped item like a rock or a toy? To measure the volume of an irregularly shaped object, a scientist fills a graduated cylinder part way with water and notes its volume. Then she carefully drops in the object to be measured and notes the new volume. The scientist can then simply subtract the first measurement from the second to determine the volume of the object. This is called the **displacement method**. For example, if the cylinder originally has 25 milliliters (ml) of water in it, and then an eraser is dropped in and the water level goes up to 47 ml, the volume of the eraser is 22 ml (47 ml minus 25 ml). Note that the original amount of water must be enough to completely cover the object that is being measured.

If an object is too big to fit in a graduated cylinder, the **overflow method** can be used. To do this, set a container that is big enough to hold the object inside a dish or tray. Then fill the container completely full of water. Carefully drop the object into the water, allowing the water to overflow into the dish. Carefully pour the water that overflowed into a graduated cylinder and measure its volume. The volume of the water that overflowed is equal to the volume of the object.

As you can see, there are many ways to measure how much room something takes up.

What did we learn?

- What is volume?
- Does air have volume?

Taking it further

- If you have a cube that is 10 centimeters on each side, what would its volume be?
- Why is volume important to a scientist?

Calculating volume

Mathematical formulas are very useful for calculating volumes of regularly shaped items. Collect the following items and use the formulas below to calculate the volume of each item. Complete the "Calculating Volume" worksheet.

- Box of crackers or other dry food
- Tennis ball
- Can of soup or other canned food
- Ice cream cone (without the ice cream)
- Die (6-sided for playing games)

Note: You can use anything that is shaped like the items above if you do not have the exact items; for example, you can make a cone out of paper if you do not have an ice cream cone available.

Formulas:
- Volume of a cube: = Side x Side x Side
- Volume of a rectangle = Length x Width x Height
- Volume of a sphere = (⅘) π Radius x Radius x Radius
- Volume of a cylinder = π Radius x Radius x Height
- Volume of a cone = (⅓) π Radius x Radius x Height
  (π = approximately 3.14)

Note: It can be difficult to accurately measure the radius of a ball. It is much easier to measure the circumference and then calculate the radius. Recall that the circumference of a circle is C = 2 x π x Radius, so the radius is equal to the circumference divided by 2π, or R = C/(2π).

8

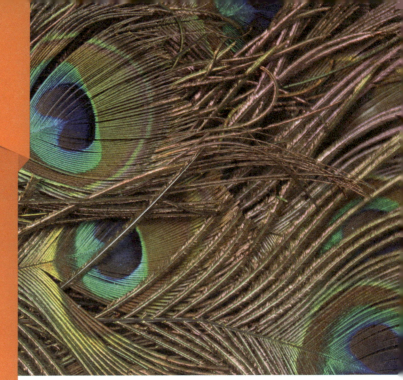

Density

Does it feel heavy?

How do we measure the density of an object?

Words to know:

density

Have you ever held a ping-pong ball in one hand and a golf ball in the other? Even though the balls have nearly the same volume—they take up about the same amount of space—the golf ball is heavier than the ping-pong ball. This is because the golf ball has more mass in the same area. It is more dense.

Density is the relationship between mass and volume. **Density** is defined as the mass divided by the volume. Here is another example to help you understand density. Which has more mass, a kilogram of feathers or a kilogram of lead? You may have answered that a kilogram of lead has more mass because we usually think of lead as being heavy and feathers as being light. But a kilogram of lead is the same amount of mass as a kilogram of feathers. However, the pile of feathers would be much larger—it would have a greater volume—than the pile of lead. Lead is very dense. A cubic centimeter of lead is 11.3 grams. But feathers are much less dense. There is less than 1 gram of feathers in a cubic centimeter of feathers. Because density is defined as mass divided by volume, its units are usually given as grams/milliliter (g/ml) or grams/cubic centimeter (g/cc).

Understanding density is important to scientists for many reasons. One use of density is to help the scientist determine what a sample of material is made of. If someone were interested in opening a mine, they would take samples of ore found in the area to a scientist to determine what metals are present. One way the scientist determines what metals are present is to measure the sample's

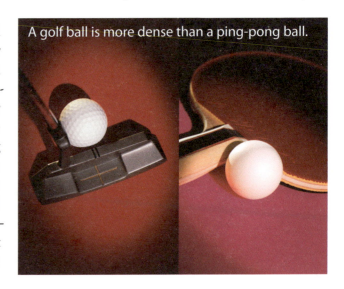

A golf ball is more dense than a ping-pong ball.

🧪 Measuring density

Purpose: To determine the density of a ping-pong ball and the density of a golf ball.

Materials: ping-pong ball, golf ball, pennies or paper clips, liquid measuring cup

Procedure:

1. Measure the mass of the ping-pong ball using your balance and pennies or paper clips.
2. Measure the volume of the ping-pong ball using the water displacement method from lesson 7. If the ball floats, carefully push the ball just under the surface of the water with your finger.
3. Divide the mass by the volume to determine the ball's density.
4. Repeat these three steps to determine the density of a golf ball.

Questions:

- How did the mass of the golf ball compare to the mass of the ping-pong ball?
- How did the volume of the golf ball compare to the volume of the ping-pong ball?
- Which ball has a higher density?

density. Different metals have different densities. Gold is very dense so it feels heavier than a similar sized piece of copper, which has a much lower density. If the ore has a density close to 9 g/cc it is likely to contain copper, and if it has a density close to 19 g/cc it is likely to contain gold.

Another reason that scientists are interested in density is to determine how a material might be used. If something is very dense it is more ideal for some uses than others. If it has a low density it may be more useful in other areas. For example, fiberglass and Styrofoam are materials that are not very dense because they have tiny air pockets trapped inside. This property makes them very good insulators. Fiberglass insulation is used to keep many houses warm in the winter and Styrofoam cups keep you from burning your hands when you hold a cup of hot chocolate. Many objects that are more dense, such as metals, conduct heat and do not make good insulators. ✳

🧠 What did we learn?

- What is the definition of density?
- If two substances with the same volume have different densities how can you tell which one is the densest?

🚀 Taking it further

- If you have two unknown substances that both appear to be silvery colored, how can you tell if they are the same material?
- If two objects have the same density and the same size what will be true about their masses?
- If you suspect that someone is trying to pass off a gold plated bar of lead as a solid gold bar, how can you test your theory?
- Why does the ping-pong ball have a lower density than the golf ball?

🏅 Density experiment

You have learned that just because two items are about the same size and shape does not mean that they have the same density. It is not always easy to tell an object's density just by looking at it or even by feeling it. It is often necessary to make volume and mass measurements to calculate an object's density. To get more practice with density, complete the "Density Experiment" worksheet by measuring the mass and volume of each object on the worksheet and then calculating the density of each object. You can use whatever method is best for measuring the mass and volume of each object.

9

Buoyancy

It floats.

What is buoyancy, and how do we measure it?

Words to know:

buoyancy

Have you ever wondered how something as heavy as a battleship can float on the water when a small rock sinks when you toss it in a lake? The ability of matter to float is called **buoyancy** and is directly related to its density. An object that is denser than its surroundings will sink while an object that is less dense will float. If the object floats it is said to be buoyant or to have buoyancy.

So why is a battleship more buoyant than a rock? The battleship is less dense than the rock. Even though the ship has much more mass than the rock, it is shaped so that is has a much greater volume. And most of the ship's volume is filled with air, which is much lighter and less dense than water. Thus the ship's overall density is less than the density of the water and so it floats. A rock, however, has a small mass and a small volume. Thus it has a higher density than the water, so it sinks. A few rocks have low densities and can float on the water.

You have probably used the principle of buoyancy when you were swimming. When you take a deep breath you are able to float because the air in your lungs decreases your overall density—increasing your buoyancy. Your body becomes less dense than the water so you are able to float.

Not all solids have the same density and not all liquids have the same density. For example, Mercury is a very dense liquid and would quickly sink to the bottom of a glass of water. You would never go swimming in a pool of mercury, but if you could, you would have no problem floating on the surface. However, if you went swimming in a pool of rubbing alcohol you would have a very difficult time keeping your head

42 • Properties of Matter   LESSON 9

🧪 Testing buoyancy

Purpose: To observe the buoyancy of various objects

Materials: modeling clay, popcorn, rubbing alcohol, vegetable oil, two cups

Solid in a liquid

Activity 1—Procedure:

1. Roll a piece of modeling clay into a solid ball.
2. Shape an equal size piece of modeling clay into a flat-bottomed boat.
3. Place both pieces of clay in a sink filled with water. Which shape floats?

Solid in a solid

Activity 2—Procedure:

1. Pop some popcorn and put it in a bowl. Be sure to include several unpopped kernels.
2. Mix the popcorn up with the kernels.
3. Gently shake the bowl for a few seconds. What did you observe happening?

Liquid in a liquid

Activity 3—Procedure:

1. Pour some water in one cup and some rubbing alcohol in a second cup.
2. Pour a small amount of vegetable oil in each cup. What happened to the oil in each cup? Why did the oil float in one cup but sink in the other?

above the surface because alcohol is less dense than water and you would no longer be buoyant.

So you can see that buoyancy is relative. It is the difference in the density of one substance compared to another. One of the most important applications of buoyancy is the fact that frozen water is less dense than liquid water so ice floats on liquid water. Water is one of the few substances that becomes less dense when it becomes a solid. This was a special design by God. If ice did not float, but instead sank to the bottom of the lake, the lakes and ponds would all freeze solid in the winter and life would quickly die in them. However, ice floats on the surface of a lake. This insulates the water below from the cold air. Even when the temperature is very cold outside, only the top few inches of most lakes will freeze allowing the fish and other animals to continue to live in the liquid water below the ice. So give thanks to God for making ice buoyant!

Fun Fact

Nuclear powered submarines must be bigger than diesel powered submarines in order to be buoyant. This is because nuclear reactors are much heavier than diesel engines. Therefore, the nuclear submarine must have a greater volume filled with air to compensate for the additional mass of the reactor in order to be buoyant in the water.

What did we learn?

- What is buoyancy?
- If something is buoyant, what does that tell you about its density compared to that of the substance in which it floats?
- Are you buoyant in water?

Taking it further

- What are some substances that are buoyant in water besides you?
- Based on what you observed, which is denser, water or alcohol?
- Why is a foam swimming tube or a foam life ring able to keep a person afloat in the water?
- Why is it important to life that ice is less dense than water?

Buoyancy questions

Now that you have an understanding of buoyancy, let's take a little quiz. If you are sitting in a boat that is floating in a lake and you have a large, heavy rock in the boat with you, what will happen to the water level of the lake if you toss the rock overboard and it sinks? Will the water level rise, stay the same, or lower?

Purpose: To test your hypothesis

Materials: sink, plastic tub, metal objects (wrenches, etc.) that fit in the tub, water, tape

Procedure:

1. Fill a sink with several inches of water.
2. Place a plastic tub in the sink to represent your boat.
3. Add wrenches or other heavy items to the boat. Add as much weight as you can but don't let the boat get low enough for any water to get into it.
4. Allow the water to become still. Place a piece of tape on the side of the sink at the water level. Or, if you don't mind the mark you can use a marker on the side of the sink to mark the water level.
5. Now, carefully remove all of the wrenches and place them in the water.
6. Allow the water to become still and observe where the water level is compared to where it was when the wrenches were in the boat. What happened to the water level? Did this match your hypothesis?

Conclusion: The water level should go down. This may seem opposite of what you think, but when something floats it pushes enough water out of the way so that the weight of the water that is displaced is equal to the weight of whatever is floating. So if something weighs about eight pounds it will push away about one gallon of water because a gallon of water weighs about eight pounds. If your rock weighs eight pounds it will move a gallon of water away from the boat and cause the water level of the lake to rise. But your rock is denser than water so when it is outside the boat it sinks. An eight pound rock may only take up as much space as a half-gallon of water so it will only push away that much water when it sinks. Outside the boat the rock displaces a smaller amount of water than it did inside the boat so the water level goes down.

Let's try another question. You are in a car and there is a helium balloon in the car with you. You, the balloon, and the car are not moving. The driver quickly accelerates. What will happen to you? What will happen to the balloon? If you have a chance, get a helium balloon and try this. You might be surprised what happens. Your body will be pushed back against the seat of the car, but the balloon will move forward. Why do you think the balloon moves forward instead of backward? The air in the car is moving backward just like you are, but the helium in the balloon is lighter than the air, so as the air moves toward the back of the car, it pushes the lighter helium forward.

44 • Properties of Matter   LESSON 9

James Clerk Maxwell

1831–1879

SPECIAL FEATURE

Math, science, nature, and Christianity; what do these things all have in common? They were all important in the life of James Clerk Maxwell. Maxwell, described as one of the outstanding mathematicians and scientists of the nineteenth century, was born in Edinburgh, Scotland. He was an only child and was home educated by his mother until her death in 1838. After spending two years with a tutor, James attended Edinburgh Academy, where he graduated at the top of his class in English and math. He then attended the University of Edinburgh and continued his studies at Trinity College.

After he completed his training, Maxwell spent some time teaching at Marischal College, where met and married Katherine Mary Dewar, the daughter of the college principal. He later became the professor of physics and astronomy at Kings College in London, and in 1871 he became the Chair of Experimental Physics at Cambridge University.

Although Maxwell spent time teaching at each of these schools, his real interest was in experimenting and testing out new ideas. Maxwell was the first to explain the kinetic theory of gases, showing that the movement of gas particles creates heat and pressure. He also tested the viscosity, or density, of gases. He was the first to suggest that the rings around Saturn were not a solid or a gas, but a collection of millions of tiny particles that orbit the planet. This idea was proven to be true when the *Voyager 1* space probe visited Saturn in 1980. It sent back pictures showing that the rings are composed of millions of pieces of ice, dust, and rocks.

Maxwell is best known for his work in electromagnetism. He worked closely with Michael Faraday and helped to mathematically describe the electromagnetic fields that generate electricity. Today, these equations are called Maxwell's Equations. Born out of his work with electricity, Maxwell advanced the idea that light was a form of electromagnetic energy.

Despite all his scientific work, the people closest to Maxwell did not describe him as a great scientist. Instead, they described him as a humble, godly man. James was an elder in his church and he fervently believed that the reason to study nature was to draw people to God so they could ultimately have a saving relationship with the Creator. Maxwell died of cancer at the age of 48. He left behind a great scientific legacy, but more importantly, he left behind a legacy of humble obedience to God.

UNIT 3

States of Matter

10 Physical & Chemical Properties
11 States of Matter
12 Solids
13 Liquids
14 Gases
15 Gas Laws

◊ **Distinguish** between physical and chemical changes.
◊ **Describe** how phase changes relate to solids, liquids, and gases.
◊ **Explain** the relationship among temperature, pressure, and volume of gases.

10

Physical & Chemical Properties

Is it something new?

How do we describe the properties of matter?

Words to know:

physical property   chemical property

As we learned in the first lesson, chemistry is the study of matter and how it reacts to the things around it. We know that scientists use tools and make measurements to study matter. When scientists, particularly chemists, study matter, they investigate both the physical and chemical properties of the matter.

A **physical property** is one that can be measured, studied, and tested without changing what kind of matter is being studied. We have already learned about some physical properties including mass, volume, density, and buoyancy. Other physical properties include color, texture, and hardness. All of these characteristics can be measured or observed without changing one substance into another substance.

This does not mean that the matter does not change form. A substance can experience a physical change without becoming a different substance. For example, when liquid water freezes it becomes ice but it is still water. Dry ice becomes a gas as it warms up but it is still carbon dioxide. Also, the shape of a substance can be changed without changing what substance it is, like when a rubber band stretches. It has a different shape, but it is still a rubber band. If a sugar cube is ground into powered sugar it has a different physical form, but it is still sugar. All of these changes in shape and state are physical changes and are one way that chemists study matter.

But another way that chemists study matter is by testing its chemical properties. **Chemical properties** describe how a substance reacts in the presence of other substances to form new substances. When you put vinegar (acetic acid) and baking soda (sodium bicarbonate) together, it foams and produces gas bubbles. This is a chemical property of the vinegar and soda. They combine to form

Sugar comes in many forms, but it is still sugar.

LESSON 10   **Properties of Matter • 47**

Physical & chemical changes

Purpose: To do some experiments to see if we can determine whether a physical change or chemical change is taking place

Materials: ice, sauce pan, stove, salt, lemon juice, baking soda, cup

Procedure:

1. Place several pieces of ice in a sauce pan and heat the pan over low heat. What do you observe happening to the ice? What did you add to the ice to cause the change to occur? When you added heat to the ice you observed the ice melting. If you add enough heat you will observe the water boiling and turning to steam. Is this a physical change or a chemical change? How do you know?

2. Remove the pan from the stove and add 1 tablespoon of salt. What do you observe happening? You should see the salt "disappear" into the water. Is this a physical or a chemical change?

3. Place the pan back on the stove and heat the water until it all boils away. Be careful not to burn the pan. What do you observe in the bottom of the pan? You should see the salt left in the bottom of the pan. Was this a chemical or a physical change? How do you know?

4. Finally, pour 2 tablespoons of lemon juice in a cup and then add 1 teaspoon of baking soda to the juice. What do you observe happening? You should observe many bubbles forming. Is this a physical or chemical change? How do you know?

Conclusion: Identifying a physical characteristic is usually easy, but identifying if a change is a physical or chemical change can sometimes be tricky. For the most part there are a few simple ways to determine if a change is physical or chemical. First, you need to identify if more than one type of material is being combined together. If there is only one type of material, such as in the ice melting experiment, then the change is probably a physical change. Also, if the change is easily undone, then it is likely to be a physical change. For example, the melted ice can be frozen again so that change is easily undone. However, in the lemon juice and baking soda experiment, the gas that escaped could not easily be changed back into lemon juice and baking soda. This indicates a chemical change.

Chemical changes are often identified because of the generation of something in a different state such as the gas generated by the lemon juice and baking soda. Also, chemical changes can be indicated by color changes, emission of light or sound, and releasing of heat. These are not hard and fast rules, but are good indicators of chemical changes.

carbon dioxide, water, and sodium acetate. Chemical reactions change the identity of the substances.

One of the most important chemical changes designed by God is photosynthesis. In photosynthesis, water and carbon dioxide are changed by sunlight into sugar and oxygen through a chemical reaction. This chemical change provides food for nearly every living thing on Earth.

So become a good chemist and learn to detect chemical and physical changes in the matter around you.

What did we learn?

- What are some physical properties of matter?
- What is a chemical change?
- Give an example of a chemical change.

Taking it further

- How can you determine if a change in matter is a physical change or a chemical change?
- Diamond and quartz appear to have very similar physical properties. They are both clear crystalline substances. However, diamond is much harder than quartz. How would this affect each one's effectiveness as tips for drill bits?

Physical or chemical?

Test your understanding of chemical and physical characteristics by completing the "Physical or Chemical Properties" worksheet.

48 • Properties of Matter   LESSON 10

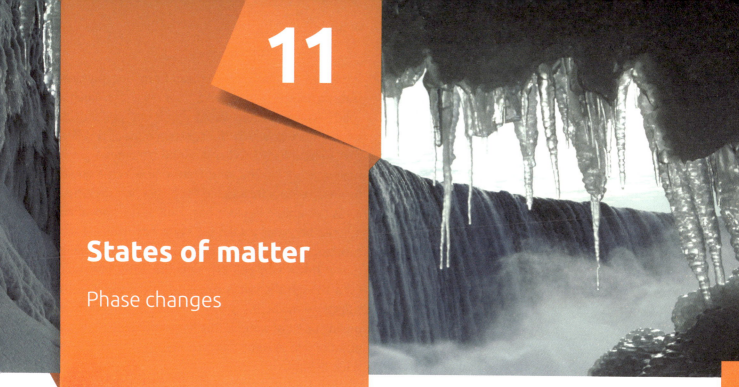

11

States of matter

Phase changes

What are the different states of matter?

Words to know:

solid
kinetic energy
liquid
gas
phase change
melt

melting point
evaporate
boiling point
condense
frozen/freeze
sublimation

All matter has physical and chemical properties. One physical property of a substance is its physical state. When we speak of physical state we are talking about whether the substance is a gas, a liquid, or a solid. A substance's physical state is determined by how tightly its molecules cling to each other and the amount of energy necessary to make the molecules move apart.

In a **solid**, the molecules are packed tightly together and do not move very much. They are strongly attracted to each other. As energy, usually in the form of heat, is added to a substance, the molecules begin to move; their kinetic energy increases. **Kinetic energy** is the energy of any moving object. As more heat is added, the molecules eventually gain enough energy to move away from each other. When the forces pulling them together equal the forces pushing them apart, the substance changes from a solid to a **liquid**. The molecules in a liquid flow over each other and move around freely. Yet they stay close together. They have more kinetic energy than they had when they were in the solid state.

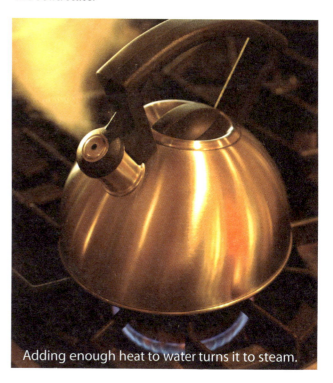

Adding enough heat to water turns it to steam.

LESSON 11   Properties of Matter • 49

Steam locomotives boil water to make steam.

If more energy is added to the liquid, the molecules move even faster. Eventually they gain enough energy to break free of the other molecules and become a gas. In a **gas**, the molecules are far apart from each other and move very quickly.

Similarly, if molecules lose energy, if heat is transferred to something else like cooler air, the molecules will slow down. A gas will become a liquid or a liquid will become a solid when it loses enough energy for the molecules to become attracted to each other again.

When a substance changes from one physical state to another it is called a **phase change**. When a solid becomes a liquid we say that is has **melted**. The temperature at which the substance melts is called the **melting point**. When a liquid becomes a gas it has **evaporated**. The temperature at which a liquid begins to evaporate is called the **boiling point**. When a gas becomes a liquid it has **condensed**, and when a liquid becomes a solid we say it has **frozen**. Each substance has a unique melting point and boiling point. This is why you can separate water from salt. Even though the salt is dissolved in the water, each substance maintains is own physical properties. The boiling point of water is much lower than the boiling point of salt, so water can be boiled away and the salt is left behind.

Some substances can change directly from a solid into a gas or directly from a gas into a solid without going through a liquid phase. This phase change is called **sublimation**. Two of the most common substances that experience sublimation are carbon dioxide and mothballs. Dry ice, shown on next page, is frozen carbon dioxide. When it is placed in a warm environment it quickly sublimates directly into a gas without leaving any liquid behind. This makes it very

🧪 Observing phase changes

Purpose: To observe phase changes

Materials: ice, saucepan, stove, hand mirror, ice tray, freezer

Procedure:

1. Observe a piece of ice. How does it feel?
2. Place a few pieces of ice in a small saucepan and melt it over medium heat just until most of the ice is melted.
3. Remove the pan from the heat and observe the liquid water. How does the liquid compare to the solid?
4. Return the pan to the stove on medium heat and watch as the water begins to boil. Be careful not to put your hand in the steam; it could burn. What did you notice as the water began to boil?
5. Place a hand mirror in the steam and watch as some of the steam condenses on the mirror. How does the water on the mirror feel?
6. Remove the pan from the stove.
7. Pour the water into an ice tray and place the tray in the freezer.
8. After 1–2 hours observe the water again. How does the water look and feel now?

Conclusion: You have now observed the phase changes of water. Review the names of each phase change:

Solid to Liquid – Melting

Liquid to Gas – Evaporation

Gas to Liquid – Condensation

Liquid to Solid – Freezing

You did not observe sublimation; the changing of a solid directly to gas or a gas directly to a solid.

popular in the food industry for keeping foods frozen while being transported.

The vast majority of substances on Earth are solids at normal temperature and pressure. Water is one very important substance that occurs in all three states, depending on the weather. And several substances, including oxygen and nitrogen, naturally occur as gases on Earth. The fact that the earth is made of substances that are solid is another example of God's provision for life on Earth. Many of the planets in our solar system are made of hydrogen and other elements that are naturally gases and would not be able to support life as we know it, but Earth was designed perfectly for life.

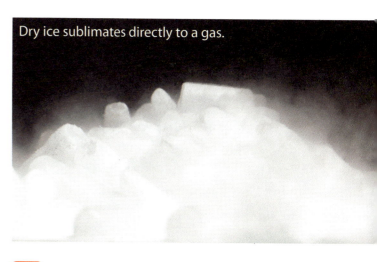

Dry ice sublimates directly to a gas.

What did we learn?

- What are the three physical states of most matter?
- What is the name for each phase change?
- What is required to bring about a phase change in a substance?

Taking it further

- Name several substances that are solid at room temperature.
- Name several substances that are liquid at room temperature.
- Name several substances that are gas at room temperature.

Water density

Kinetic energy plays a vital role in determining the phase of a substance. As you learned, when molecules have very little kinetic energy, they are closely attracted to each other and form a solid. As molecules gain energy they begin to move apart from each other and become a liquid. When enough energy is added to a substance the molecules break completely free of each other and become a gas.

With this in mind, would you expect a solid to be more or less dense than a liquid of the same substance? Recall that density is mass per volume. You would expect that since the molecules of a solid are more attracted to each other than the molecules in a liquid, the solid would have a higher density than the liquid of the same substance. In nearly every case you would be right. Liquid gold is not as dense as solid gold. Liquid carbon dioxide is not as dense as solid carbon dioxide. Almost every known substance is denser as a solid than as a liquid. There is one important exception.

Do you know which substance becomes less dense when it freezes than when it is a liquid? Water is one of the few substances that expands as it

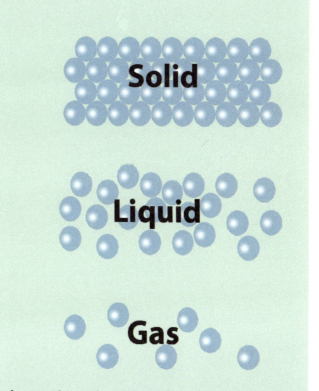

freezes. This is why ice cubes float in a glass of water; they are less dense than the liquid water.

LESSON 11 **Properties of Matter • 51**

Purpose: To demonstrate the density of ice

Materials: glass jar, water, marker, freezer

Procedure:

1. Fill a glass jar half full of water.
2. Use a marker to mark the water level on the outside of the jar. Do not place a lid on the jar.
3. Place the jar in the freezer for several hours.
4. After the water has frozen, compare the level of the ice with the level of the water. You will see that the water has expanded and has become a solid.
5. Now place the jar of ice on a counter and allow the ice to melt. You will see the level of water and ice go down as the ice becomes liquid water again. Since the liquid water takes up less space than the ice, its volume is less for the same amount of mass so its density is greater.

Conclusion: As mentioned before, this unusual characteristic of water is vitally important to keeping everything alive in lakes and ponds during the winter. But the fact that water expands as it freezes has other effects as well. Water flows into cracks in rocks, buildings, and sidewalks. As it freezes it pushes the cracks wider. This allows more water to flow in when it warms up and more water freezes the next time. After several cycles of freezing and melting, rocks can be broken and sidewalks and buildings may need to be repaired. This freezing and thawing cycle also affects farmers' fields. Water flows into the ground. As it freezes it expands. If the water is under a rock, it pushes the rock up slightly. When the water melts more water flows under the rock. When it freezes it pushes the rock up a little more. After weeks or months of freezing and thawing, many rocks are pushed up to the surface of the field and must be removed before the next crop is planted. The power of freezing water is great. If you had filled your jar completely full of water and placed the lid on it, the jar would have been broken by the force of the expanding water.

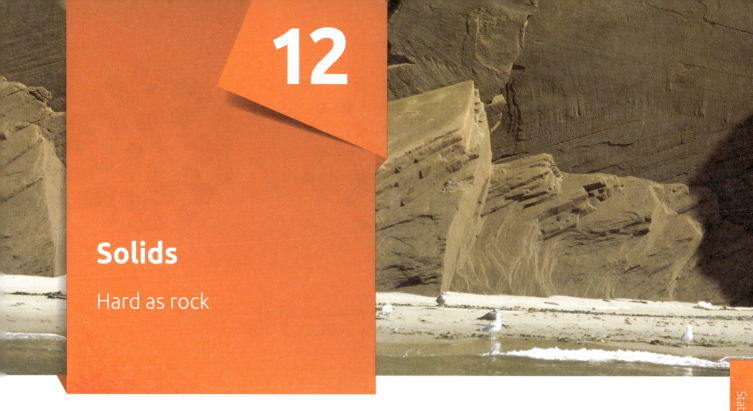

12

Solids

Hard as rock

What are the properties of a solid?

Challenge words:

crystalline solid   amorphous solid

Matter on Earth almost always exists in one of three states: solid, liquid, or gas. When a substance is a solid, its molecules have relatively low kinetic energy compared to other phases, and the molecules are strongly attracted to each other. Because of this the molecules are not free to move around, but this does not mean they are not moving at all. All matter moves. But in a solid the molecules are only able to vibrate. They vibrate so quickly that we cannot actually see or feel them moving.

There are several other characteristics that are common to all solids. If you pick up something, how do you know if it is a solid, a liquid, or a gas? You might say it is a solid if it is hard or if it is heavy. However, not all solids are hard and not all solids are heavy. A piece of paper is a solid but it is not hard or heavy. So what makes a solid different from a liquid or a gas?

First, solids are substances that have a definite shape. Unless an outside force causes the shape to change, a solid will stay the same shape. Even a rubber band, which can be stretched, will keep the same shape if it is left alone.

Second, solids have a definite volume. They take up the same amount of room or space. Your desk does not take up more space today than it did yesterday. A book has the same length, width, and height all the time, so its volume does not change.

Third, solids have a high density. Because the molecules are packed closely together there is more mass in a given volume in a solid than in a liquid or a gas. (Water is an exception.) Different types of solids have different densities as we saw in lesson 8; however, solids are almost always more dense than liquids and gases. And many items that appear to have low densities, like a ping-pong ball, are actually filled with a gas such as air, making them less dense.

Solids are formed when a liquid is cooled. Often, if a liquid is cooled slowly the molecules have time to line up in regular patterns and form crystals. If a liquid is cooled quickly, the molecules are frozen in random order so crystals cannot form. Because of this, some solids have different characteristics even if they are made of the same substance. For example, graphite and diamond are both solids comprised of carbon atoms. Graphite is formed when carbon atoms cool slowly. It is a very soft slippery solid because the carbon atoms align in sheets, that slip past one

LESSON 12 **Properties of Matter** • 53

Diamond crystal

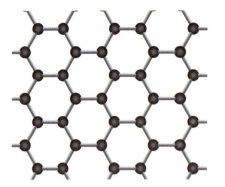
Graphene sheet

Buckyball

another. Diamond is formed when carbon atoms bond together under elevated temperatures and pressures, and then are brought to normal temperatures rapidly. The carbon atoms in diamond align in a lattice pattern, making diamond the hardest substance on Earth. In recent years researchers have discovered other forms of solid carbon as well. Graphene is composed of sheets of carbon that are one atom thick and arranged in hexagon shapes. Fullerenes, nicknamed Buckyballs, are spheres of carbon that are hexagons connected together to form balls in a similar way to soccer balls. All of these are forms of solid carbon, yet they have different properties because their atoms line up in different ways. Similarly, crystallized sulfur is a hard rock-like substance, but sulfur that is cooled quickly forms a rubbery substance.

The majority of elements, the pure building blocks that all other substances on Earth are made from, are solids. Only two elements, mercury and bromine, are liquids at room temperature. Eleven elements are gases, and all the other elements are solid at room temperature. So look around and you will find solids everywhere.

What did we learn?

- What are three characteristics of solids?
- How do large crystals form in solids?
- What state is the most common for the basic elements?

Taking it further

- Is gelatin a solid or a liquid?

Testing for solids

Purpose: To identify which substances are solids and which are not

Materials: a wooden block, a rock, a metal spoon, Silly Putty®, some honey

Procedure: For each substance ask the following questions.

1. Does it take up the same amount of room if you put it in another container?
2. Is it denser than air, or does it float in air?
3. Does its shape stay the same if I move it or put it in another container?
4. Is it a solid?

Conclusion: You probably had no problem determining that honey is a liquid. But Silly Putty® is trickier. The putty is such a thick liquid that it takes a long time for it to flow and change its shape. Nevertheless, Silly Putty® is not a solid because its shape will change over time.

Glass: solid or liquid?

Based on what you have learned about solids, would you describe glass as a solid or a liquid? Glass appears to hold its shape. The pane of glass in a window stays in place and does not flow out of the frame. Things made of glass have a definite volume. If the window pane does not move or change shape, then its volume does not change either. Finally, glass is dense. Because of its flat shape, you may be able to get a flat piece of glass to float on the surface of the water, but if a marble is made of glass or the glass is made into some other shape, it is dense enough to sink. Therefore, we would classify glass as a solid.

However, there are many sources that would classify glass as a liquid. Why the controversy? Glass has a few characteristics that do not fit well with other solids. First, glass does not have a specific melting point. Ice always melts at the same temperature,: 0 degrees Celsius (at sea level). Other solids also have specified melting points, but different samples of glass can melt at different temperatures.

Glass also does not have a crystalline structure. Most solids are **crystalline solids**. This means that their molecules line up in regular patterns that repeat. Pure metals form regular patterns and are thus crystalline solids. Water forms crystals as it freezes. Sometimes these crystals take on the shape of beautiful snowflakes and other times these crystals form large blocks of ice, but on a molecular level, the molecules are lined up in regular patterns. Glass molecules, however, are totally random. A solid with random molecules is called an **amorphous solid**. So if glass is a solid, it is an amorphous solid.

Finally, samples of very old glass, such as glass in thousand-year-old cathedrals, are thicker at the bottom than at the top. Many people believe this

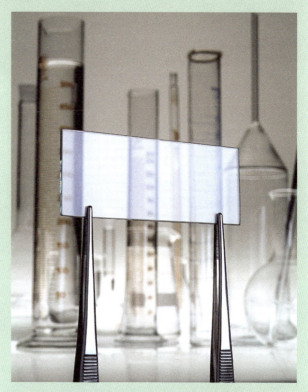

shows that glass flows very slowly and is thus a very thick liquid and not a solid. Other people believe that the glass in these cathedrals is thicker at the bottom because of the process used to make the glass. They believe that the panes of glass were thicker at the bottom when they were installed and not because they have flowed over time. No one really knows for sure which is true.

If you look in different science books and read many articles about glass, you find that some people say that glass is a solid, others call it an amorphous solid. Some people call it a liquid or a super-cooled liquid. There does not seem to be a consensus on whether glass is a solid or not. Do some research on this topic. Do you think glass is a solid or a liquid?

13

Liquids

Can you pour it?

What are the properties of a liquid?

Words to know:

viscosity

We have seen that matter can take on one of three different states: solid, liquid, or gas. When a substance has enough energy that the molecules can easily slide over one another, but not enough energy to easily move away from one another, it is called a liquid. The molecules are held close to each other, yet they move in a random order. A liquid has more kinetic energy than a solid of the same substance.

Because the molecules in a liquid can move, a liquid does not have a definite shape like a solid does. A liquid will take on the shape of its container. Also, a liquid has a definite volume that can be measured. The molecules are close together so liquids are more dense than gases. Liquid water is denser than ice, but in general, liquids are less dense than solids, and gases are much less dense than liquids and solids.

Have you noticed that some liquids are thicker than others? If the molecules have a strong attraction for each other, they do not flow easily. A thick liquid is said to be viscous or to have high viscosity. **Viscosity** is a measurement of how strongly a liquid's molecules are attracted to each other. When we think of liquids we usually think of water. Water is the most common liquid on Earth. Water has a relatively low viscosity. It flows easily. You can easily pour it from one container to another. What other liquids can you think of that have a low viscosity? Milk, soft drinks, and many other drinks are mostly water and therefore have a low viscosity. Other liquids have a high viscosity as you will see in the following activity.

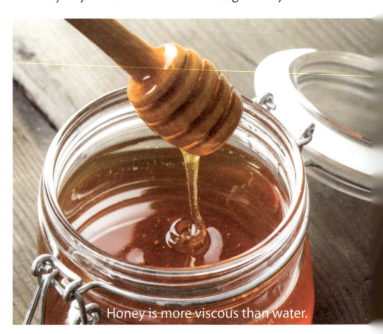

Honey is more viscous than water.

What did we learn?

- Which has more kinetic energy, a solid or a liquid?
- What shape does a liquid have?
- What is viscosity?

Taking it further

- How is a liquid similar to a solid?
- How is a liquid different from a solid?
- How would you change a solid into a liquid?

Observing viscosity

Complete the "Viscosity" worksheet.

Evaporation

When you think of a liquid changing into a gas, you generally think of adding heat so that the molecules gain enough energy to break away from each other. However, if you spill water on the counter and do not clean it up, the water will eventually go away. What happened to the water molecules? They changed from a liquid into a gas, but the water was not heated to the boiling point.

When a liquid changes into a gas it is called evaporation. Evaporation occurs very quickly when a liquid reaches its boiling point; however, evaporation can take place even when the liquid is not near its boiling point. Not all molecules in a liquid have the exact same amount of kinetic energy. Some molecules have more energy than others. When molecules with higher energy reach the surface of the liquid, they can escape into the air and become gas molecules—they evaporate.

Evaporation occurs from the surface of every lake, river, and ocean and is what keeps the water cycle going. Evaporation also occurs when your body sweats. The liquid on your skin evaporates into the air. As it does, it uses energy to change from liquid to a gas. This cools down your body.

The rate of evaporation is affected by three main things. First, the temperature of the liquid affects its evaporation rate. Do you think increasing the temperature will increase or decrease the rate of evaporation? Since it takes energy to change a liquid into a gas, increasing the temperature will increase the rate of evaporation. This is why it is

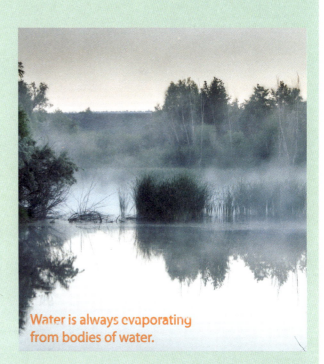

Water is always evaporating from bodies of water.

better to water your lawn early in the morning than in the middle of the afternoon. In the morning the air is cooler and less water will evaporate and more water will stay on the ground to be used by your grass than in the heat of the day.

Evaporation rate is also affected by the humidity of the air above the surface of the liquid. Do you expect a liquid to evaporate more quickly when the air is humid or when the air is dry? When there is less water vapor in the air, the water molecules have an easier time moving from the surface of the liquid into

LESSON 13 **Properties of Matter** • 57

the air because there are fewer water vapor molecules for them to collide with. If wind is blowing across the surface of a lake much more water will evaporate than if the air is still.

Finally, surface area affects the evaporation rate. Do you think that more water will evaporate from a tall narrow cup or from a short wide cup? The greater the surface area, the more opportunities molecules have to escape so water will evaporate more quickly from a wide cup than from a narrow cup.

Purpose: To feel the effects of wind on evaporation

Materials: cup of water

Procedure:

1. Dip your index finger from each hand into a cup of water.
2. Hold one finger in front of your mouth and gently blow on your finger.
3. At the same time, hold the other finger up but away from your mouth.

Questions:

- Which finger becomes dry faster?
- How does the finger you blow on feel compared to the finger you did not blow on?

Conclusion: The finger you blow on feels cool because energy is being used to evaporate the water on it.

14

Gases

Lighter than air?

What are the properties of a gas?

Words to know:

gas pressure   atmospheric pressure

Challenge words:

diffusion

What is the most important gas you can think of? You probably said air or oxygen. If you don't have air to breathe you will quickly die. God designed our atmosphere with just the right amount of oxygen to support all life on the planet. We know that air is made up of gases, but what exactly is a gas?

A substance becomes a gas when it has enough energy that the molecules can freely move away from each other. These molecules have overcome the attractive forces that were holding them close to each other. Because the molecules in a gas do not attract each other, a gas does not have a definite shape. The molecules move randomly to fill whatever container the gas is in. Also, for this reason, the gas does not have a definite volume. Its volume expands to fill its container. Because the gas molecules move randomly, they become evenly distributed within the container.

Because it takes a large amount of energy to overcome the attractive forces of most molecules, gas molecules move at very high speeds. They have very high kinetic energy. Air molecules are usually moving at about 1,100 miles per hour (490 meters per second). This is fast enough to cross the United States in about three hours! However, the molecules do not move all the way across the United States because they move in a straight line until they hit something. When they hit objects they change their direction. Air molecules experience millions of collisions per second. They collide with other air molecules and they collide with any object in their path. This means that billions of atoms hit each square centimeter of everything on Earth each second.

Inside a container, the air molecules collide with each other and with the inside of the container. If only a few molecules were hitting the inside of the container no one would ever notice it. But because there are billions of molecules inside a container such as a ball, the constant collisions against the inside of the ball create pressure on the inside of the ball. This is called **gas pressure**. Fortunately, there is gas on the outside of the ball that is also pushing against the ball. This is called **atmospheric pressure**. At sea level, the amount of pressure applied by the air to one

LESSON 14 **Properties of Matter** • 59

Observing air pressure

Why does a ball bounce? The air molecules inside the ball are constantly moving and pushing outward. When the ball hits the floor these molecules push against the floor causing the ball to move upward. When a ball is "flat" it does not have as many air molecules inside as when it is "pumped up." A flat ball does not bounce as well because there are fewer molecules to push back against the floor when you drop it. Another thing that can affect how well a ball bounces is temperature. Do you think gas molecules will move faster when they are warm or when they are cold? You are going to do an experiment to see how temperature affects the way a ball bounces.

Purpose: To see how temperature affects air pressure

Materials: two tennis balls, freezer

Procedure:

1. Place one tennis ball in the freezer for at least 30 minutes, and leave the other one at room temperature.
2. Make a guess as to which ball will bounce higher: a warm tennis ball or a cold tennis ball.
3. Remove the tennis ball from the freezer.
4. Drop both balls together from the same height onto the floor.

Questions:

- Which ball bounced highest? Why?

Conclusion: Warm gas molecules have more energy and are moving more quickly than cold molecules. This means that the air molecules inside the warm tennis ball are hitting the inside of the ball more frequently than the colder ones. This causes the air pressure inside the warm ball to be higher. Thus more molecules will push against the floor at any one time when you drop the ball so it bounces higher than the colder ball.

---

square centimeter is called one atmosphere. As you go up in altitude the air molecules are farther apart so there are fewer molecules per square inch and the atmospheric pressure is lower.

Even though the air all around us is constantly pressing against our bodies we do not notice it. God designed our bodies perfectly for our environment. Our bodies naturally exert an outward pressure equal to about 1 atmosphere. Our bodies are also designed to adjust to changes in atmospheric pressure. Have you ever gone up in the mountains or on an airplane and had your ears "pop"? That was your body adjusting to a change in pressure. Changes in air pressure cause the air pocket inside the ear to expand when the plane climbs and contract when it descends. God created the air we need to breathe and designed our bodies to adapt to the air pressure. ✻

What did we learn?

- When is a substance called a gas?
- What is the shape of a gas?
- In which state of matter are the molecules moving the fastest?
- What is atmospheric pressure?

Taking it further

- How is a gas similar to a liquid?
- How is a gas different from a liquid?
- Why is it necessary that a spacesuit be pressurized in outer space?

Diffusion

Gas molecules are free to move about. This movement causes gas molecules to mix with each other. The molecules move from an area of higher concentration to an area of lower concentration. This movement is called **diffusion**.

Your body uses diffusion to gain the oxygen that it needs. We already mentioned that you get oxygen from the air you breathe, but how does the oxygen get from the air into your body? When you inhale a breath of air it contains about 21% oxygen. The amount of oxygen in your blood as it flows through your lungs is much lower, so the oxygen molecules in the air diffuse into your blood, moving from an area of higher concentration to an area of lower concentration. Similarly, the carbon dioxide gas in your blood is at a much higher concentration than the carbon dioxide in the air so the carbon dioxide diffuses out of your blood into the air. This is God's design for providing you a way to obtain oxygen and get rid of carbon dioxide.

Another example of diffusion is the movement of scent particles in flowers. The scent particles are very light gas molecules. The concentration of the scent molecules is much higher in the flower than in the air so the scent molecules move from the flower into the air. When some of those particles enter your nose, you can detect the scent.

Purpose: To demonstrate the diffusion of scent molecules in the air

Materials: perfume, cup, step ladder

Procedure:

1. Pour a small amount of perfume into a small cup.
2. You are going to place the cup in a small closet, but before you do, smell the air in the closet. Do you detect the smell of perfume? You should not be able to smell any perfume in the closet.
3. Place the cup with the perfume on the floor of the closet and close the door.
4. After 5 minutes open the door.
5. Stand on a step ladder and smell the air high up in the closet, then smell the air near the floor. Which air has the stronger scent of perfume?
6. Close the closet door and wait 5 more minutes.
7. Again, smell the air in various parts of the closet. Is it more difficult to tell a difference in the air at the top of the closet from the air at the bottom? If there is still a strong difference in the smell at the top of the closet and the bottom of the closet, close the door and wait 1 hour and smell the air again.

Conclusion: The scent molecules will diffuse through the air molecules in the closet, moving from the area of higher concentration near the floor to lower concentration near the ceiling. Many air fresheners use diffusion to spread their scents throughout a room.

15

Gas Laws

Rules to live by

How do gases behave?

Words to know:

Boyle's law
Charles's law

Understanding how matter behaves under different conditions has always been a goal of scientists. Many scientists make observations and conduct experiments to try to predict what will happen under various conditions. They cool materials down or heat them up to see how they react. They put them under pressure and see what happens. They combine different substances together to try to make new materials. All of these experiments have revealed important properties and have led to many important discoveries. Two of the most important discoveries related to gases are called gas laws because they describe how all gases behave, no matter what the gas is made of.

The first gas law was discovered in 1660 by a scientist named Robert Boyle and is called **Boyle's law**. Boyle discovered that as pressure on a gas increases, the volume of the gas decreases. This should make sense because when you increase the pressure, the gas molecules are pushed closer together so they take up less room. Conversely, as the pressure goes down, the volume of the gas goes up. He also discovered that this relationship is proportional. This is true because the gas molecules are free to move farther apart when there is lower pressure so they can fill a greater space. That means that if the pressure is cut in half the volume doubles and if the pressure is three times what is was initially, the volume will be one-third of its original volume. This relationship between the volume of a gas and the pressure is true for all gases, as long as the temperature of the gas does not change.

The temperature of a gas can affect its volume as well. This relationship was studied about 100 years later by a scientist named Jacque Charles and

The temperature of a gas can affect its volume.

is called **Charles's law**. Charles discovered that the higher the temperature, the greater the volume of the gas. As the temperature is increased the molecules have more energy and can move faster. If the pressure remains the same, these more energetic molecules can push outward with more force and thus take up more space. If the temperature of the gas doubles, the volume of the gas will double as well. You saw the effect of this relationship in the last lesson when you tested how high a cold ball would bounce. As the temperature of the gas decreases, the molecules have less energy so they do not move around as much and will not take up as much room. This is true only if the pressure on the gas remains the same.

So you see that pressure and temperature both affect the volume of a gas. The gas laws say that if temperature does not change, the volume goes up as the pressure goes down; and if the pressure does not change, the volume goes up as the temperature goes up.

Fun Fact

Weather balloons are very large helium-filled balloons that are used to carry weather instruments through the atmosphere to measure such things as temperature, pressure, and wind at high altitudes. When a weather balloon is released, it is only partially inflated. This is because as the balloon rises the pressure drops and the volume of gas expands inside the balloon. If the balloon were fully inflated before it was released, it would explode at a lower altitude and the weatherman would not get the needed data.

What did we learn?

- If temperature remains constant, what happens to the volume of a gas when the pressure is increased?
- If pressure remains constant, what happens to the volume of a gas when the temperature is increased?
- What are two different ways to increase the volume of a gas?

Taking it further

- Why might you need to check the air in your bike tires before you go for a ride on a cold day?
- Why do you think increasing pressure decreases the volume of a gas?
- Why do you think increasing temperature increases the volume of a gas?
- What might happen to the volume of a gas when the pressure is increased and the temperature is increased at the same time?

LESSON 15 **Properties of Matter** • 63

Hot and cold gas

It is somewhat difficult to test the effects of changing pressure on gas at home because changing the pressure requires a vacuum chamber. However, it is very easy to change the temperature of the gas while keeping the pressure the same and see the change in volume. You can test this two different ways.

Purpose: To observe the effect of temperature on gas volume

Materials: balloon, cloth tape measure, freezer, empty plastic gallon milk carton, microwave oven, oven mitts

Activity 1—Procedure:

1. Fill a balloon with air and tie the end of the balloon.
2. Measure and record the circumference of the balloon.
3. Place the balloon in the freezer for 15 minutes.
4. Remove the cold balloon from the freezer and measure its circumference.

Conclusion: The circumference of the balloon should be smaller because the air molecules are cooler and therefore take up less room inside the balloon.

Activity 2—Procedure:

1. Fill a sink half full with cold water.
2. Place ½ cup of water inside an empty plastic gallon milk carton. Do not place the cap on the carton.
3. Place the carton in a microwave oven and heat for 30 seconds.
4. Use oven mitts to remove the carton from the microwave oven.
5. Immediately place the cap on the carton and set the carton in the sink of cold water.

Conclusion: As the gas cools down it will have less volume and the milk carton will "shrink."

Charles's law

Recall that Charles's law states that as the temperature of a gas increases the volume also increases and as the temperature decreases the volume of the gas decreases. You can demonstrate Charles's law with another simple experiment.

Purpose: To demonstrate Charles's law

Materials: dish soap, two small bowls, small plastic bottle

Procedure:

1. Mix one teaspoon of dish soap with two teaspoons of water to make a bubble solution.
2. Fill one small bowl with hot water and another small bowl with cold water.
3. Dip the mouth of a small plastic bottle in the bubble solution to form a film across the mouth of the bottle.
4. Place your hands around the bottle but do not squeeze. What do you see happening at the mouth of the bottle? You should see a bubble forming. Why does a bubble form?
5. For the remaining parts of the experiment, dip the bottle in the bubble solution as often as necessary to replace the soap film across the mouth of the bottle.
6. Next, place the bottle in the bowl of hot water. What happens? Why?
7. Move the bottle to the cold water. What do you see happening? Why does this happen?

Conclusion: As the air inside the bottle heats up, either from the heat in your hands or from the heat in the water, the volume of the air expands. This expanding air pushes on the soap film and forms a bubble. As the air inside the bottle cools, the volume of the air decreases and sucks the soap film into the bottle.

Robert Boyle
1627–1691

SPECIAL FEATURE

Do you have fourteen brothers and sisters? Robert Boyle, a very famous chemist, did. Robert was the fourteenth of fifteen children in his family. He was born in Munster, Ireland in 1627. Robert's father was the Earl of Cork and he was not only blessed with many children, but he was considered the wealthiest man in all of Great Britain.

The Boyle family had a deep faith in God and Robert learned to spend time every day reading his Bible. He was a devout Christian. In fact, throughout his life, Boyle was offered many important positions of leadership in the Anglican Church. However, he refused, believing that God had called him to other areas of ministry.

The most important area of ministry that Boyle had was the pursuit of biblical science. When Boyle began his scientific studies, chemistry as we know it did not exist. Instead, the study was called alchemy and it was a strange mix of science and mysticism. Robert Boyle, however, believed that there were definite scientific properties to elements that did not need mystical explanations. With this in mind, he performed many experiments. He was the first to define an element as a substance that cannot be broken down by ordinary chemical means. Boyle recognized numerous elements when many other people only recognized four types of matter. Boyle was also the first scientist to distinguish between mixtures and compounds. In fact, Boyle has been described as the leading chemist of the seventeenth century.

Boyle viewed his work in science as his ministry and God blessed his work. In many lectures, Boyle taught that Christianity could be defended intellectually and was more reasonable to believe than other philosophies of the day.

In addition to his scientific work, Robert Boyle was also a great supporter of world missions. He used his own wealth to support missionaries to Ireland, Scotland, Wales, India, and North America. He also paid to have the gospels and the Book of Acts translated into Turkish, Arabic, and Malay, and to have the entire Bible translated into Irish.

Robert Boyle is best remembered for his gas laws and other contributions to chemistry; however, Christians should also remember him for his service to God in defending Christianity to the scientific world of his day, and for his work in spreading the gospel around the world.

Properties of Matter

UNIT 4

Classifying Matter

16 Elements
17 Compounds
18 Water
19 Mixtures
20 Air
21 Milk & Cream

◊ **Distinguish** between elements, compounds, and mixtures using examples.
◊ **Identify** unique characteristics of water and air and their importance for life.
◊ **Describe** how pasteurization and homogenization are used to process milk.

66 • Properties of Matter

16

Elements

The basic building blocks

What are the basic building blocks of matter?

Words to know:

classification
element
atom
compound
mixture
diatomic molecule

In order to understand any complex subject, it is necessary to break it into smaller pieces. One way to do this is by classification. **Classification** is grouping things together according to their similar characteristics. For example, most living things are classified as either plants or animals. Animals are then classified as either vertebrates or invertebrates, depending on whether or not they have a backbone.

In a similar way, scientists have classified all matter by analyzing what it is made of and then grouping similar types of matter together. Matter is classified into three groups: elements, compounds, and mixtures. An **element** is defined as a substance that cannot be broken down into simpler substances by ordinary chemical means. Elements, also called **atoms**, are the simplest kinds of matter. Some common elements that you are familiar with are oxygen, helium, gold, silver, and carbon. All matter on Earth is made up of basic elements.

Compounds are substances that are formed when two or more elements combine chemically to make a new substance. Common compounds include water (a combination of oxygen and hydrogen) and carbon dioxide (a combination of oxygen and carbon). Many substances around you are compounds. Elements were designed to easily combine to form new and wonderful things. Salt, sugar, baking soda, gasoline, acids, plastic, and many other substances are compounds.

A **mixture** is formed when two or more substances are mixed together but do not chemically combine to form a different substance. Common mixtures include milk and salt water. When salt is dissolved in water the salt molecules spread out in the water but do not combine with the water molecules to form a new substance. Similarly, sweet tea is a mixture of water, tea, and sugar. These substances are all in the same cup and mixed up but do not react with each other. A bowl of cereal and milk is also a mixture. Both elements and compounds are considered pure substances, but mixtures are not pure because they contain more than one kind of substance.

There are 92 elements that occur naturally and more than 20 additional elements that have been made by man. All substances on Earth are made

LESSON 16   Properties of Matter • 67

from some combination of the elements. However, although everything is made from elements, elements rarely occur by themselves as single atoms in nature. Most atoms join with other atoms, either of the same kind or different kinds, to form molecules. For example, the air we breathe is mostly nitrogen and oxygen. However, it is very unlikely to find a single nitrogen atom or a single oxygen atom. Most nitrogen and oxygen atoms combine in pairs. Each molecule is made of two nitrogen or two oxygen atoms connected together. A molecule consisting of two atoms of the same element is called a **diatomic molecule**.

Each element has a unique symbol derived from its name. The symbol is usually the first letter or two of the name. Some symbols include O for oxygen, H for hydrogen, He for helium, and Si for silicon. A few symbols come from the Latin name

🧪 Classification exercise

When scientists first started trying to figure out what things were made of they had a difficult task. The ancient Greeks believed that everything was made of either fire, air, earth, or water. Today, we know that things are much more complex than just four elements.

Purpose: To appreciate the process of classification

Materials: jigsaw puzzle

Procedure:

1. Dump the pieces of a jigsaw puzzle in the middle of the table. This mixed up pile represents what early scientists understood about the basic building blocks of matter.

2. Put all of the edge pieces in one pile and all of the inside pieces in another pile. As scientists began to learn more about matter, they discovered that some types of matter had similar characteristics and began to group them.

3. Now sort the edge pieces into piles by their similar shapes. Some pieces may have an indent on each side, some may have no indents at all.

4. Sort the interior pieces by similar shapes such as the ones shown here. Your puzzle pieces may be shaped differently. Find common shapes and sort the pieces according to the common shapes. This is how scientists began classifying matter—by looking at what was the same and what was different.

5. Once you have the pieces sorted, with each shape representing an element, see if you can make a small picture with a few of the pieces from the different piles. This is like making a compound. A few elements put together in a certain way make a new substance, just as a few pieces put together in a certain way make a new shape or picture.

Conclusion: When you put the whole puzzle together, think about when God put all of the elements together and created the beautiful world we live in.

68 • Properties of Matter   LESSON 16

for the element. For example, Ag is the symbol for silver whose Latin name is *argentum* and Fe is the symbol for iron whose Latin name is *ferrum*.

Metal names often end in "um" or "ium," and nonmetals usually end in "n" or "ine." Many man-made elements are named for people or places such as Einsteinium or Californium. Newly discovered elements are called by their electron structures until they are given official names: pentium, herium, ununbium, ununtrium, etc.

We are very familiar with many of the elements around us. Nearly everyone knows what gold, silver, aluminum, oxygen, carbon, hydrogen, and helium are. These familiar building blocks are used to form every other substance around us. As we learn about each element, we can thank God for His wonderful design of each substance.

What did we learn?

- What is an element?
- What is a compound?
- What is a mixture?

Taking it further

- If a new element were discovered and named *newmaterialium*, would you expect it to be a metal or a nonmetal?
- Is salt an element, a compound, or a mixture?
- Is a soft drink an element, a compound, or a mixture?

Periodic table

In order to better understand the elements, scientists have placed them in a chart called the periodic table of the elements (see next page). The elements are arranged according to their atomic numbers. The atomic numbers are shown at the top of each box on the periodic table. The periodic table also lists an element's name, atomic mass, and symbol. Many periodic tables list additional information, but they all list these four items.

Elements have been divided into three main groups. The first group includes the elements that are metals. Most elements are metals and can be found on the left side and in the middle of the table. Metals on this periodic table are colored pink, purple, yellow, and light green. Most metals are solids and are good conductors of electricity.

On the right side of the table are the elements that are nonmetals. These are shown in light blue and dark blue. Hydrogen, colored red, is also a nonmetal even though it is placed on the far left side of the table. There are only a few nonmetals. Most of the nonmetals are gases at room temperature, but a few are liquids or solids. Nonmetals are poor conductors of electricity.

Separating the metals and nonmetals are seven elements that have some characteristics of metals and some characteristics of nonmetals. These elements have been given the name of *metalloids* and are sometimes referred to as semiconductors. The metalloids are dark green on the periodic table. The semiconducting properties of these elements make them ideal for computer and other electronic applications.

To learn more about the elements, complete the "Learning about Elements" worksheet.

Periodic Table of the Elements

Legend:
- Alkali metals
- Alkali-earth metals
- Transition metals
- Poor metals
- Metalloids
- Nonmetals
- Noble gases
- Hydrogen nonmetal

Key:
- Atomic number: 12
- Symbol: Mg
- Atomic Mass: 24.31
- Name: Magnesium
- Electron structure by energy level: 2,8,2

	IA	IIA	IIIB	IVB	VB	VIB	VIIB	VIII	VIII	VIII	IB	IIB	IIIA	IVA	VA	VIA	VIIA	VIIIA
1	1 **H** 1.008 Hydrogen 1																	2 **He** 4.0026 Helium 2
2	3 **Li** 6.941 Lithium 2,1	4 **Be** 9.012 Beryllium 2,2											5 **B** 10.81 Boron 2,3	6 **C** 12.01 Carbon 2,4	7 **N** 14.01 Nitrogen 2,5	8 **O** 16 Oxygen 2,6	9 **F** 19 Fluorine 2,7	10 **Ne** 20.18 Neon 2,8
3	11 **Na** 22.99 Sodium 2,8,1	12 **Mg** 24.31 Magnesium 2,8,2											13 **Al** 26.98 Aluminum 2,8,3	14 **Si** 28.09 Silicon 2,8,4	15 **P** 30.97 Phosphorus 2,8,5	16 **S** 32.07 Sulfur 2,8,6	17 **Cl** 35.45 Chlorine 2,8,7	18 **Ar** 39.95 Argon 2,8,8
4	19 **K** 39.1 Potassium 2,8,8,1	20 **Ca** 40.08 Calcium 2,8,8,2	21 **Sc** 44.96 Scandium 2,8,9,2	22 **Ti** 47.9 Titanium 2,8,10,2	23 **V** 50.94 Vanadium 2,8,11,2	24 **Cr** 52 Chromium 2,8,13,1	25 **Mn** 54.94 Manganese 2,8,13,2	26 **Fe** 55.85 Iron 2,8,14,2	27 **Co** 58.93 Cobalt 2,8,15,2	28 **Ni** 58.69 Nickel 2,8,16,2	29 **Cu** 63.55 Copper 2,8,18,1	30 **Zn** 65.39 Zinc 2,8,18,2	31 **Ga** 69.72 Gallium 2,8,18,3	32 **Ge** 72.59 Germanium 2,8,18,4	33 **As** 74.92 Arsenic 2,8,18,5	34 **Se** 78.96 Selenium 2,8,18,6	35 **Br** 79.9 Bromine 2,8,18,7	36 **Kr** 83.8 Krypton 2,8,18,8
5	37 **Rb** 85.47 Rubidium 2,8,18,8,1	38 **Sr** 87.62 Strontium 2,8,18,8,2	39 **Y** 88.91 Yttrium 2,8,18,9,2	40 **Zr** 91.22 Zirconium 2,8,18,10,2	41 **Nb** 92.91 Niobium 2,8,18,12,1	42 **Mo** 95.94 Molybdenum 2,8,18,13,1	43 **Tc** 99 Technetium 2,8,18,14,1	44 **Ru** 101.1 Ruthenium 2,8,18,15,1	45 **Rh** 102.9 Rhodium 2,8,18,16,1	46 **Pd** 106.4 Palladium 2,8,18,17,1	47 **Ag** 107.9 Silver 2,8,18,18,1	48 **Cd** 112.4 Cadmium 2,8,18,18,2	49 **In** 114.8 Indium 2,8,18,18,3	50 **Sn** 118.7 Tin 2,8,18,18,4	51 **Sb** 121.8 Antimony 2,8,18,18,5	52 **Te** 127.6 Tellurium 2,8,18,18,6	53 **I** 126.9 Iodine 2,8,18,18,7	54 **Xe** 131.3 Xenon 2,8,18,18,8
6	55 **Cs** 132.9 Cesium -18,18,8,1	56 **Ba** 137.3 Barium -18,18,8,2	57 **La** 138.9 Lanthanum -18,18,9,2	72 **Hf** 178.5 Hafnium -18,32,10,2	73 **Ta** 180.9 Tantalum -18,32,11,2	74 **W** 183.9 Tungsten -18,32,12,2	75 **Re** 186.2 Rhenium -18,32,13,2	76 **Os** 190.2 Osmium -18,32,14,2	77 **Ir** 192.2 Iridium -18,32,15,2	78 **Pt** 195.1 Platinum -18,32,17,1	79 **Au** 197 Gold -18,32,18,1	80 **Hg** 200.5 Mercury -18,32,18,2	81 **Tl** 204.4 Thallium -18,32,18,3	82 **Pb** 207.2 Lead -18,32,18,4	83 **Bi** 209 Bismuth -18,32,18,5	84 **Po** (209) Polonium -18,32,18,6	85 **At** (210) Astatine -18,32,18,7	86 **Rn** (222) Radon -18,32,18,8
7	87 **Fr** (223) Francium -18,32,18,8,1	88 **Ra** (226) Radium -18,32,18,8,2	89 **Ac** (227) Actinium -18,32,18,9,2	104 **Rf** (261) Rutherfordium	105 **Db** (262) Dubnium	106 **Sg** 262.94 Seaborgium	107 **Bh** (264) Bohrium	108 **Hs** (265) Hassium	109 **Mt** (266) Meitnerium	110 **Ds** (271) Darmstadtium	111 **Rg** (280) Roentgenium	112 **Cn** (285) Copernicium	113 **Nh** (284) Nihonium	114 **Fl** (289) Flerovium	115 **Mc** (289) Moscovium	116 **Lv** (293) Livermorium	117 **Ts** (294) Tennessine	118 **Og** (294) Oganesson

Lanthanides (row 6):

58 **Ce** 140.1 Cerium -18,20,8,2	59 **Pr** 140.9 Praseodymium -18,21,8,2	60 **Nd** 144.2 Neodymium -18,22,8,2	61 **Pm** (145) Promethium -18,23,8,2	62 **Sm** 150.4 Samarium -18,24,8,2	63 **Eu** 152 Europium -18,25,8,2	64 **Gd** 157.3 Gadolinium -18,25,9,2	65 **Tb** 158.9 Terbium -18,27,8,2	66 **Dy** 162.5 Dysprosium -18,28,8,2	67 **Ho** 164.9 Holmium -18,29,8,2	68 **Er** 167.3 Erbium -18,30,8,2	69 **Tm** 168.9 Thulium -18,31,8,2	70 **Yb** 173 Ytterbium -18,32,8,2	71 **Lu** 175 Lutetium -18,32,9,2

Actinides (row 7):

90 **Th** 232 Thorium -18,32,18,10,2	91 **Pa** 233 Protactinium -18,32,20,9,2	92 **U** 238 Uranium -18,32,21,9,2	93 **Np** (237) Neptunium -18,32,22,9,2	94 **Pu** (244) Plutonium -18,32,24,8,2	95 **Am** (243) Americium -18,32,25,8,2	96 **Cm** (247) Curium -18,32,25,9,2	97 **Bk** (247) Berkelium -18,32,26,9,2	98 **Cf** (251) Californium -18,32,28,8,2	99 **Es** (252) Einsteinium -18,32,29,8,2	100 **Fm** (257) Fermium -18,32,30,8,2	101 **Md** (258) Mendelevium -18,32,31,8,2	102 **No** (259) Nobelium -18,32,32,8,2	103 **Lr** (262) Lawrencium -18,32,32,9,2

Note: the lowest electron levels are not shown for rows 6 and 7, instead they are indicated by a -, which means 2, 8.

William Prout

1785–1850

SPECIAL FEATURE

A doctor, a chemist, and a father of seven, William Prout was a meticulous man who was born on January 15, 1785, in the town of Horton, in Gloucestershire, England. He was the oldest of three brothers born to a tenant farmer named John Prout. Even though he was not physically strong, he left school at the age of thirteen to work on his father's farm.

He was aware of his lack of education, so when he was seventeen he left home to change this. He went to a school in Wiltshire to learn Latin and Greek. Three years later he entered a classical seminary in Redland to further his studies. To cover the costs of his schooling, he taught the younger students. This is when he became interested in chemistry. Because of this interest, he was encouraged to study medicine, which he did at Edinburgh University. He graduated in 1811 with his M.D. (doctorate in medicine).

William moved to London to complete his studies and become a licensed physician. In 1813 he set up a medical practice. At that time he also married Agnes Adam, oldest daughter of Dr. Alexander Adam. They had seven children together.

In addition to practicing medicine, Prout began to study chemistry in many forms. Using the test results from other chemists relating to the atomic weights of different elements, he suggested that all other elements were whole number multiples of the atomic weight of hydrogen. This suggestion was highly controversial. It changed the notion of the indivisible atom. His suggestion was rejected by some of the top chemists of his day; however, other experiments throughout the nineteenth century eventually proved this to be true. In one of his experiments, he showed that one volume of oxygen and two volumes of hydrogen would form

two volumes of steam. From this he stated that water was made up of two hydrogen atoms and one oxygen atom. Today this seems obvious, but in the early nineteenth century this was astounding.

He later turned his attention to developing new techniques for studying organic chemistry. Sparing no expense in his search for more accuracy, he developed many of his own instruments, including equipment to remove moisture from chemicals. He performed a multitude of experiments, but he was only willing to publish a few of his results as he did not want to put any information out until he was sure his measurements were completely accurate. Because of his work and drive for accuracy, by 1827 he had become the leading physiological chemist in England.

Prout devoted much of his research to understanding the chemicals in the bodies of animals and humans. His work included developing new methods for studying and separating compounds. He often used himself as the test subject. For example, he measured the carbon dioxide in his breath at different times of the day and night, before and after eating, and in various emotional and physical states.

In addition to studying the chemicals in his breath, Prout also worked to find the properties and origins of blood. Based on his study of blood and the respiratory system, he suggested that blood flowed through the lung to remove carbon, in the form of carbon dioxide, from the system. He also studied the components that make up urine in animals and men. He found that the differences between healthy and diseased urine could be readily explained and understood.

He studied animal digestive systems to see the chemical changes that took place. He examined the contents of rabbits' stomachs at various times after they had eaten and found that acids where introduced into the system at certain locations. This led to the discovery that the acid found in the stomach is hydrochloric acid. He also felt certain that the respiration and the assimilation of food and the production of heat in animals were all linked in the maintenance of life for warm-blooded animals; he just was not sure of the mechanisms that linked them. William also felt that the foods a person eats could have an effect on the person's mental abilities.

Prout made great strides in understanding chemicals and especially in understanding their effects on human and animal bodies. However, he did not live to see the results of many of his discoveries. In his youth he had suffered many intense earaches and in the 1830s he became deaf. At this time he withdrew from the scientific society. He died in 1850.

17

Compounds

Making new substances

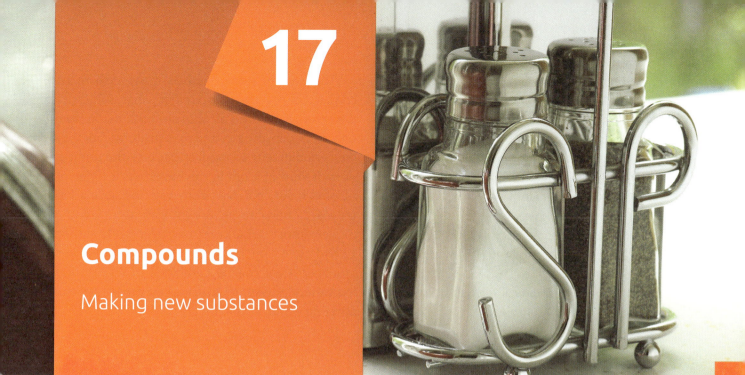

What are compounds, and how are they made?

Words to know:

chemical formula

Challenge words:

mineral
silicate
carbonate
halide
halogen

sulfide
phosphate
oxide
native mineral
native element

Everything around us is made up of atoms. These atoms sometimes occur by themselves, but usually they combine with other atoms to form new kinds of matter. When atoms combine together they form molecules. If two atoms of the same type of element combine they form what is called a diatomic molecule, but if two or more different types of elements combine to form a molecule it is called a compound. The most familiar compound is water. Water is formed when two hydrogen atoms and one oxygen atom combine.

When atoms of different elements combine to form a new substance, the new substance usually has completely different characteristics from those of the original atoms. For example, hydrogen and oxygen are both gases at room temperature, but when they combine to form water, they become a substance that is a liquid at room temperature. Similarly, sodium is a soft metal and chlorine is a poisonous gas, but when they combine chemically they form a solid that you eat every day—table salt.

There are more than 110 known elements and they can be combined in innumerable ways. Scientists have identified over three million different compounds. A few of the very familiar compounds include salt, sugar, water, starch, alcohol, carbon dioxide, and chlorophyll.

Some elements are very reactive and combine easily with other elements. Others are very stable and do not combine easily with any other substance. Hydrogen is very reactive, but nitrogen is very stable. God designed the earth's atmosphere to have a large percentage of nitrogen specifically because it does not easily react with other substances. If the atmosphere was 100% oxygen, fires would burn out of control, but the nitrogen in the atmosphere dilutes the oxygen and keeps life on Earth safe. You can learn more about why some elements are more

LESSON 17 **Properties of Matter** • 73

reactive than others in the upcoming unit: *Properties of Atoms and Molecules*.

Scientists use symbols called **chemical formulas** to describe what a compound is made of. The symbol for each kind of atom is followed by a subscript number showing how many of that type of atom are in the molecule. Water has two hydrogen atoms and one oxygen atom so its chemical formula is H_2O. Sugar (sucrose) has 12 carbon atoms, 22 hydrogen atoms, and 11 oxygen atoms so its chemical formula is $C_{12}H_{22}O_{11}$. All other compounds have similar chemical formulas. Some are simple, others are very complex.

The kinds of atoms and how many of them are connected together determines what substance is formed. Sugar and alcohol are both made from only carbon, hydrogen, and oxygen atoms, but they are very different substances because the number of each kind of atom in the molecules is different. Alcohol is very different from sugar because it has only 4 carbon atoms, 8 hydrogen atoms, and 1 oxygen atom (C_4H_8O), whereas sugar has 12 carbon, 22 hydrogen, and 11 oxygen atoms. God designed atoms so they can be combined in an astounding variety of ways so that we can experience this amazing world around us. ✳

Electrolysis of water

A water molecule is made when two hydrogen atoms bond with an oxygen atom. The bond between these atoms is fairly strong but can be broken by an electrical charge. As the electrons flow through the water, the molecules break apart and the hydrogen and oxygen return to their gas forms.

Purpose: To break water apart into its elements

Materials: two small jars or test tubes, sink, copper wire, battery, baking soda

Procedure:

1. Fill a sink with water and place two small jars or test tubes in the water so they are completely filled with water.
2. Cut two pieces of copper wire long enough to reach from a battery on the counter into the jars.
3. Strip off an inch of insulation from the ends of each wire and attach one end of a wire to the positive terminal and one end of the other wire to the negative terminal of the battery.
4. Place the end of one wire in one jar and the other wire in the second jar.
5. Hold the jars inverted in the water so that the mouth of each jar is below the surface of the water and the jar remains filled with water (see illustration).
6. Sprinkle a teaspoon of baking soda in the water and stir to help it dissolve. The baking soda encourages the electricity to flow through the water.

Conclusion: After a short time you should notice bubbles forming on the ends of each of the wires. These bubbles will slowly push the water out of the jars. After a few minutes it should become obvious that one jar is being filled with gas faster than the other jar. Note: You may notice a bluish/green substance forming on the end of one of the wires. This is a reaction between copper atoms and the oxygen atoms.

Questions:

- What do you think is in each jar?
- Which jar do you think has the hydrogen in it?
- Why do you think the battery is needed to separate the atoms?

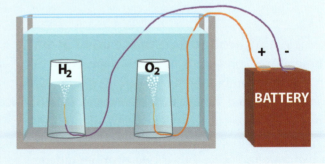

What did we learn?

- What is a compound?
- What is another name for an element?
- What is another name for a compound?
- Do compounds behave the same way as the atoms that they are made from?

Taking it further

- The symbol for carbon dioxide is CO_2. What atoms combine to form this molecule?
- The air consists of nitrogen and oxygen molecules. Is air a compound? Why or why not?

Minerals

The most abundant element in the universe is hydrogen. Most of the outer planets including Jupiter and Saturn are believed to be made primarily of hydrogen gas. But God designed the earth differently. Earth is a solid planet composed primarily of five elements: oxygen, silicon, aluminum, iron, and calcium. Over 90% of the earth's crust is made up of these five elements. However, as you just learned, these elements are seldom found by themselves. They are chemically combined with other elements to form compounds. The compounds that make up the earth's crust are called **minerals**.

Minerals are grouped according to the elements that they are made from. The largest group of minerals is the **silicates**, which are formed when different elements bond with silicon and oxygen. The next most abundant minerals are the carbonates. **Carbonates** are formed when elements bond with carbon and oxygen.

Other mineral compounds include the **halides**. These are compounds containing **halogens**, which are elements in column VIIA of the periodic table, headed by fluorine. **Sulfides** are compounds containing sulfur, and **phosphates** are compounds containing phosphorus and oxygen. As you can see, most of the mineral compounds contain oxygen. There are many other smaller groups of minerals that are called **oxides** because they also form when elements combine with oxygen.

A few minerals are not compounds. A few elements are found in the earth's crust in their pure form. Minerals that are comprised of just a single element are called **native minerals** or **native elements**. These include silver, gold, and carbon (in the form of diamond). These elements are considered more valuable because they are rare compared to the many other compounds that are found in the earth's crust. As you learn more about compounds, don't forget about the important compounds in the ground under your feet.

18

Water
God's compound for life

What are the properties of water?

Words to know:
universal solvent

Challenge words:
percolate   transpiration

Over 70% of the surface of the earth is covered with water. This fact, more than any other, is why life is able to exist on Earth and not on any other planet in our solar system. God designed water to be the perfect compound to sustain life.

Water is nearly a **universal solvent**. This means that it dissolves nearly everything it touches. Obviously, some things dissolve more easily than others, but it is very difficult to obtain completely pure water because water even dissolves very tiny amounts of glass or metal from its container. This is a very important quality of water that most other liquids do not possess. Because so many substances dissolve in water, it is used to transport nutrients throughout plants. And because blood is comprised of mostly water, it can transport all the chemicals you need throughout your body.

One reason that water is such a good solvent is its unique shape. The hydrogen atoms attach to the oxygen atom at a 105° angle giving the molecule a lopsided shape. This means that the hydrogen side of the molecule is slightly positive and the oxygen side is slightly negative, which allows the water to break apart most other substances and hold the atoms away from each other. This process will be explained more in later lessons.

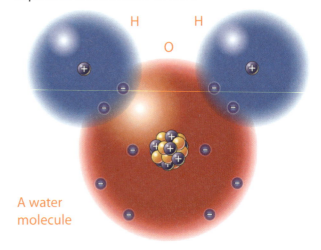

A water molecule

Water has no flavor or smell, but it is in nearly everything we eat or drink. It is found throughout all plants and animals. Water is necessary for nearly every bodily function. This is why it is so important to drink enough water every day.

Besides being necessary for life, water is very useful in many natural and man-made processes. Water is one of the main ingredients needed for photosynthesis. In photosynthesis, plants use water and carbon dioxide to produce sugar. Water is important in farming to produce the food our world needs. Water is also a key ingredient in many manufacturing processes. Water is necessary for the production of paper, electronic circuits, food, beverages, and many other manufactured goods.

Water is also used for recreation. Water sports are some of the most popular activities. Also, water is used for decoration. Fountains and waterfalls are a common sight in many areas. In addition, water is needed for many daily activities such as bathing, laundry, cooking, and gardening.

With so many uses for water, you might be concerned that the world will run out of water. However, as you learned in lesson 6, God designed the world so that water is constantly recycled. Even though plants and animals use water in their normal growing processes, they also release water back into the atmosphere. Also, when a plant or animal dies, the water in its body is released. Water is also a by-product of many chemical reactions. For example, many spaceships use liquid hydrogen and liquid oxygen for fuel. The fuel molecules combine to form large clouds of steam that billow from the ship on take off. Also, some new cars are now using hydrogen as fuel and they produce water as the end product instead of carbon monoxide. So water is continually being added to the earth at about the same rate it is being used.

The next time you watch the rain or drink a glass of water, thank God for creating water so life could exist on Earth. To learn more about water, go to the Answers in Genesis website for information and articles.

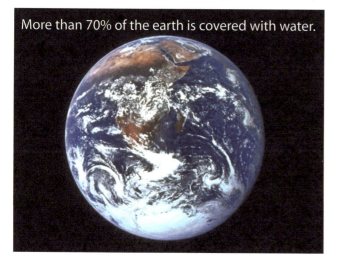

More than 70% of the earth is covered with water.

🧪 Hunting for water

Using a copy of the "Water, Water Everywhere" worksheet, record all of the uses of water in your home for the past day.

🧠 What did we learn?

- What two kinds of atoms combine to form water?
- Why is water called a universal solvent?
- What is unique about the water molecule that makes it able to dissolve so many substances?

🚀 Taking it further

- What would happen to your body if oxygen could not be dissolved in water?
- Is water truly a universal solvent?
- Why is it important for mothers with nursing babies to drink lots of water?

🏅 Water recycling

Where does water come from to water plants that are not in a yard or park, but in a natural area? If you said the water comes from rain, you are correct. But rain water is not the only source of water for many plants. A tree can continue to grow even when it does not rain for an extended period of time. Where does it get water from? Trees, and many other plants, have roots that grow down deep into the soil. Water from under the ground **percolates** or wicks up from underground rivers and the plants can access this water.

LESSON 18 **Properties of Matter** • 77

After plants use this water, how does more water get into the air and into the ground? Plants release excess water through their leaves into the air. This process is called **transpiration**. As much as 70% to 90% of the water that is absorbed by a plant's roots is released through transpiration without being used for photosynthesis. This water can be recycled and reused later by the plants.

Purpose: To demonstrate how water is recycled in nature.

Materials: three 2-liter soft drink bottles, string, scissors, potting soil, bean seeds, ice water

Procedure:

1. Cut the top off of a 2-liter soft drink bottle.
2. Pour ⅔ cup of water into the bottle.
3. Cut the bottom off of a second bottle.
4. Punch a hole into the bottle cap and thread a 14-inch long piece of string through the hole so a loop sticks up into the second bottle, and the ends of the string stick down into the water of chamber A when the second bottle is inserted into the first.
5. Place 3–4 inches of potting soil into the second bottle (chamber B). Be sure that the loop of string is standing up in the middle of the soil and not trapped against the side of the bottle.
6. Plant several bean seeds in the soil. Keep the soil moist until the bean plants begin to grow. Once you have several plants growing, you can complete your experiment.
7. Cut the bottom off of a third soft drink bottle. Leave the cap on the mouth of the bottle. Place this bottle on top of the second bottle. (See diagram)
8. Pour 1 cup of ice water into the top bottle (chamber C).
9. Carefully observe what is happening to the water in each chamber of the model.

Conclusion: What do you observe happening in each chamber? The water from the bottom bottle is being sucked up into the second bottle just as ground water percolates up into the ground. Is water from the top bottle entering the second bottle? No. The water in the top bottle is only cooling the air in the second bottle, allowing the water vapor in the second bottle to condense and form rain that falls on the soil to be used by the plants.

Leave your bottles in a sunny location and watch the water being recycled as your plants use the water to grow and also release water into the air to be reused.

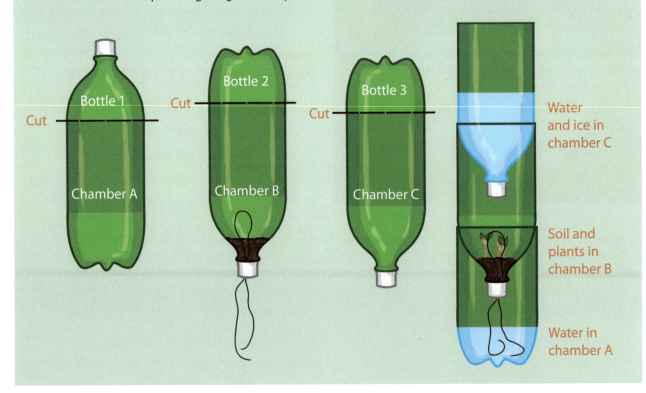

78 • Properties of Matter   LESSON 18

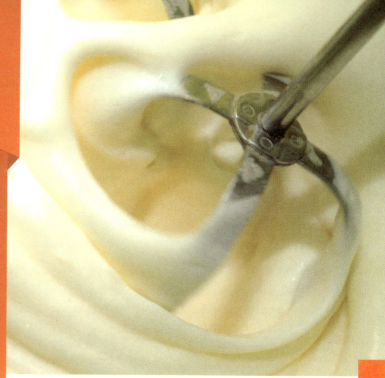

19

Mixtures

All mixed up

What is a mixture, and how is it made?

Words to know:

homogeneous mixture   heterogeneous mixture

Challenge words:

nonsoluble   chromatography
decantation   centrifuge
distillation

Elements and compounds are pure substances. They each consist of only one kind of material. Yet most things in the universe are not pure substances. Most items are a combination of two or more elements or compounds mixed together. These substances are called mixtures.

All mixtures have the following characteristics:

1. Mixtures are comprised of two or more pure substances.
2. Each substance keeps its own properties and does not create a new substance.
3. The different substances can be separated by physical means. This means that the different substances can be separated by filtering, boiling, sorting, or some other non-chemical method.
4. The substances in a mixture can be found in any proportion. In a compound such as carbon dioxide, the ratio of atoms is always the same; there are always 2 oxygen atoms to each carbon atom. But in a mixture such as salt water there could be 1 part salt to 1 part water, or 1 part salt to 100 parts water, or any other proportion, and it would still be salt water.

One of the most abundant mixtures on Earth is seawater. Seawater consists of water, salt, and a variety of other minerals, as well as tiny bits of sand, plants, and other materials. Each of these substances has its own physical properties such as boiling point, density, hardness, etc. And each item can be separated from the water if the water is boiled away. If you add more salt to seawater it is still seawater. And if you filter out the sand and other debris, you still have seawater. Other common mixtures include air (which is a mixture of gases), milk (which is a mixture of liquids and solids), and stainless steel (which is a mixture of iron, carbon, and chromium).

Mixtures can be classified by how the materials are distributed throughout it. If the substances in the mixture are evenly distributed it is called

🧪 Separating a mixture

Purpose: To see how a mixture is different from a compound

Materials: funnel, coffee filter, cup, orange juice

Procedure:

1. Line a funnel with a coffee filter.
2. Place the funnel so the contents will drip into a cup or bottle.
3. Shake up a container of orange juice, then slowly pour some orange juice through the filter. Reserve some orange juice for the end of the experiment. The water will flow through the filter and the orange pulp will be trapped in the filter. This may take several hours depending on the filter you use. Do not squeeze the filter.
4. After most of the liquid has passed through the filter, compare the liquid to some of the original orange juice. The liquid should be much clearer. Some of the orange pulp may be small enough to pass through the filter, but much of it will be trapped by the filter.

Conclusion: Orange juice is a mixture of orange pulp and water. The water is still water no matter how much orange pulp is mixed with it. The orange pulp is still orange pulp even when it is mixed into the water. So orange juice is not a new substance, just a mixture of substances.

a **homogeneous mixture**. Air and seawater are homogeneous mixtures. A mixture in which the substances are not evenly distributed is called a **heterogeneous mixture**. Granite is an example of a heterogeneous mixture. Granite is made from a combination of the minerals feldspar, quartz, and mica. These minerals can be found in any proportion and are not evenly mixed. One piece of granite may be mostly feldspar and another may have a large chunk of quartz with bits of mica spread around in it.

Although pure substances (such as elements like oxygen and compounds like water) are vitally important, most things around you are a combination of elements and compounds mixed together and are thus mixtures.

🧠 What did we learn?

- What are two differences between a compound and a mixture?
- What is a homogeneous mixture?
- What is a heterogeneous mixture?
- Name three common mixtures.

🚀 Taking it further

- If a soft metal is combined with a gas to form a hard solid that doesn't look or act like either of the original substances, is the resulting substance a mixture or a compound?
- How might you separate the salt from the sand and water in a sample of seawater?

🎖 Separating compounds

Substances in a mixture can be separated by physical means. You have just demonstrated one method for separating various compounds in a mixture. Filtering is an important way to separate nonsoluble compounds from a mixture. **Nonsoluble** compounds are those that do not dissolve in water or other liquids. Filtering is used in many applications.

One of the most important uses of filtering is in water treatment. Filters are used to screen out all but the smallest particles in our drinking water.

There is an even easier way to separate nonsoluble compounds. What happens to a bottle of orange juice if you just let is sit for a while? The pulp will settle to the bottom. This method of letting layers settle out is called

decantation. Decantation can be used in situations that do not require complete separation. When you pour off the liquid, a small amount of the solid is likely to move with it, but decantation can be a quick way to separate out most of the solids from the liquids. Filtering may be needed to remove the fine particles that are left.

Although filtering and decantation have many uses, there are several other ways to separate compounds in a mixture. What other method have you demonstrated? Evaporation is another important way to separate compounds in a mixture. Evaporation is used to separate compounds that have been dissolved. You used evaporation to separate the water from the salt that had been dissolved in it in lesson 10. Evaporation usually uses heat to turn a liquid, often water, into a gas while other compounds remain behind.

Distillation (shown at right) is closely related to evaporation. If you want to obtain a sample of pure water, you can evaporate the water, but instead of letting the water vapor dissipate into the air, you can collect the water vapor and allow it to condense into another container. This is the process of distillation. The evaporation separates the water from the other materials that might be dissolved in it and then collects that water vapor in a separate container.

Chromatography is another method for separating compounds. If various compounds are dissolved in a liquid, they can be separated by placing some of the liquid on a piece of paper. As the liquid spreads through the paper, the various compounds will move at different speeds through the paper so different compounds will end up at different locations on the paper. This is easily demonstrated with water soluble ink. Ink, particularly black ink, consists of several different pigments mixed together. The various pigments are dissolved in water. These pigments will separate as the water moves through a piece of paper.

A fourth method for separating various compounds in a mixture is the use of a centrifuge. A **centrifuge** is an instrument that spins very quickly. If a mixture that contains solids and liquids is placed in a centrifuge, the solid particles will collect on the sides of the container, allowing them to be separated from the liquid. This process is often used to separate blood cells from the plasma in which they are suspended.

There are other methods for separating mixtures as well. Can you think of an easy way to separate metal filings that have been mixed in with sand? You could use a magnet to collect the metal filings. Think about the various physical characteristics of the compounds in a mixture. This will help you decide the best method for separating the compounds. For each mixture below, decide what method you think would be best for separating the various compounds.

- Cream and milk
- Mud and water
- Mixture of food colors
- Lemon juice and water
- Oil and water
- Coffee and coffee grounds

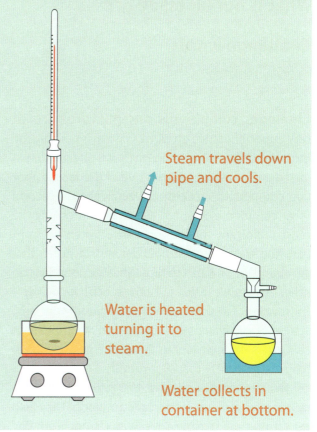

Steam travels down pipe and cools.

Water is heated turning it to steam.

Water collects in container at bottom.

LESSON 19 **Properties of Matter** • 81

20

Air
What we breathe

What is our air made of?

Challenge words:

fractional distillation

Elements, compounds, and mixtures are all very important to life on Earth. Oxygen is possibly the most important element on Earth. It is one of the two ingredients in water, which is one of the most important compounds on Earth. Oxygen is also the second most common element in air which is probably the most important mixture on Earth. When we compare the elements, compounds, and mixtures that are abundant on Earth to those that are abundant on other planets, we realize just how special our planet is. The mixture of air that we breathe is unique in our solar system. Some planets do not have an atmosphere at all. Others have poisonous atmospheres. Only Earth has an atmosphere that supports life.

Because air is a mixture, the ratios of the components of air vary from place to place. They also vary from time to time; they even change from day to night. But on average, air consists of about 78% nitrogen, 21% oxygen, and 1% other gases. The other gases in air include carbon dioxide, argon, helium, and various pollutants including carbon monoxide and sulfur dioxide. There are also solid particles mixed into the air including soot and pollen. The air also contains water vapor. All of these together comprise the wonderful mixture we call air.

Even though nitrogen is the most abundant element in air, it is not very reactive and its main purpose is to dilute the oxygen in the air. Oxygen is really the most important component of air. Oxygen is necessary for most of the processes that occur in human and animal bodies. In order to break down food into usable energy, we need oxygen. God provided oxygen in the air.

However, it would be dangerous to have too much oxygen. If a person breathes pure oxygen for prolonged periods of time, he can experience serious physical problems. Life could not occur on Earth if the air was 100% oxygen. Also, as the amount of oxygen in the air increases, the chance of natural forest fires increases as well. Since oxygen is needed for combustion, the more oxygen there is, the faster things will burn. Thankfully, God designed the earth's atmosphere to be mostly nitrogen. Nitrogen does not react easily with most things. It does not react with the human body, so even though we breathe it in with every breath, we breathe it right out again without it affecting us in any way. The air we breathe is perfectly designed for life on Earth.

In addition to providing oxygen for us to

🧪 Importance of oxygen

Purpose: To demonstrate the importance of oxygen in the air

Materials: small candle, jar, baking soda, vinegar, bottle

Procedure:

1. Light a small candle. It burns because it is able to use the oxygen in the air for combustion.
2. Place a jar over the candle. What happens after a few seconds? The candle goes out because it has used up the oxygen in the air. Without a fresh supply of air, there is no oxygen for combustion and the flame goes out.
3. Remove the jar and relight the candle.
4. Mix a small amount of baking soda and vinegar in a bottle.
5. Hold the mouth of the bottle near the candle. What do you observe happening? The flame again goes out. Why does the flame go out? The chemical reaction between the baking soda and vinegar produces carbon dioxide gas. The carbon dioxide is heavier than the oxygen in the air so it pushes the oxygen molecules away from the flame. Without oxygen, combustion cannot take place and the flame goes out.

Conclusion: As you can see, oxygen is very important for combustion. Breaking down food into energy is a form of combustion that takes place inside your body, so you need oxygen to keep your body functioning; that oxygen comes from the air you breathe.

breathe, the atmosphere also protects us from harmful radiation from the sun. Although the sun provides us with light to see by and heats the planet, it also produces very high energy waves that can damage living tissue. Fortunately, God has designed our atmosphere with just the right mixture of molecules to protect life from these dangerous rays. Nitrogen is a diatomic molecule, which means there are two nitrogen atoms bonded together. The bond between these two atoms is very strong. But when gamma rays from the sun hit our upper atmosphere they encounter a large amount of nitrogen. Most of the energy in these gamma rays is used up breaking nitrogen atoms apart. This keeps most of that harmful energy from reaching the earth. Similarly, energy from harmful x-rays from the sun is used up breaking apart oxygen molecules in the upper atmosphere and energy from ultra-violet rays is used up breaking apart ozone molecules. Thus the mixture of gases in our upper atmosphere is just right to protect the plants, animals, and people on the surface.

It is also important that ozone is found only in the upper atmosphere and not in the air near the surface because ozone is poisonous to people and animals. So God designed the atmosphere to have ozone only up high where it can counteract ultra-violet rays without harming the life on Earth. Air is indeed an amazing mixture. ✳

🧠 What did we learn?

- What is likely the most important element on Earth?
- What is likely the most important compound on Earth?
- What is likely the most important mixture on Earth?
- What are the main components of air?

🚀 Taking it further

- Why is nitrogen necessary in the air?
- Why is oxygen necessary in air?
- How does the composition of air show God's provision for life?

Fractional distillation

We know that air is a mixture of several different gases. But how did scientists separate out these various gases to determine what air was made of? They used a special type of distillation called **fractional distillation**.

To begin, a sample of air is cooled and compressed until all of the gases in the mixture become liquids. This occurs at very low temperatures. The liquid mixture is then slowly heated. The various elements and compounds boil at different temperatures. As each compound begins to boil, the vapor is collected and stored in a separate container. After the final liquid is boiled and then collected, each container has a fraction of the mixture and these fractions can be compared to find the whole composition of the mixture, thus the name fractional distillation.

Fractional distillation is used for separating many mixtures besides air. For example, a mixture of ethanol and water can be separated by fractional distillation. Ethanol boils at 78.5°C while water boils at 100°C, so a mixture of the two can be heated to 78.5°C and kept at that temperature until the ethanol boils away. The ethanol vapor can be trapped and allowed to condense in a separate container. This leaves the water behind. The temperature can then be raised to 100°C to boil away the water. The water vapor can then be trapped and allowed to condense in another container. This separates the water from any other impurities.

One of the most common uses of fractional distillation is to separate the components of crude oil during the refinement of petroleum. Crude oil is heated and the various hydrocarbons and nonhydrocarbons are distilled at different temperatures. Petroleum processing is somewhat more complex than separating ethanol and water. Instead of boiling the substances at different temperatures, the crude oil is heated to about 340°C. The gases are then piped into a special tower called a fractionating column. As the gases rise up the tower they cool and condense. The heavier molecules condense first and are collected at the bottom of the tower. The lighter molecules have lower boiling points and thus rise higher and condense at higher levels in the tower. This process allows the crude oil to be continuously fed into the distiller and the various components to be continuously siphoned off allowing the process to go very quickly. The diagram illustrates the various substances that are separated in the distillation process.

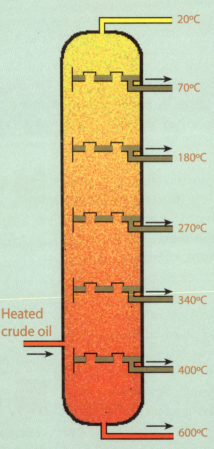

20°C — Refinery gases
1–4 carbon atoms per molecule. For heating and cooking (e.g., propane).

70°C — Gasoline compounds
5–12 carbon atoms per molecule. Gasoline for cars and for making plastics and chemicals.

180°C — Kerosene compounds
9–15 carbon atoms per molecule. Used for heating, lighting, and jet fuels.

270°C — Diesel oils
12–25 carbon atoms per molecule. Fuel for trucks and trains.

340°C — Lubricants
20–50 carbon atoms per molecule. Used for heating oil, candle waxes, polishes.

400°C — Fuel oils
20–70 carbon atoms per molecule. Fuels for ships, factories, and central heating.

600°C — Residue compounds
>70 carbon atoms per molecule. Road surfaces and roofing.

21

Milk & Cream
"Udderly" delicious

How are milk and cream processed?

Words to know:
pasteurization
homogenization
foam
structural integrity

Challenge words:
curds
whey
enzyme

One of the most common mixtures found inside refrigerators around the world is milk. In some parts of the world, people drink goat's milk; in other parts of the world, people drink cow's milk. Cow's milk is a mixture that is approximately 87% water, 5% lactose (a type of sugar), 4% fat, 3% protein, and 1% ash. The protein, vitamins, and minerals in milk make this a popular drink and a popular ingredient in many foods.

At one time, milk had to be consumed shortly after it came from the cow because it quickly spoiled. But after a scientist named Louis Pasteur discovered that bacteria were the cause of most food spoilage, and with the invention of the modern refrigerator, milk can now be stored for many days before going bad.

After the cow is milked, the milk undergoes several processes before it reaches the supermarket

Most dairy products consumed around the world originate from cow milk.

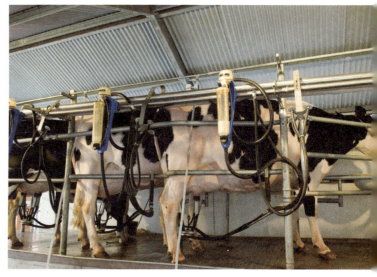

Fun Fact
In 1923 the federal government passed a law requiring that butter must contain at least 80% milk fat.

LESSON 21 **Properties of Matter** • 85

How strong is your whipped cream?

The ability of the cream molecules to trap and hold air is called its **structural integrity**.

Purpose: To see which kind of cream has more structural integrity

Materials: whipping cream, sugar, vanilla, plate, canned spray whipped cream

Procedure:

1. First, make homemade whipped cream by whipping 1 cup of whipping cream with an electric mixer on high speed until peaks form.
2. Gently stir in 1 tablespoon of sugar and 1 teaspoon of vanilla extract.
3. Place a large spoonful of this whipped cream on a plate.
4. Next to this, spray an equal amount of canned spray whipped cream. Be sure the canned whipped cream is made from real cream. Many of the products sold are made from other substances such as vegetable or soybean oil.
5. Place the plate in the refrigerator for one hour.
6. Check to see which sample looks the most like whipped cream after one hour.

Conclusion: Eventually the cream molecules will lose their ability to keep the air molecules trapped. When the air starts to escape, the cream begins to "weep" and becomes runny. Which type of cream would be best to use if you want to make a dessert ahead of time and store it until your dinner guests have finished their dinner?

shelves. The first process is called **pasteurization**, named for Louis Pasteur. In this process the milk is heated to 145°F (63°C) for 30 minutes, or to 161°F (72°C) for 15 minutes, to kill the bacteria. It is then cooled and stored at temperatures below 45°F (7°C). Most milk that is packaged for drinking experiences a second process called **homogenization**. This process breaks up the fat molecules into tiny bundles so they stay suspended in the liquid of the milk instead of floating to the top.

Butter, cheese, milk, yogurt, cottage cheese, and sour cream are all dairy products.

Some milk, however, is not homogenized. If the milk is not homogenized, the cream (the part containing the fat molecules) will rise to the top of the milk. In the past, people would skim the cream off the top of the milk and use it to make butter, cheese, and other products. Today, instead of just letting the cream rise to the top of the vat, the milk is placed in a machine called a separator, which spins the milk and separates the cream from the milk. The cream is then used to make many products including butter, cheese, ice cream, and whipped cream. The milk that remains is then sold as skim milk.

Butter is one product that is made from cream. Butter is made by churning or vigorously shaking the cream. This churning causes the fat molecules to combine together in large clumps. These clumps are then removed and sometimes a small amount of salt and coloring are added before the butter is formed

Fun Fact

Some milk is pasteurized at a temperature of 280°F (138°C), for two seconds. This kills all bacteria and allows the milk to be stored at room temperature until it is opened. This is called ultra-high temperature sterilization or UHT.

🧪 Making butter

While you are waiting for your whipped cream to break down, you can make some butter.

Purpose: To make butter

Materials: 1 cup of whipping cream, jar with lid

Procedure:

1. Place one cup of cream in a clean jar and screw the lid on tightly.
2. Shake the jar vigorously until you become tired. Then ask someone else to shake it for a while. After several minutes you will begin to notice that the cream is becoming foamy.
3. Keep shaking until you notice several clumps in a thinner liquid. This is the butter in the buttermilk.
4. Drain the buttermilk into another container.
5. Collect all of the pieces of butter and rinse them with cold water.
6. Press the pieces together or press them in a small mold. You now have fresh butter to enjoy on a piece of bread or a cracker. You can use the buttermilk to make your favorite pancakes or biscuits.

into cubes and packaged to be sold. The liquid left over from the butter making process is called buttermilk. Buttermilk is sold separately and is also used in many products such as pancakes or biscuits.

Another popular product made from cream is whipped cream. Whipped cream is a foam. A **foam** is a liquid with air trapped inside it. Have you ever seen sea foam? The movement of the waves against the shore causes air bubbles to become trapped between the water molecules. Similarly, if cream is whipped, air molecules become trapped in between the fat molecules, causing the cream to be very fluffy. Usually sugar and vanilla are added to improve the flavor of the whipped cream, and make it a delightful topping to any sweet treat.

Finally, one of the most popular cream products is ice cream. Ice cream is a mixture of cream, milk, sugar, and other flavors that are frozen into a smooth delicious dessert.

🧠 What did we learn?

- Is milk an element, a compound, or a mixture?
- What is pasteurization, and why is it done to milk?
- What is homogenization, and why is it done to milk?
- What is a foam?

🚀 Taking it further

- Why does whipped cream begin to "weep"?
- Why must cream be churned in order to make butter?

LESSON 21 **Properties of Matter** • 87

Cheese

One of the most useful and popular milk products is cheese. Cheese is formed when an acid reacts with the sugar in milk, causing the milk to curdle. The curdled milk separates into solid chunks called **curds** and a watery liquid called **whey**. Thus, little Miss Muffet, who was eating curds and whey, was actually eating curdled milk, or cheese.

Cheese can be made from any kind of milk, but most cheese is made from cow's milk or goat's milk. A specific kind of acid-producing bacterium is added to the milk to begin the curdling process. An **enzyme**, which is a biological substance that helps speed up a chemical reaction, is then added to complete the process. The enzyme that is most often used is called *rennet*. Rennet is an enzyme found in the lining of calves' stomachs. It is extracted and condensed for use in commercial cheese making. Because there is a limited amount of calf rennet available, scientists have developed genetically modified bacteria that produce the active ingredient in rennet. Today about 60% of the cheese produced is made with synthetic rennet.

There are hundreds of varieties of cheese. Each variety depends on the milk that is used, the bacteria that are used, the length of time the curds are aged, and various flavorings that are added. Some soft cheeses are made by using only the acid-producing bacteria to produce soft curds. Most cheeses include rennet to solidify the curds, giving them a stiffer structure. Hard cheeses are formed when the curds are pressed for extended periods of time.

Cheese has been eaten since early Bible times. Cheese has nearly the same nutrition as milk, but because it has gone through a chemical change, it is much less likely to spoil. Some cheese can stay fresh for months, especially if refrigerated. This makes it a more portable food than milk.

Purpose: To see the separation of curds and whey

Materials: whole milk, pan, stove, vinegar

Procedure:

1. Heat 1 cup of whole milk just to simmering. Do not allow it to boil.
2. Remove the pan from the heat and add 1 tablespoon of vinegar. You should see the milk curdling before your eyes.
3. Use a colander to drain off the whey. The curds will be left behind.

Conclusion: This is a very simple cheese making process that yields a substance similar to cottage cheese. The acid in the vinegar caused the milk to curdle and separate into curds and whey. Some recipes use vinegar or lemon juice to make soft cheese, but most cheese is made using acid-producing bacteria.

UNIT 5

Solutions

22 Solutions
23 Suspensions
24 Solubility
25 Soft Drinks
26 Concentration
27 Seawater
28 Water Treatment

◊ **Distinguish** between solutions and suspensions using examples.
◊ **Identify** the most common solvent.
◊ **Describe** the relationship between solubility and concentration.

Properties of Matter • 89

Solutions

Not just an answer to a problem

What are solutions, and how are they made?

Words to know:
solution
solute
solvent
solubility
saturated

We have seen that all materials in the universe can be classified into three categories: elements, compounds, and mixtures. Mixtures can be further classified as solutions or suspensions. You may think that a solution is an answer to a problem, but in chemistry a **solution** is a homogeneous mixture where one substance has been dissolved in another substance. Salt water is a common example of a solution.

In a solution, the substance that is dissolved is called the **solute** and the substance in which it is dissolved is called the **solvent**. In the example of salt water, the salt is the solute and the water is the solvent. In a true solution, the solute stays dissolved in the solvent and does not easily come out. If the solute eventually settles to the bottom of the container it is a suspension instead of a true solution. We will discuss suspensions more in the next lesson.

When a solute is dissolved in a solvent, the solvent works to separate the molecules. As we discussed earlier, water molecules are shaped so that they easily break apart other molecules. A salt crystal is easily broken apart by the movement of the water molecules around it. The tiny salt molecules are then surrounded by water molecules and held away from each other so they cannot recombine. This is what is happening when something is dissolved—its molecules are separated from each other and held apart by the solvent.

Solubility is the measure of how much solute can be dissolved in a given amount of solvent. For example, how much salt can be dissolved in a cup of water? You can easily dissolve a teaspoon of salt in a cup of water. But if you keep adding salt, eventually there are no longer enough water molecules to surround the salt molecules and some of the salt will not dissolve, but will instead settle on the bottom of the cup. The water is then said to be **saturated** because it can no longer dissolve any more salt. This does not mean it is no longer a solution; it is just a saturated solution.

How well one substance dissolves in another substance depends on what the substances are. Although many substances dissolve in water, others dissolve only a small amount or not at all. This is because a solvent can only dissolve a solute that

90 • Properties of Matter   LESSON 22

has a similar molecular shape or structure. Often people say that "like dissolves like." For example, water does not easily get rid of grease on dirty dishes because the water and oil molecules are very different so oil will not dissolve in water. However, soap molecules are shaped on one end like a fat or oil molecule and thus are able to break apart the grease molecules on the dishes, while the other end of the soap molecule is similar in structure to a water molecule and can thus be dissolved in the water and rinsed away. This makes soap a very useful substance.

Even if you have two substances that can form a solution, the speed at which a solute dissolves in a solvent can be affected by several things including temperature and surface area. When a liquid is warmed up its molecules move more quickly. So they can more easily move in between the molecules of the solute. In general the warmer the solution the faster the solute will dissolve. Also, the amount of surface area exposed to the solvent will affect how quickly the substance dissolves. A sugar cube will dissolve more slowly in a glass of water than a teaspoon of granulated sugar will, and the same amount of powdered sugar will dissolve even faster because more of the sugar molecules are exposed to the water at one time.

What did we learn?

- What is a solution?
- Is a solution a homogeneous or heterogeneous mixture?
- In a solution, what is the name for the substance being dissolved?
- In a solution, what is the substance called in which the solute is dissolved?
- What is solubility?

Taking it further

- Why can more salt be dissolved in hot water than in cold water?
- If you want sweet iced tea, would it be better to add the sugar before or after you cool the tea?

Understanding solutions

To help understand how solvents and solutes work together to make a solution, we will dissolve some Life Savers candies in water. Follow the instructions on the "Solutions Experiments" worksheet.

Like dissolves like

You have learned that molecules with similar shapes more easily dissolve each other than molecules with very different shapes. You also learned that water and oil have very different molecular structures. So, would you expect substances that dissolve in water to also dissolve in oil?

Complete the experiment on the "Like Dissolves Like" worksheet.

LESSON 22 **Properties of Matter • 91**

23

Suspensions

And we don't mean getting kicked out of school.

What is a suspension, and how is it made?

Words to know:

immiscible
emulsifier
suspension
colloid

Have you ever heard the expression, "Oil and water don't mix"? This usually means that two very different people may not get along or two very different ideas may not work well together. This saying comes from the fact that even though many different substances dissolve in water, oil does not dissolve in water. Oil molecules are not easily broken apart by water, so even if you stir the oil and water together they quickly separate. Liquids that do not mix, such as oil and water, are said to be **immiscible**.

When a mixture is made of one substance that does not dissolve in the other, that mixture is called a **suspension** instead of a solution. The orange juice mixture you tested was a suspension. In a suspension, the molecules of one substance are held apart for a while, but eventually they come back together and settle out. The molecules cannot be broken down small enough for them to stay dissolved. If you leave a container of orange juice alone for a while, much of the pulp will settle to the bottom. That is why most juice bottles say to shake before opening. In the case of oil and water, the oil is lighter than the water so it floats to the top. Either way, the water is not able to keep the molecules in suspension indefinitely.

Even though some substances are immiscible, it is often desirable to have the particles stay in suspension. This can sometimes be accomplished by adding an **emulsifier**—a substance that helps

Orange juice is a suspension.

to keep the suspended particles from coming back together. When baking bread, it is desirable that the oil be evenly distributed throughout the dough, so an emulsifier is often added to help break up the oil and hold it in suspension in the water. Lecithin, derived from soybeans or other plants, is a common emulsifier added to bread dough. Lecithin can also be derived from egg yolks, so egg yolks are often used as emulsifiers. A suspension with very tiny particles throughout the liquid is called a **colloid**.

A common suspension found in most homes is milk. Milk is mostly water with fat and protein molecules suspended in it. Have you ever seen spoiled milk? As it ages the milk is no longer able to hold the particles in suspension and the solids separate from the liquid. Italian salad dressing is another suspension. It contains water, oil, herbs, and spices that must be shaken together before being used on your salad.

Fun Fact
Emulsifiers are often used in ice cream to give it a more creamy texture.

Fun Fact
Many products with oil in them list *partially hydrogenated vegetable oil* as an ingredient. Hydrogenation is a process where hydrogen is added to the oil to change the molecular structure of the oil to make it a solid or semi-solid fat. This is a different way to keep oil from separating out of many products.

What did we learn?

- What is a suspension?
- What does immiscible mean?
- What is an emulsifier?
- What is a colloid?

Taking it further

- What would happen to the mayonnaise if the egg yolk were left out of the recipe?
- How is a suspension different from a true solution?

Making your own suspension

Another common suspension is mayonnaise. The main ingredients in mayonnaise are vegetable oil, vinegar, and lemon juice. However, if you just mixed those ingredients together you would not get the creamy white spread that so many people love on their sandwiches. In order to make the creamy texture, the oil molecules must be broken into tiny particles and held in suspension in the vinegar and lemon juice. And to keep the molecules from separating back out, an emulsifier must be used. In this case that emulsifier is egg yolk.

Purpose: To make your own homemade mayonnaise

Materials: small mixing bowl, dry mustard, paprika, salt, one egg, vinegar, vegetable oil, lemon juice

Procedure:

1. In a small mixing bowl combine ¼ teaspoon dry mustard, ⅛ teaspoon paprika, and ½ teaspoon salt.
2. Add 1 egg yolk and 1 tablespoon vinegar.
3. Beat the mixture at medium speed with an electric mixer until blended.
4. Begin adding vegetable oil, 1 teaspoon at a time, beating constantly. Continue adding oil, 1 teaspoon at a time, until ¼ cup oil has been added.
5. While continuing to beat, add ¾ cup more oil in a thin, steady stream.
6. Add 1 tablespoon of lemon juice.
7. Now you can eat this tasty suspension on your lunch. The rest should be refrigerated for up to 4 weeks.

Cake

One very fun type of suspension is a cake. You may not normally think of a cake as a science experiment, but a cake is full of chemistry. What are the main ingredients in making a cake? It really depends on your recipe, but most cakes contain flour, sugar, oil, eggs, water, and baking powder. If you use a cake mix, many of these ingredients are already measured out for you and all you have to add is oil, water, and eggs.

The first important step, from a chemistry point of view, is mixing the oil and the sugar. The fat molecules in the oil are creamed together with the sugar. In the process of creaming, air molecules are trapped between the fat molecules. This is what gives the cake its soft, spongy texture.

The cake batter is considered an oil in water emulsion. The sugar is dissolved in the water, and then the oil molecules are suspended in the sugar water. Eggs are added as an emulsifier to keep the fat molecules suspended. The flour molecules also help to add stabilization to the emulsion as the cake rises and bakes in the oven.

Now that you see how much chemistry is involved in making a cake, you can bake your favorite cake and enjoy the sweet suspension.

24

Solubility

How well does it dissolve?

What affects the solubility of a substance?

Words to know:

Henry's law
precipitate
precipitation

When you think of solutions you probably think of a solid solute being dissolved in a liquid solvent like sugar dissolved in water. And many solutions are made this way. But there are other kinds of solutions. Some types of liquids can be dissolved in other liquids, and gases can be dissolved in liquids as well. There are also a few solutions where a gas or liquid solute is dissolved into a solid solvent, but these are much less common.

The maximum amount of solute that can be dissolved in a given amount of solvent is called solubility. You have already learned that solubility is determined by what the substances are. The amount of salt that can be dissolved in water is different from the amount of salt that can be dissolved in rubbing alcohol. Also the amount of salt that can be dissolved in water is different from the amount of sugar that can be dissolved in the same amount of water. So, the identities of the solute and the solvent affect the solubility of the solution.

Temperature and pressure can also affect the solubility of the solution. How temperature and pressure affect the solution depends on the type of solution you have. If you have a solid solute dissolved in a liquid solvent, like salt water, the solubility increases with temperature. You already learned that the solute will dissolve faster at higher temperatures because the molecules are moving more quickly, but this also means that more solute can be dissolved when the molecules are moving more quickly. Think of the water molecules as being like a juggler. A juggler can easily keep two or three balls moving through the air. But as more balls are added the juggler must exert more energy to keep them all up. Similarly, in a salt water solution, the water can keep more salt dissolved if it is warmer and has more energy.

The temperature of the solvent does not significantly affect the solubility of the solution when both the solvent and solute are liquids. This is because the molecules of both liquids move freely to begin with so more energy is not needed to make one dissolve in the other. However, the temperature of the solution greatly affects the amount of gas

LESSON 24 Properties of Matter • 95

that can be dissolved in a liquid. In this situation the molecules of the liquid are trying to keep the gas molecules from escaping. So, increasing the temperature makes the solvent molecules move farther apart so more gas molecules can escape. Thus, for gases dissolved in liquids, the solubility decreases as temperature increases.

A change in pressure can also affect solubility of gases in liquids. Pressure change has little effect on solids and liquids because their molecules are already close together. But gas molecules are more spread out, and increasing the pressure pushes them closer together. If they are closer together they are more likely to stay dissolved, so increasing pressure increases solubility for a gas dissolved in a liquid. This property was discovered by William Henry in 1801 and is called **Henry's Law**.

When a solvent can no longer dissolve any more solute the solution is said to be saturated. If more solute is added it will settle on the bottom of the container if it is a solid, or will escape into the air if it is a gas. Also, if conditions change, for example if the temperature drops, the solubility of the solution may change and the solvent may no longer be able to hold as much of the solute as it did before, and some of the solute will come out of the solution. This is called **precipitation**, and the material that comes out of the solution is called a **precipitate**. You may have heard the term precipitation referring to rain or snow. This

Warm and cold solutions

Purpose: To see how temperature relates to gas solubility

Materials: two canned soft drinks, two clear cups

Procedure:

1. Open two soft drink cans: one that has been refrigerated and one that is at room temperature.
2. Pour from each can into a clear cup.
3. Observe the cups without touching them and answer the following questions.
   - What are the main ingredients in the solution you are observing?
   - Which cup appears to be more bubbly?
   - Does gas escape more easily from a warm solution or a cold solution?
   - Which cup contains the colder liquid?
4. Taste the solutions and determine if your hypothesis was correct.

Conclusion: The water molecules in the cold soft drink are moving more slowly than the ones in the warm soft drink so the cold one is better able to keep the gas molecules trapped.

is the same idea. When the clouds can no longer hold all of the water that is in them, some of the water leaves the atmosphere as rain or snow. This is call precipitation. When a solution can no longer hold all of its solvent, some of it leaves the solution; this is also called precipitation.

What did we learn?

- What is solubility?
- What are the three factors that most affect solubility?
- What is the name given to particles that come out of a saturated solution?

Taking it further

- Why are soft drinks canned or bottled at low temperatures and high pressure?
- Why do soft drinks eventually go flat once opened?
- If no additional sugar has been added to a saturated solution of sugar water, what can you conclude about the temperature and/or pressure if you notice sugar beginning to settle on the bottom of the cup?

Solubility of various substances

Solutions are made when one substance dissolves in another. In lesson 22 you did an experiment where candy, primarily sugar, was dissolved in water. What other substances can you think of that dissolve in water? So many substances can be dissolved in water that water is often called a universal solvent. However, not all solutes dissolve equally in water. Even though table salt and sugar look very similar and are both soluble in water, their solubilities are not the same. Similarly, different salts have different solubilities in water.

Purpose: To measure the solubilities of various substances

Materials: four clear cups, table salt, potassium salt, baking soda, sugar, "Solubility of Various Substances" worksheet

Procedure:

1. Write your hypothesis about which substance will best dissolve in water on a copy of the "Solubility of Various Substances" worksheet.
2. Pour ½ cup of room-temperature water into each of the four clear cups.
3. In the first cup, add table salt ¼ teaspoon at a time until no more salt will dissolve. Stir with a spoon after each addition. Record the amount of salt that you were able to dissolve on the worksheet.
4. Repeat step 3 with potassium salt, adding ¼ teaspoon at a time to the second cup. Record the amount you were able to dissolve on the worksheet.
5. Repeat step 3 with baking soda.
6. Repeat step 3 with sugar.
7. Answer the questions on the worksheet.

LESSON 24 **Properties of Matter** • 97

25

Soft Drinks

America's (second) favorite drink

How are soft drinks made?

Americans love soft drinks, also called soda, pop, soda pop, coke, seltzer, mineral, lolly water, or carbonated beverages. In fact, after water, soft drinks are the most popular drink in the United States. People around the world drink more than 1 billion 8-ounce soft drinks every day. That is equal to 790 gallons per second! What is it about soft drinks that makes them so popular? Not only are they sweet and flavorful, but they have a tangy bubbly taste that results from the carbon dioxide gas that is dissolved in the liquid.

Let's take a look at the process that creates America's second favorite drink. Soft drinks are 90% water. So the process starts with purifying the water. Even though the water coming into the manufacturing plant is clean, chemicals are added to the water to remove any flavor from chlorine or other minerals that may be in the water. This is necessary to ensure that the soft drink made at one plant will taste exactly like the one made at another plant. The water is tested carefully to make sure it is pure, and then it is passed through a microscopic filter to get rid of any remaining impurities.

The second step in the process is to add a sweetener to the water. Seventy percent of all soft drinks are sweetened with corn syrup. The other 30%, diet soft drinks, are sweetened with aspartame or another artificial sweetener. Corn syrup is the cheapest sweetener, so it is the most popular, although in recent years there has been an increasing trend toward using regular sugar in some sodas.

Corn syrup is produced through an interesting chemical process. Dried corn is soaked in water and sulfuric acid. Then the softened corn is crushed and the endosperm, the center of the seed that contains the starch, is separated from the rest of the corn. An enzyme, a special chemical, is added to liquefy the cornstarch. A second enzyme is added to the liquid to turn it into a sugar called dextrose. But this is not sweet enough for soft drinks, so a third enzyme is added to turn the dextrose into a sweeter sugar called fructose. The fructose is boiled to remove the water and the result is a very sweet liquid called high fructose corn syrup. This is the sweetener used in many foods including soft drinks.

Fun Fact

Water is the most popular drink in the world. And although soft drinks are second in the United States, it is believed that tea is the second most consumed beverage worldwide.

Diet soft drinks are sweetened with artificial sweeteners that have significantly fewer calories than corn syrup. The most commonly used artificial sweetener is a chemical called aspartame. Aspartame is a combination of two amino acids and is 200 times sweeter than sugar.

After the water is sweetened, the third step in the process is to add flavor and coloring to the liquid. Every brand and flavor of soft drink has a secret recipe that is carefully guarded by each company. Each flavor is a special blend of herbs, spices, oils, extracts, and acids. Cola products are flavored with an extract from the kola nut and are the most popular flavor of soft drink. In fact, 70% of all soft drinks sold are colas. Lemon-lime is the second most popular flavor, followed by pepper, root beer, and orange. All other flavors account for only 2–3% of all soft drink sales.

In addition to flavorings, color is also added to the liquid. Color is an important step in the process because color affects how we think about the drink. We expect orange flavored soft drinks to have an orange color. We expect cola flavored soft drinks to be brown. So coloring is added to enhance the perceived taste of the product. Once the flavor and color have been added, the liquid is now a finished syrup.

The fourth step in the process is to add the carbonation. The finished syrup is cooled and then sprayed into a pressurized cooler of carbon dioxide. The water in the syrup dissolves the carbon dioxide gas. The carbonated liquid is then put into bottles or cans. The containers are X-rayed for fullness and then they are warmed up and dried before being packaged in cartons or boxes. For soft drinks dispensed from fountain machines, a flavored concentrate is combined with carbonated water at the moment the drink is dispensed.

Although soft drinks are very popular in America, there are many concerns with the amount consumed. Regular soft drinks contain a large amount of sugar, and consuming too much sugar can lead to

Fun Fact

Some firsts for soft drinks:
- The first soft drink was recorded in the seventeenth century in France. It was made from lemon, honey, and water. It was similar to today's lemonade.
- Carbonation was first added to soft drinks in the eighteenth century when Jacob Schweppe added carbonation to water in 1794.
- John Pemberton first produced Coca-cola in 1886. Pemberton, a chemist living in Atlanta, Georgia, first made Coca-cola as a brain tonic.

Making your own soft drink

Although you will not be able to recreate your favorite soft drink at home, you can make some very tasty soft drinks of your own. Following are two recipes for homemade soft drinks.

Cola

In a large glass, combine the following ingredients:

- 8 ounces club soda
- 8 teaspoons sugar or corn syrup
- 1 teaspoon vanilla extract
- 1 teaspoon lemon juice
- ¼ teaspoon orange juice
- dash nutmeg
- dash cinnamon
- 4 drops yellow food color
- 2 drops red food color
- 2 drops blue food color

Orange (has a somewhat salty taste but can be good)

In a large glass, combine the following ingredients:

- 6 ounces orange juice
- 2 ounces water
- 3 teaspoons sugar or corn syrup
- 1 teaspoon baking soda (to produce carbon dioxide bubbles)

many physical problems including obesity, heart disease, and diabetes. Also, in addition to the sugar, soft drinks contain a lot of salt and caffeine, all of which can actually cause you to feel thirsty because they remove water from your system rather than quenching your thirst. So the next time you are thirsty, drink America's number one drink—water.

Fun Fact

The sweetener aspartame was discovered by accident. A scientist had been working with phenylalanine and aspartic acid and some of each ingredient got on his fingers. When he licked his finger to turn the page of a book, he noticed that it tasted sweet, and thus aspartame was born.

What did we learn?

- What are the main ingredients of soft drinks?
- What is the most popular drink in the United States? The second most popular?
- What are the two most popular sweeteners used in soft drinks?

Taking it further

- Why are soft drink cans warmed and dried before they are boxed?
- Why are recipes for soft drinks considered top secret?
- Why would the finished syrup be tested before adding the carbonation?

Regular or diet?

Do you drink regular or diet soft drinks? Most people prefer regular, but some people prefer diet because it has fewer calories. So what is the difference between regular and diet soft drinks? Regular soft drinks are sweetened with high fructose corn syrup. Most diet soft drinks are sweetened with aspartame or saccharine, which are artificial sweeteners that do not have the calories of corn syrup. If you were given two identical soft drink cans and were told to decide which one was regular and which one was diet without tasting them, could you do it? Let's try an experiment to find out.

Purpose: To determine if a soft drink is regular or diet

Materials: large bucket, cans of regular and diet soft drinks of the same brand and flavor

Procedure:

1. Fill a sink or large bucket with water.
2. Place unopened cans of regular and diet soft drinks in the water. They must be the same brand and flavor.

Questions:

- What do you observe happening?
- Which can sinks? Which can floats?
- Why do you think this happens?
- What are the differences in the ingredients between the two cans?

Conclusion: The main difference between the regular and diet drinks is the sweetener that is used. Recall from lesson 9 that buoyancy is related to the density of an object. If something is denser than water it will sink; if something is less dense than water it will float. So the densities of the two drinks must be different. Sugar is denser than water and the amount of sugar dissolved in the regular drink causes the whole can to be denser than water, so it sinks. The artificial sweetener is about 200 times sweeter than sugar so less is required to obtain the same sweetness as sugar or corn syrup. Therefore, the liquid in the diet drink is about the same density or slightly less dense than the water, so it floats.

26

Concentration

Is your lemonade weak?

How do we measure the concentration of a solution?

Words to know:

dilute   concentration

concentrated

Have you ever tasted Kool-Aid® that was too weak or lemonade that was too strong? Both of these drinks are solutions and as you learned, a solution could have a very small amount of solute dissolved in it or it could have enough solute to saturate the solution. If there is a relatively small amount of solute in the solution it is said to be **dilute**. This would be the Kool-Aid® that tastes watered down. If it has a relatively large amount of solute, the solution is said to be **concentrated**. This would be the very strong lemonade.

Dilute and concentrated are both qualitative descriptions of the concentration of solute in a solution. As we learned in earlier lessons, qualitative descriptions can be helpful, but quantitative measurements are often more helpful. The quantitative measurement of the **concentration** of a solution measures the number of grams of solute dissolved in 100 grams of solution. For example, if 5 grams of sugar are dissolved in 95 grams of water the resulting 100 grams of sugar water is said to be a 5% sugar solution. This quantitative measurement is necessary to enable scientists to be able to repeat results of experiments.

Other than getting the right concentration in your soft drink, there are many reasons for changing the concentration of a solution. Increasing the concentration of a liquid solution generally raises its boiling point and lowers its freezing point. The boiling point of the solution is higher than pure water because the dissolved molecules get in the way of the water molecules and make it harder for the water molecules to reach the surface and escape into the air. Thus it takes more energy to boil the solution. This can be very useful in an engine where you want to cool the engine without boiling away the liquid that is absorbing the heat. In order to make the engine in your car work better, antifreeze is added to the water in the radiator. This allows the solution of water and antifreeze to absorb heat from the engine and not boil over.

Similarly, antifreeze keeps the water in your radiator from freezing, even when the temperature is below 32°F—the freezing point of water. In order for water to freeze, the molecules must line up and form crystals. In a solution, the solute keeps the water molecules separated and makes it harder

LESSON 26   **Properties of Matter • 101**

🧪 Making ice cream

One very fun application of a salt water solution is used in the making of ice cream. Salt is added to the ice that surrounds the ice cream container. The salt lowers the freezing point of the water and thus allows it to absorb more heat from the ice cream mixture, causing the cream mixture to freeze more quickly.

Purpose: To observe the effects of salt on ice by making ice cream

Materials: small and large plastic zipper bags, milk, sugar, vanilla, ice cubes, salt

Procedure:

1. In a small sandwich-sized plastic zipper bag combine:

    ½ cup milk

    1 tablespoon sugar

    $\frac{1}{8}$ teaspoon vanilla extract

2. Zip the bag shut and make sure it is sealed.
3. Place the small bag inside a larger zipper bag.
4. Add 2 cups of ice cubes and 2 teaspoons of salt into the large bag.
5. Zip the larger bag closed and shake the bags for several minutes until the milk mixture freezes. If it does not seem to be freezing after several minutes, add another teaspoon of salt to the ice mixture.
6. When the milk mixture is frozen, remove the small bag.
7. Carefully wipe off the salt water from the outside of the bag, and then open it and enjoy your sweet dessert!

for the water to freeze. Therefore, the temperature must get much colder before the solution will freeze. This helps your car engine keep running even in very cold weather. Another important application of solutions is the sprinkling of salt on sidewalks in the snow. The salt molecules become dissolved in the water and this lowers the temperature at which the water freezes, so it is less likely that ice will form on the sidewalk. ✳

Rock salt melts snow because it lowers the temperature at which water freezes.

What did we learn?

- What is a dilute solution?
- What is a concentrated solution?
- How does the concentration of a solution affect its boiling point?
- How does the concentration of a solution affect its freezing point?

Taking it further

- Why is a quantitative observation for concentration usually more useful than a qualitative observation?
- If a little antifreeze helps an engine run better, would it be better to add straight antifreeze to the radiator? Why or why not?

Salt water

We have learned that salt changes the freezing and boiling point of water, but how much does it change? Let's find out.

Purpose: To see how much salt affects the freezing and boiling point of water

Materials: salt, water, freezer, sauce pan, stove, two cups, thermometer, "Salt's Effect on the Freezing and Boiling Point of Water" worksheet

Freezing point—Procedure:

1. Dissolve as much salt as you can into 1 cup of water.
2. Pour ½ cup of the salt water into a cup and place it in the freezer.
3. Pour ½ cup of plain water into a second cup and place it in the freezer as well.
4. Measure the temperature of the liquid in each cup every 2 minutes until the liquids begin to freeze. Record your measurements on the "Salt's Effect on the Freezing and Boiling Point of Water" worksheet.
5. Complete the questions on the worksheet.

Boiling point—Procedure:

1. Pour ½ cup of plain water into a sauce pan.
2. Heat the water on medium heat.
3. Measure the temperature every minute until the water begins to boil. Record your measurements on the worksheet.
4. Remove the pan from the heat and pour the water out of the pan.
5. Pour ½ cup of the salt water from the previous experiment into the pan.
6. Heat the mixture over medium heat.
7. Measure the temperature every minute until the saltwater begins to boil. Record your measurements on the worksheet.
8. Remove the pan from the heat.
9. Answer the questions on the worksheet.

27

Seawater

The world's most common solution

What is seawater composed of?

Challenge words:

salinity   conductivity

As we have already learned, a solution is formed when one substance is dissolved in another substance. We have also learned that water is one of the best solvents around. Because of this, when rain falls and water flows over the surface of the earth, the water dissolves many minerals that are found on the ground. These minerals are carried into streams and rivers, and eventually end up in the oceans. Therefore, seawater, which covers about 71% of the surface of the earth, is the largest and most common solution in the world.

The most common mineral dissolved in seawater is salt. But many other minerals are found there as well. Magnesium and bromide, along with sodium chloride (table salt), are available in high enough concentrations in seawater to be removed and sold commercially. All together, there are 55 different elements that have been identified in seawater. In 1,000 grams of water there are about 35 grams of minerals. Of the minerals in the seawater, about 75% is salt, although the salt concentration varies from place to place. This gives seawater a salt concentration of about 3%: three grams of salt in one hundred grams of seawater.

In addition to minerals, gases such as oxygen, nitrogen, and carbon dioxide are dissolved in the water. The concentration of oxygen is highest near the surface of the ocean. Some oxygen dissolves from the air into the water, but most of it is produced by algae that grow near the surface of the water.

The ocean is a wonderful solution of water, minerals, and gases that God has provided to support life on Earth.

What did we learn?

- What is the most common solution on Earth?
- What are the main substances found in the ocean besides water?
- How does salt get into the ocean?
- Name one gas that is dissolved in the ocean water.

Taking it further

- Why is seawater saltier than water in rivers and lakes?
- Why is there more oxygen near the surface of the ocean than in deeper parts?

Making seawater

Activity 1

Purpose: To make your own seawater solution

Materials: salt, straw, water, cup

Procedure:

1. Add 1¼ teaspoons of salt to 1 cup of water.
2. Stir until the salt is dissolved. This makes a solution that is about 3% salt water.
3. Next, dissolve gases in the water by gently blowing air bubbles through a straw into the water for about 15 seconds. The air coming out of your lungs contains oxygen, nitrogen, and carbon dioxide—the three most common gases that are found in seawater.

Conclusion: You now have a solution that is close to the composition of seawater. Actual seawater has other minerals in it, but they are usually found in small quantities, so your solution is close to actual seawater.

Seawater is more dense than freshwater so you will be more buoyant in the ocean than in a swimming pool. In fact, many people like to swim in the Great Salt Lake in Utah because it is even saltier than the ocean, thus making it easy to float in.

Activity 2

Purpose: To demonstrate density of seawater

Materials: cup, egg, water, seawater solution

Procedure:

1. Carefully place an egg in a cup of fresh water. What happens? (It sinks to the bottom.)
2. Now place the egg in your seawater. Does it float above the bottom of the cup? The egg should float. If it does not, remove the egg and stir in an additional teaspoon of salt, then try it again.

Salinity

The amounts of salt and other substances that are dissolved in seawater are fairly constant; however, there are several factors that affect the actual composition of the mixture in any particular location at any particular time. The amounts given earlier in the lesson are averages; the amount of salt, gases, and other substances will be different at different locations. **Salinity** is a measure of the amount of salts dissolved in water. The salinity of seawater varies from place to place and varies with temperature. The varying salinity affects many of the properties of the ocean. How do you think salinity affects boiling point and freezing point of the seawater? The higher the salinity, the higher the boiling point and the lower the freezing point will be. This is what you demonstrated in the previous lesson.

Salinity also affects the electrical **conductivity** of the water. Pure water does not conduct electricity at all. However, as the salinity rises, the conductivity of the seawater increases. In fact, scientists often calculate the sea's salinity by measuring its conductivity.

The amounts of various gases in a sample of seawater are also dependent on the salinity of the water. How do you think salinity affects the amount of gas that will dissolve in the seawater? The more salt there is in the water, the fewer water molecules there are available to dissolve and hold the gases in the solution. Therefore, as salinity increases, the concentration of gases goes down. Temperature also affects the amount of gas that is dissolved. As you learned earlier, the warmer the solvent, the less gas it can dissolve, so as water temperature goes up, there is less gas dissolved in the seawater.

Salinity and temperature also affect the density of the seawater. You saw that the saltier the water was, the better able it was to support an egg. This is because the higher the salinity, the denser the seawater becomes. How do you expect temperature to affect the density of seawater? The colder the water is, the closer the molecules are together so the denser it is. Therefore, you may find a layer of less salty warm water on top of a layer of saltier, colder water. The composition of the seawater may not be constant as you go down in depth.

Another factor affecting the composition of seawater at various depths is the animal and plant life in the water. Since plants can only grow where there is sunlight, you will not find plants more than a few hundred feet down. Therefore, the gases that are produced by the plants, primarily oxygen, will be found in higher concentrations near the surface of the ocean than in deeper areas.

Rain or other fresh water entering the ocean will also affect its composition. If there is a river nearby, the salinity will be less than it is farther from the source of the river. If a hurricane or series of storms dumps a large amount of rain, the seawater will become diluted in that area for a time.

Even though there are a great number of factors affecting the composition of seawater, the water is always moving so there is constant mixing of the water. This helps to keep the concentration of salt and other minerals fairly constant throughout the oceans.

It's easy to float in the Dead Sea because of the very high salinity of the water.

Desalination of Water

SPECIAL FEATURE

Desalination is the process of removing salt from ocean water to produce fresh water. Many places in the world do not have enough fresh water for drinking, irrigating, and manufacturing. While other areas have more than enough water, the cost of transporting it is too great. Therefore, many people are trying to find ways to take salt water from the ocean and turn it into fresh water.

One way this is done is through distillation. Distillation is a process in which water is heated until it evaporates, leaving the salt and other minerals behind. The steam is then allowed to condense in another container to form fresh water. One of the earliest recorded uses of distillation occurred around 50 B.C. when the armies of Julius Caesar made distillation devices using the sun's heat to evaporate water from the Mediterranean Sea. These units were small and therefore could only make small amounts of fresh water.

In addition to distillation, man is exploring reverse osmosis, electrodialysis, and vacuum freezing as other desalination processes. Reverse osmosis is the most common of these. Reverse osmosis forces water through a membrane that only allows water molecules to pass and blocks all the larger molecules. One osmosis process uses hollow membrane spheres that are pulled under the surface of the ocean. The water pressure of the ocean forces water into the sphere, while the membrane blocks the salt and other minerals, leaving them outside the sphere.

One group of people that has always been interested in desalination is sailors. To combat the fear of dying at sea for lack of fresh water, many different ideas were developed to turn salt water into fresh water. In 1791 Thomas Jefferson, then Secretary of State, had a technical report printed on the back of all papers on board the ships describing how to use distillation as a source of fresh water, in case of an emergency. Later, steam ships used distillation units to make the fresh water from seawater for their boilers. This reduced the cost of running steam ships by allowing the cargo bays to carry cargo instead of fresh water. By the time of World War II, mobile desalination units were in wide use. Because of many fresh water shortages after World War II, governments and private parties around the world began working to find ways to reduce the costs of desalination, while increasing production.

Saudi Arabia and other countries in the Middle East have led the way in developing desalination plants. Some of the largest desalination plants include the Ashkelon sea water reverse osmosis plant in Israel, which opened in 2005, and the Ras Al Khair desalination plant in Saudi Arabia, which came online in 2014. California is also building a very large desalination plant in Carlsbad which is projected to be able to produce over 50 million gallons of fresh water per day. Over 40% of the domestic water used in Israel is produced by desalination. There are over 16,000 desalination plants worldwide which produce enough water for more than 300 million people, and that number is increasing every year.

The Ras Al Khair desalination plant in Saudi Arabia

28

Water Treatment

Making it clean

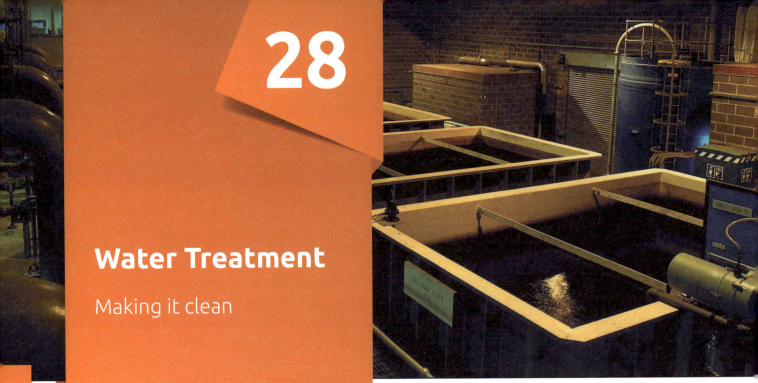

How can we treat drinking water to make it safe?

Words to know:

sedimentation

Challenge words:

hard water   water softener

limescale

Because water is a nearly universal sol-vent, many substances become dissolved in it. This can be very useful if you are trying to wash away dirt or other substances. However, this can be very bad if you are trying to obtain good drinking water. Most water we use today comes from rivers, lakes, and streams. This water is not clean enough to be safe to drink because of the substances dissolved in it as well as the particles suspended in it. Therefore, most towns and cities have freshwater treatment facilities that clean up the water and remove unwanted and harmful substances before the water gets to your house.

The first stage in water treatment is called **sedimentation**. The water is placed in a large holding tank and large particles are allowed to settle to the bottom. These particles are not actually dissolved in the water but are suspended by the movement of the water. When the water sits still for a length of time, these particles fall to the bottom of the tank and are removed.

The water is then sent through a filter that removes most smaller particles that are still suspended in the water. The water then goes to a mixing basin where chemicals are added to the water. Chlorine is added to kill bacteria and to eliminate odors in the water. Fluoride is added to the water in many cities to help strengthen teeth. And alum is added because it causes any remaining particles to clump together. This water is sent to another settling basin where the remaining particles precipitate out and fall to the bottom of the tank and are scraped away.

Fun Fact

Some homes are not connected to a sewer system. Instead, they have septic systems that filter out large particles and then allow the wastewater from the home to flow into the ground where the natural filter of the soil, sand, and rocks purifies the water before it enters the water table.

🧪 Cleaning our water

Purpose: To make your own mini-water treatment plant

Materials: 2-liter soft drink bottle, cotton balls, sand, charcoal briquettes, plastic zipper bag, goggles, hammer, small gravel or pebbles, dirt, two cups, alum

Procedure:

1. Have an adult cut off the bottom of a 2-liter soft drink bottle.
2. Tightly stuff the top or mouth of the bottle with cotton balls.
3. Turn the bottle upside down and add ½ cup of sand to the bottle.
4. Place 3 charcoal briquettes in a plastic zipper bag.
5. While wearing goggles or other eye protection, use a hammer to break the charcoal into very small pieces.
6. Place the charcoal pieces in the bottle on top of the sand.
7. Add a 2-inch thick layer of small gravel or pebbles.
8. Make some dirty water by adding a small amount of dirt to a cup of water and stir thoroughly.
9. Pour ¼ of the water into another cup for later comparison.
10. Allow the rest of the water to settle for 5 minutes and notice that much of the dirt settles to the bottom.
11. Pour this water into another cup; be sure not to pour the sediment into the cup.
12. Add 1 teaspoon of alum to the water.
13. Stir for thirty seconds and again allow the water to settle for 5 minutes. Notice that more of the particles have clumped together and settled to the bottom.
14. Have someone hold your bottle filter over a dish or cup. Carefully pour most of the water into the bottle. Be sure not to pour any of the particles from the bottom of the container into the filter.
15. Allow the water to drip through the filter.

Conclusion: Compare the water that comes out of the filter with the original water you saved. The filtered water should be much cleaner. Even though this water looks clean, do **not** drink it. It still may contain harmful bacteria.

Finally, the water is sent through one last filter to eliminate any remaining unwanted particles. The filter usually consists of a layer of gravel, then a layer of charcoal, then a layer of sand. In some treatment facilities a final amount of chlorine is added to the water and then it is released into the main water pipes where it flows to homes and other buildings for people to use.

After the water is used in homes and offices, the water in most areas goes into the sewage system where it flows to the wastewater treatment facility. Because used water often has bacteria and other harmful substances in it, it needs to be cleaned up before it is returned to the streams and rivers.

At a wastewater treatment facility, the water first flows through a series of bars that trap large objects and remove them from the water. Next, the water is placed in a holding tank where smaller particles settle to the bottom. The water is then drained away leaving behind a slimy layer on the bottom of the tank called sludge. The sludge is treated with a helpful type of bacteria that kills the harmful bacteria and is then spread on fields and used as fertilizer. The water is also treated with a type of bacteria that eats harmful bacteria and dirt. The water is then sent to another holding tank and then released into a river or stream.

This cleaned water is then safe for wildlife, although it may not be safe for human consumption. It is important that we keep our water clean so that we do not harm wildlife or people. Water can become polluted when people dump harmful chemicals, oil, or other dangerous substances in the water. These items need to be disposed of in a responsible manner so that we are good stewards of the world God has given us.

Fun Fact

The first recorded case of man trying to improve his water supply is in Exodus, when the Israelites were at Marah. God had Moses throw a tree branch into the water to make the bitter water sweet. Other examples include drawings estimated to be from around 1400 BC showing Egyptians using sedimentation methods to improve their drinking water. Around 500 BC Hippocrates invented a special cloth to strain rainwater called the *Hippocrates Sleeve*.

What did we learn?

- Why do we need water treatment plants?
- What are the three main things that are done to water to make it clean enough for human consumption?
- Why is it important not to dump harmful chemicals into rivers and lakes?

Taking it further

- How is the filter you built similar to God's design for cleaning the water?

Hard water

Not all water treatment takes place at a water treatment plant. Some people have water treatment systems in their homes. Although water treatment plants remove harmful substances such as bacteria, and often add fluoride to the water, the water treatment process does not remove every element that is dissolved in the water. Therefore, some areas have what is called hard water. **Hard water** is water that contains dissolved minerals, particularly calcium and magnesium.

Hard water is not a health hazard. The minerals in the water do not harm people. However, these minerals can cause problems that people wish to avoid. First, the minerals in hard water react with soap to prevent the soap from lathering. Instead, the soap and minerals form a soap scum, which is hard to rinse off of your dishes or skin. The minerals in hard water also react with laundry soap and can cause your clothes to feel rough and scratchy and to look dingy.

Another problem with hard water is that the calcium and magnesium precipitate out of the water and can form a scaly substance called **limescale**. Limescale can build up in pipes and machines that use water and eventually clog the pipes or ruin the machinery. This is especially a problem with boiler systems and water heaters that heat water since heat causes the chemical reactions to occur faster and thus increases the problem.

There are two main ways to eliminate hard water. One of the most common methods is to install a water softener. A **water softener** is a system that water flows through as it enters your house. The system contains a water softening agent, usually sodium carbonate, which reacts with the calcium and magnesium to form salts that do not cause these problems. Other water softening systems contain a material called zeolite, which is sodium aluminum silicate. When the water passes over this material, the calcium and magnesium ions are replaced with salt and the water becomes soft.

A second method for treating hard water is to install a water filter that removes some of the minerals as the water passes through the filter. Water filters are sometimes preferred to the chemical water softeners because the water is not salty after treatment. Water filtering is a more expensive method so is not as widely used. Regardless of the method used, water softening is a form of water treatment that occurs in many homes.

UNIT 6

Food Chemistry

29 Food Chemistry
30 Chemical Analysis of Food
31 Flavors
32 Additives
33 Bread
34 Identification of Unknown Substances—Final Project
35 Conclusion

◊ **Describe** how chemistry is involved in the food we eat.
◊ **Identify** the use of food additives.
◊ **Describe** the importance of chemical reactions in making bread.

Properties of Matter • 111

29

Food Chemistry

You are what you eat.

What chemicals are in our food?

Words to know:
carbohydrate fat
protein

One of the most interesting places to learn about chemistry is in the kitchen. Our bodies are made up of a very complex collection of chemicals, and the food we eat is a collection of chemicals. So there are a lot of chemicals to be found in the kitchen. The first place to begin looking for chemicals is on food labels. Foods consist of three major types of chemicals: carbohydrates, proteins, and fats. So these three items will be listed on most food labels.

Carbohydrates are molecules made from carbon, oxygen, and hydrogen. There are two types of carbohydrates: sugars and starches. Carbohydrates occur naturally in many fruits, vegetables and grains. Starches are found in wheat, rice, oats, other grains, and potatoes. Good places to find foods with carbohydrates are in a loaf of bread and a box of cereal. Natural sugar is found in many foods containing fruits and vegetables, but sugar is also added to many other foods as well, including many breakfast cereals. When you look at the list of ingredients don't be fooled. Sugars are called by many names, so even if the list does not include the word "sugar" the food may still have carbohydrates from sugar. You can look for words like sucrose, fructose, or high fructose corn syrup. These are different types of sugar.

To find proteins you can look for foods that contain meat, nuts, and many types of beans such as navy, pinto, or black beans. Proteins are also found in many dairy products such as milk, cheese, and ice cream. **Proteins** are made from molecules called amino acids that consist of carbon, hydrogen, oxygen, and nitrogen atoms.

Fats are very long molecules that are found in many foods. Fats are found in nuts, meats, and many dairy products. Fats can also be found in many vegetables. Fats are also found in snack foods and anything made with oil, margarine, or butter.

Food chemistry does not end with the carbohydrates, proteins, and fats. There are many other chemicals in our food. First, foods also contain vitamins and minerals, which are all chemical compounds your body needs. Second, many foods have chemicals added to them such as preservatives, flavor enhancers, and color enhancers. Finally, some chemicals are added to foods to change their texture. For example, baking soda and baking powder are added to cakes, cookies, and other baked items to make them light and fluffy.

The naturally occurring chemicals are needed by your body. Carbohydrates are your body's main source of energy. During photosynthesis plants lock the sun's energy in the carbohydrates they produce, then your body releases and uses that energy when it digests the sugars and starches in the food you eat. Proteins are used as building blocks for most of the tissues in your body so it is important to eat enough protein every day. Fats, vitamins, and minerals are also necessary for your body to function correctly. So make sure you eat a wide variety of healthy food so that the great chemical lab called your body will have all of the chemicals it needs to help you grow strong and healthy.

What did we learn?

- What are the three main types of chemicals that naturally occur in food?
- What kinds of chemicals are often added to foods?
- Why is the kitchen a great place to look for chemicals?

Taking it further

- If you eat a peanut butter and jelly sandwich, which part of the sandwich will be providing the most carbohydrates? The most fat? The most protein?

Baking soda—a vital chemical

Purpose: To demonstrate the importance of chemical reactions in our foods

Materials: cookie recipe and necessary ingredients

Procedure:

1. Make a half batch of your favorite cookie dough following the recipe, except leave out the baking soda.
2. Next, make a half batch of the same cookie dough with the baking soda included.
3. Bake several cookies from each batch of dough. Be sure to keep track of which cookies have baking soda and which do not.
4. After baking the cookies, observe the cookies and then taste them.

Questions:

- How do the cookies with baking soda look compared to the cookies without baking soda?
- Are there differences in color, shape, or texture?
- What differences do you notice in the taste? Which cookies taste better?

Conclusion: Baking soda is a chemical called sodium bicarbonate. Baking soda is a base and it reacts with an acid. The cookie dough, when moistened, is acidic enough to cause a bubbly reaction. That reaction releases carbon dioxide, which becomes trapped in the dough and makes the cookies light and fluffy.

Food research

As you learned, there are a wide variety of chemicals in food. Do some research and answer the questions on the "Food Chemicals" worksheet.

Genetically Modified Foods

SPECIAL FEATURE

For centuries, farmers and scientists have worked to provide the best food at the best prices. Since the 1850s when Gregor Mendel first discovered that traits were passed down from parent plants to children plants, scientists have used selective breeding to grow plants that produce more and bigger fruit, resist insects and diseases, and better survive the trip from the farm to the store. However, recently scientists have discovered a new way to try to improve crops. This relatively new method is called genetic modification (GM).

Scientists can change specific genes in a seed's DNA to make it produce a plant with desired characteristics, or suppress an undesirable trait. Corn has been developed that kills insects that try to eat it. Genetically modified soybeans are able to survive even when sprayed with weed killer. Some GM tomatoes can stay firm even after they are ripe so they don't become damaged when they are shipped to the store. And a type of rice is being developed that will have a higher amount of vitamin A so that it has more nutrition than regular rice.

The stated goals by those companies developing these genetically modified crops is to produce better food at a lower cost, and to find ways to grow crops in otherwise unfavorable conditions, such as in very dry regions. If these goals are met, everyone wins, they say. The world can produce more food and people will spend less money buying it.

However, there are concerns about GM crops and many people claim that these crops do not meet these goals. For instance, many of the GM crops do not produce more food than non-modified versions, and they are not less expensive than traditional crops. In fact, the GM tomato

One advantage of GM fruits and vegetables could be a longer shelf life.

was taken off the market because tomatoes were developed through selective breeding that had the same characteristics at a lower price.

There are also health concerns about GM crops. For example, in several animal studies those animals fed GM foods had fewer and smaller babies. Some rats that were fed GM foods showed excessive cell growth, which could lead to cancer. Some animals developed liver and kidney damage after just three months of being fed GM corn. And other animal studies have linked GM foods with increased toxins in the body, allergic reactions, and intestinal damage.

GM crops could harm the environment as well. Insects that don't eat the GM crops could be killed when they eat nearby plants that have pollen on them from the GM crops. Some weeds might become resistant to weed killers when the modified genes get passed from the GM crops to

Seeds from GM crops are known to spread into non-GM farms, which could cause an organic farm to lose its certification.

the weeds that grow near them, producing "super weeds." And seeds from GM crops are known to spread into non-GM farms, which could cause an organic farm to lose its certification.

On the other hand, the FDA (Food and Drug Administration), which is responsible for certifying that the food sold in the United States is safe does not require labeling of GM foods because testing has shown GM foods to be essentially identical to non-GM foods. Nearly all of the corn and soybeans grown in the U.S. are genetically modified varieties and have been for many years. So many people claim that GM foods are safe for people to eat. Supporters of GM foods argue that those who are fearful of GM products have not proven there are any health risks while the producers have shown their products to be safe.

Genetic modification is a new science and there are many problems that need to be addressed. It is possible that in the future scientists and farmers may decide that GM crops are not worth it, and will choose to concentrate on breeding better crops, as we have done for thousands of years. It is also possible that scientists may work out many of these problems and develop crops that are beneficial to us all, without many of the harmful side effects. So watch the news and see what happens with GM crops in the next few years. There are many websites and books about genetically modified organisms to help you become informed about the issue.

30

Chemical Analysis of Food

How do I know what I'm eating?

How do we analyze the food we eat?

Words to know:
translucent   indicator

Challenge words:
calorie   calorimeter
food calories

The main chemicals that naturally occur in food are carbohydrates, proteins, and fats. Foods also contain water, vitamins, and minerals. If you look at the labels on most foods, they will tell you how many grams of carbohydrates, protein, and fat are in each serving. The label will probably also tell you which vitamins and minerals are present in the food. This information can be very helpful in deciding which foods to eat and which foods to avoid. You may want to limit the amount of salt or fat in your diet, and food labels can help you do that. But how did the manufacturer of the food determine the chemical composition of the food?

Scientists have designed many tests that help determine the chemical composition of food. Some of these can be done easily at home. Others require special chemicals and equipment that are not readily available. A simple way to detect the presence of fats, particularly fats that are in the form of oil, is called the brown paper test. If you place a sample of food that contains oil on a piece of brown paper, like a shopping bag or lunch bag, the oil will cause the paper to become **translucent**—it will allow light to pass through it. If the sample only contains water, the paper will not be translucent, but will only look darker. It's simple to test for the presence of fats.

116 • Properties of Matter   LESSON 30

🧪 Analyzing your food

Purpose: To test for the presence of oils and starch

Materials: bread, water, vegetable oil, peanut butter, potato or tortilla chips, apple slice, flour, brown paper, iodine, "Chemical Analysis" worksheet

Testing for Oils—Procedure:

1. Place a sample of each of the foods listed on the worksheet on a piece of brown paper. Write the name of the sample next to it on the paper. Allow the sample to sit for 5 minutes.

2. While you wait, record your prediction about which samples you think will contain fat/oil in the first column of the chart on the "Chemical Analysis" worksheet.

3. Wipe off any excess sample that has not evaporated.

4. Hold the paper up to the light and notice which samples caused the paper to become translucent (allow light to pass through). Write your observations on the worksheet.

5. Check the food label (if available) to see if oil is listed in the ingredients. Write your answers on the worksheet.

6. Write "yes" in the last column if your test shows that the sample contains oil. Compare this with your prediction.

Testing for Starch—Procedure:

1. Predict which samples you think will contain starch. Write your predictions in the first column of the chart on the worksheet.

2. Test each of the foods listed on the worksheet for the presence of starch by placing a drop of iodine on each sample.

3. Record the color of the iodine after it combines with the food.

4. Check the food label (if available) to see if grains such as wheat, rice, or barley are listed in the ingredients. Write your answers on the worksheet.

5. If the iodine turned blue or greenish it indicates the presence of starch. Write "yes" in the last column if starch was detected. Compare this result with your prediction.

Testing for carbohydrates and proteins is a little more difficult. There are two kinds of carbohydrates in foods: sugars and starches. The element iodine will chemically combine with starch molecules to form a substance that changes from the red/orange of the iodine to a blue color. This is a quick test for starch. A substance that is used to detect the presence of another substance, usually by changing color, is called an **indicator**. Iodine is an indicator used to test for starch. Other indicators are used to test for other chemicals. Benedict's solution is an indicator used to test for glucose or sugar. Biuret solution is used to test for the presence of protein, and indophenol solution is used to test for vitamin C. You probably don't have most of these indicators in your home, so it is more difficult to test for some of these chemicals. Chemists use many other chemicals to test for various substances. These tests enable scientists to make food labels that tell us what is in our food so we can make good choices about the foods we eat. ✳

🧠 What did we learn?

- What are the main chemicals listed on food labels?
- How do food manufacturers know what to put on their labels?
- What is one way to test if a food contains oil?
- What is an indicator?

🚀 Taking it further

- How do you suppose indicators work?
- Why is it important to know what chemicals are in our food?

Calories

One important property of food that is found on most food labels is the number of calories the food contains. The calorie is a measure of the amount of energy in the food. In the metric system, a **calorie** is defined as the amount of energy needed to raise the temperature of one gram (or one cubic centimeter) of water by one degree Celsius.

Food contains a large amount of energy, so when talking about the energy in food we are really talking about a kilocalorie or 1,000 calories. This unit of measurement is often written with an uppercase C instead of a lowercase c, but in common usage the energy in food is just called calories. But you need to keep in mind that **food calories** are equal to 1,000 metric calories.

So how does a scientist determine the amount of energy in a sample of food? A special device called a **calorimeter** is used (see illustration below). A sample of the food is placed in the center of the device, in what is called the bomb. This is not an explosive device; it is just a chamber with oxygen where the food is burned. Surrounding the bomb is a container of water. As the food burns, it releases heat, which in turn raises the temperature of the water. Often, a calorimeter will contain a small mixer that will move the water around the bomb container so that the temperature is consistent throughout the sample of water.

The temperature of the water is measured before the food is burned and after the food is completely burned. Using the volume of the water and the change in temperature, a scientist can calculate the number of calories that were released from the food. Scientists carefully perform these experiments on many samples of food to determine the number of calories in various foods.

It has been determined that one gram of protein contains four calories of energy, one gram of carbohydrates contains four calories of energy, and one gram of fat contains nine calories of energy. Thus, foods that have a high amount of fat have more calories than foods that are low in fat for the same serving size. We all need calories every day. Our body uses the energy in our food to keep our bodies moving and growing. However, when you eat food with more calories than your body needs, the excess energy is stored in the form of fat in your body. Your goal should be to eat a variety of foods every day to get the amount of energy your body needs, without eating more calories than you need.

Fill out the "How Many Calories Did I Eat?" chart. For 24 hours write down everything you eat or drink and how many calories were in it. Use the calories on the food label and the serving size to determine how many calories you took in. If you eat something that does not have a nutrition label, you can use the calorie chart to estimate the calories in your food. At the end of 24 hours, calculate your total calorie intake. It is recommended that teen girls take in about 2,200 calories per day and that teen boys take in about 2,800 calories per day. The amount of energy you actually need may be different depending on your size, age, and activity level.

Calorimeter

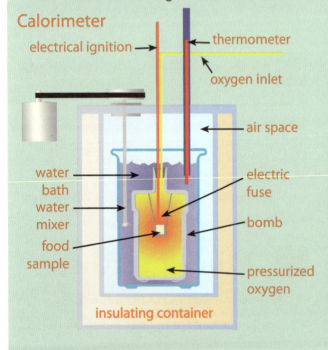

31

Flavors

Chocolate or vanilla?

How are our foods flavored?

What is your favorite flavor of ice cream?

For many Americans it is chocolate or vanilla. Flavor is a combination of taste and smell. Flavors are chemical compounds that stimulate the taste buds in your mouth and the olfactory (smell) receptors in your nose. When you take a bite of ice cream or a swallow of juice, food molecules touch your taste buds and gas molecules go into your nasal cavity to allow you to fully enjoy the flavor.

God has created a nearly endless variety of flavors. Many are very pleasant and others are very unpleasant. The chemicals that produce the flavors we like may be present in only certain parts of a plant or may only be present when the fruit is ripe. Some plants must be heated to develop the flavor.

Herbs are plant leaves that are used to add flavor to our cooking. Some common herbs you may find in your kitchen are basil, oregano, parsley, and sage. Herbs can be used fresh, but often the leaves are dried and crushed so they can be stored and used as needed.

Spices are flavors that come from roots, stems, and seeds of plants. Some common spices include pepper, cinnamon, and nutmeg. In the past, spices and other natural flavors have been very expensive. In fact, finding a new trade route to India and China, where most of the spices were grown, was one reason Columbus sailed west across the Atlantic Ocean. However, most flavors are much more affordable today than in the middle ages. Scientists have learned to isolate the desired chemical compounds and concentrate them. You can buy vanilla or almond extract and have access to these flavors much more easily today than you could have several hundred years ago.

In addition to natural flavors, scientists have learned to recreate many flavors by copying the chemical composition of particular flavors. They have also learned to make new chemical combinations that have flavors without any natural counterpart, including bubble gum, spice, and smoke flavors. Artificial flavors are usually less expensive

Parsley

LESSON 31 **Properties of Matter • 119**

🧪 Enjoying your favorite flavor

You can enjoy your favorite flavor by mixing your favorite flavor of instant pudding with milk and then eating it together.

🧪 Taste and smell

Purpose: To see how sense of smell affects flavor

Materials: Life Savers candy

Procedure:

1. Plug your nose and have someone place a Life Savers candy in your mouth without telling you what flavor it is.
2. Keep your nose plugged and try to figure out what flavor the candy is.
3. Do not bite the candy. Keep concentrating on the flavor. Does the flavor change as the candy dissolves in your mouth?
4. Repeat the experiment without plugging your nose. Was it easier to determine the flavor of the candy?

Conclusion: You should find that without the use of your sense of smell you can only detect sweetness and possibly some sour, but it is difficult to determine the flavor of the candy. It is possible that after a period of time the flavor became more distinct as some of the scent molecules were able to enter your nasal passage from your throat, allowing you to smell the candy as well. It should be much easier to determine the flavor of the candy when your nose is not plugged.

Food Chemistry

than the natural flavors they are replacing. However, even though the artificial flavor may have the same chemical formula as the natural flavor, the experienced tongue can usually tell the difference.

Your kitchen is full of various flavors. Take a few minutes to smell and taste the different herbs and spices and other delicious flavors found there.

Fun Fact

There are about 200 natural flavors and 750 artificial flavors used in the food industry today.

🧠 What did we learn?

- What two parts of your body are needed in order to fully enjoy the flavor of your food?
- What is the difference between an herb and a spice?
- What is the difference between a natural flavor and an artificial flavor?

🚀 Taking it further

- Why might a cook prefer to use fresh herbs rather than dried herbs?
- Why do you think artificial vanilla tastes different than natural vanilla even though they may have the same chemical formula?

Essential oils

The chemistry of herbs and spices does not end in the kitchen. Leaves, flowers, seed, bark, and other parts of plants are used in many other applications besides flavoring food. One of these applications is to make essential oils. Essential oil does not mean that the oil is necessary or essential for a person's well-being as the name might suggest. Instead, the name comes from the word essence, since the oils are the essence of the plant's fragrance.

In order to make an essential oil, the part of the plant containing the desired fragrance must be processed to remove the oil. This is most often done by distillation. The plant is placed in an apparatus that forces steam through the plant. This causes the oils containing the fragrance to vaporize. The vapor is then allowed to condense in a separate container. This process is very similar to the fractional distillation described in lesson 20 since the oil condenses at a different temperature from the steam. Some oils are removed by pressing the plant. This is similar to how olive oil is removed from olives. And a few oils are removed when chemicals are added to the plant to react with the oils and draw them out.

Once the oils are removed, they can be used for many different purposes. They are used to make perfumes and are added to cosmetics, soaps, and incense. As we mentioned earlier some of these oils such as vanilla and almond extract are used to flavor foods. And some essential oils are used as cleaning supplies or insect repellents.

In recent years the use of essential oils for aromatherapy and other medical uses has increased. Many people claim that inhaling certain essential oils can have a calming effect or can help with headaches or other medical problems. Other people apply small amounts of essential oils to their skin where it is quickly absorbed into the blood stream to help with various medical problems.

There is much controversy over whether essential oils really help in most of these medical situations. The supporters of essential oils claim that people have been using these remedies for hundreds of years and they have been proven to be effective. Opponents claim that most of these remedies have never been scientifically tested and do not really provide the cures that many claim they do. This is an area in which more scientific study is needed. But even without the medical uses for essential oils, the chemistry of producing them and the different ways in which they are used in other industries is very interesting.

Chocolate & Vanilla

SPECIAL FEATURE

Although Americans love chocolate they eat only about half as much chocolate as their European cousins. Americans eat an average of 11 pounds of chocolate per person per year, but Europeans eat more than 22 pounds of chocolate per person per year.

Two of the most popular flavors in the world are chocolate and vanilla. Because cacao trees and vanilla plants only grow in tropical areas, both of these flavors were known to the inhabitants of Mexico and South America long before the Europeans discovered their wonderful properties. But with the Spanish discovery of the New World, these flavors were taken to the Old World where they quickly became extremely popular.

The word *vanilla* comes from the Spanish word for *little pod* and was so named by Cortez in 1519. Vanilla flavoring is widely used in sweets, cough syrups, medicines, and is almost always found in chocolate flavored items. The Aztecs are credited with being the first to combine the delicious flavors of chocolate and vanilla. And the Spanish took these flavors back to Europe.

In 1602 a doctor named Hugh Morgan made a medicine for Queen Elizabeth that was flavored with vanilla. This is believed to be the first European use of vanilla by itself, without chocolate, and led to the use of vanilla as a separate flavor in its own right. Queen Elizabeth was so taken with vanilla that it is recorded that later in life she refused to eat foods that were not flavored with vanilla. Today, vanilla can be found in nearly every American and European household. America imports about three million pounds of vanilla every year.

Vanilla flavoring comes from the pods of the vanilla plant. These 5–10 inch long pods are long and slender, and look somewhat like green beans when they are on the plant. The vanilla pods are the fruit of the orchid-like flowers on the plant. On vanilla plantations, the flowers are hand pollinated to ensure the best results.

Once the pods are ready to be picked, the process required to fully develop the desired flavor takes about six months. This process is called fermentation and requires several steps. First, the pods are placed in the sun and heated to start the fermentation process. Next, they are folded in blankets and allowed to sweat over night. Sometimes this heating and sweating process is repeated. Next, the pods are removed from the blankets and allowed to dry in the sun for several weeks. When completely dry, the pods are placed in boxes and aged for 2–3 months.

Once the pods are aged, they are sorted, packaged, and shipped to various countries where they

Vanilla pods

are sold to bakeries and food processing plants. Most often the pods are soaked in alcohol to concentrate the flavor into vanilla extract, which is then used in many recipes.

Even more popular than vanilla is chocolate, which comes from the cacao beans that grow on the cacao tree. The name *cacao* comes from the Aztec word *cacahautl*, which was a chocolate drink enjoyed by the Aztecs. But unlike the hot cocoa of today, this drink was spiced with vanilla and peppers and did not contain sugar.

Cacao trees grow only in the tropics. They cannot grow at altitudes above 300 feet or in temperatures below 60°F (15.5°C). The trees also require at least 50 inches (127 cm) of rainfall each year, as well as an atmosphere with high humidity. These conditions only occur in a very limited part of the world, so cacao trees are only grown in a few places.

Cacao flowers are pollinated by the midge, a small mosquito-like fly. Once pollinated, the flowers produce football-shaped pods that are about 10 inches long and 3–4 inches around. Each pod contains 20–40 seeds. Once the pods are ripe, they are picked and then experience a process very much like the fermentation process for the vanilla pods. However, the process of turning cacao beans into cocoa is more complicated.

First, the pods are cut open and the flesh and seeds are scooped into a box that is placed in the sun and allowed to ferment for 5–7 days. The fermentation causes the flavor to begin to develop in the seeds. During this process the pulp turns to a liquid and drains away and the seeds begin to dry.

After curing, the seeds are roasted at a temperature between 225–300°F (105–150°C) for 15–20 minutes. Next, the seeds are cracked open and the seed coats are removed. The seeds are crushed producing a cocoa butter solution. When cooled, this

Cacao pods

becomes chocolate liquor, which is baking chocolate and is very bitter. If the cocoa butter is removed from the baking chocolate, powdered cocoa is produced.

Scientists have identified over 200 chemical compounds that contribute to the flavor of chocolate. Therefore, it is very difficult to make an artificial chocolate flavor. So most chocolate flavored items are made using cocoa powder or baking chocolate. The most common imitation chocolate flavor is made from carob, which is the fruit of a locust tree. However, most people will tell you that carob does not come close to the delicious flavor of the real thing. Similarly, scientists have developed an imitation vanilla that is produced from a sugar of wood fiber. It has the same chemical formula as the vanilla flavor, but has a bitter aftertaste that some people dislike.

Overall, most people, especially in Europe and America, are willing to pay for their favorite flavors of vanilla and chocolate. Are you hungry for a chocolate bar now?

32

Additives

What's really in your food?

What things are added to our food?

Words to know:

preservative
antioxidant
flavor enhancer
Food and Drug Administration (FDA)

Unless you eat only fresh fruits and veg-etables that are grown on organic farms, you probably have eaten chemical additives in your food. Chemicals are added to foods for many reasons. Some are added to keep the food from spoiling; others are added to improve the food's look, taste, or texture.

For centuries people have added sugar or salt to their foods to keep them from spoiling. For example, sugar is added to fruit to make jams and jellies. Salt is added to meat to make jerky, bacon, or ham. Sugar and salt work as preservatives because they absorb much of the moisture in the food. Bacteria need moisture to grow, so sugar and salt help to reduce bacteria growth.

Today, there are over 3,000 different additives that are used in food products. These additives include preservatives, antioxidants, emulsifiers, stabilizers, flavorings, and colorings. Vitamins and minerals are also added to many foods to improve their nutritional value.

Preservatives are added to keep food fresh and to prevent mold and bacteria from growing in the food. One common preservative is acid. Vinegar is added to cucumbers to make pickles. Vinegar is an acid that prevents bacteria from growing. In a similar way, **antioxidants** are chemicals that prevent food from reacting with oxygen. BHA and BHT are antioxidants that are often added to oils in very small amounts to prevent them from reacting with the oxygen and becoming spoiled.

Emulsifiers and stabilizers are chemicals that are added to change the texture of a food, usually to thicken it. Emulsifiers are often used to help prevent oil and water from separating. Natural emulsifiers include eggs, tree sap, and seaweed. Artificial chemical emulsifiers include polysorbate and propylene glycol.

Flavors and flavor enhancers are added to improve the flavor of the food. You just learned

Fun Fact

Americans eat an average of 150 pounds of food additives each year.

Food Chemistry

124 • Properties of Matter   LESSON 32

Fun Fact

Fifty percent of the products on grocery store shelves did not exist only ten years ago. This is primarily due to the extensive research into chemical additives. These additives are necessary ingredients in many products today. Of course, this does not mean that many of the products are more nutritious, just that they are more convenient.

about what flavors are and how chocolate and vanilla flavors are produced, but many other flavors are added to foods as well. Sugars or other sweeteners are often added to improve the flavor of a product. Hundreds of other natural and artificial flavors are used as well. **Flavor enhancers** are used to make the flavor stronger rather than to change it. A small amount of salt can enhance the flavor of a food. Another common flavor enhancer is MSG (monosodium glutamate), a compound originally made from seaweed, but today made from molasses.

Finally, colorings are added to many foods. Colorings do not have any nutritional value and do not change the flavor of the food, but can make the food more appealing. Many foods change color when they are heated. So frequently, colorings are added to processed foods to make them look more like we expect them to look. For example, strawberries turn brown when they are heated, so red food coloring is added to strawberry jam to make it more appealing. Certain shapes and colors suggest flavors and help us enjoy our food more.

The federal **Food and Drug Administration (FDA)** oversees the use of additives in food. Each manufacturer is required to perform tests to show that the additives they use are safe for humans. These findings are submitted to the FDA. Some additives are not added by the food manufacturer, however. Some chemicals get into the food when it is grown. This could include pesticides and herbicides as well as hormones given to animals. These items would not appear on a list of ingredients.

The most nutritious food is always the fresh, unprocessed food that God created; however, it is not always possible to eat fresh foods, and preservatives keep foods from spoiling and allow us more convenience.

What did we learn?

- What is a food additive?
- Name three different kinds of additives.
- Why are preservatives sometimes added to foods?
- What compound has been used as a preservative for thousands of years?
- Why are emulsifiers sometimes added to foods?

Taking it further

- Why are vitamins and minerals added to foods?
- Why does homemade bread spoil faster than store bought bread?

LESSON 32 **Properties of Matter** • 125

🧪 Preserving our food

Although many food additives do not add to the nutritional value of the food, some additives, such as preservatives, are very useful.

Purpose: To demonstrate the use of preservatives

Materials: apple, knife, lemon juice

Procedure:

1. Slice an apple into quarters.
2. Spread lemon juice over all exposed surfaces on two of the pieces.
3. Allow all of the pieces to sit exposed to the air for one hour.
4. After one hour, compare the appearance of the pieces.

Conclusion: The lemon juice is an antioxidant, which means it prevents the molecules on the surface of the apple from reacting with the oxygen in the air. Therefore, the pieces covered with lemon juice remained white while the pieces without the lemon juice reacted with the oxygen to form molecules that turn brown. These molecules form a protective barrier that prevents the rest of the apple from spoiling, but gives the apple a brown appearance. Antioxidants in many foods keep the foods from spoiling.

🎖 Food additive checklist

There are thousands of additives in the food you buy. Look around your kitchen and read the food labels in your cupboards, refrigerator, and freezer. In the right hand column of the "Food Additives Checklist" write down which foods contain each of the additives you find. Add any other additives that you find to the end of the list.

33

Bread

Why is it light and fluffy?

How is bread made?

Words to know:

fermentation beta-starch
gluten alpha-starch

If you mix some flour, water, and yeast with a little sugar, butter, and salt, you can make a loaf of bread. This delicious food is an important part of most diets around the world. The ancient Egyptians are credited with discovering that adding yeast to dough makes the bread rise and become fluffy. Today, fluffy bread is a staple in most American households.

Most bread in America is made from wheat flour. Wheat contains a protein called gluten. Gluten plays a very important part in the formation of bread. The other major ingredient in the bread we enjoy is yeast. Yeast is a type of fungi. Yeast reacts chemically with the sugar and starch in the bread dough and produces carbon dioxide gas. This reaction is called **fermentation**. The **gluten** in the flour allows the dough to stretch, and traps the carbon dioxide gas bubbles, thus allowing the dough to rise. When the bread dough is baked, the heat kills the yeast and the carbon dioxide escapes, leaving behind tiny air pockets that make the bread fluffy.

Bread contains many important chemical compounds. Bread mostly contains carbohydrates in

Yeast reacts chemically when activated with sugar and starch releasing carbon dioxide.

Food Chemistry

LESSON 33   Properties of Matter • 127

Wheat is the most common grain used to make bread, and a bread's texture is partly determined by how finely ground and processed its flour is.

the form of starch and a small amount of sugar. Bread also contains water, protein, fat, fiber, vitamins, and minerals. In general, whole wheat bread is more nutritious than white bread because much of the wheat kernel is removed to make white flour.

Mixing the dough and allowing it to rise are important steps in the bread making process. But baking the dough is also very important. The type of starch found in flour is called beta-starch. **Beta-starch** is formed from long rigid chains of glucose that are bound tightly together. The beta-starch molecules have a crystalline structure that is hard to digest. So eating raw bread dough would not be a good idea. When the dough is baked, the heat causes the starch molecules to break down, and water molecules get in between the starch molecules. These smaller molecules are called **alpha-starch** molecules.

Alpha-starch gives the bread its pleasing smell and soft consistency. Also, the enzymes in your stomach more easily digest these smaller molecules. So baking is necessary to change bread dough into something wonderful to eat. As the bread gets old, the water evaporates from the bread, and the alpha-starch molecules begin to revert back to beta-starch. This is why we don't enjoy eating stale bread. But if you enjoy a sandwich with nice soft bread, thank God for creating wheat with gluten and yeast to make the bread rise.

What did we learn?

- If you want fluffy bread, what are the two most important ingredients?
- Why is gluten important for fluffy bread?
- Why does bread have to be baked before you eat it?
- Why is whole wheat bread more nutritious than white bread?

Taking it further

- What would happen if you did not put any sugar in your bread dough?
- Can bread be made without yeast?

🧪 Baking bread

Purpose: To bake your own loaf of bread

Materials: large bowl, flour, sugar, yeast, salt, small bowl, milk, butter, cooking oil, baking pan

Procedure:

1. In a large bowl, combine 1 cup of flour, 2 tablespoons sugar, 1 package of active dry yeast (2 ¼ teaspoons), and 1 teaspoon salt.
2. In a separate bowl or measuring cup, heat ¾ cup water, ¼ cup milk, and 1 tablespoon butter or margarine until very warm (120–130°F).
3. Gradually add the liquid to the dry ingredients, beating on low speed with an electric mixer until all the liquid is added. Then beat at high speed for 2 minutes, occasionally scraping the sides of the bowl.
4. With a spoon, stir in enough remaining flour to make a soft dough that is not sticky.
5. Knead the dough on a lightly floured surface for 8–10 minutes until the dough is smooth and elastic, adding flour as necessary to form a smooth dough.
6. Spray a large bowl with cooking oil, place the dough in the bowl, and turn the dough over so the top of the dough is greased.
7. Cover and let rise in a warm area until it doubles in size, about 30–60 minutes.
8. On a floured surface, roll the dough into a 12 x 7 inch rectangle.
9. Beginning at the short end, roll the dough tightly and place seam side down in a greased baking pan.
10. Cover and allow to rise until it doubles in size, about 1 hour. Put the loaf in the oven and bake at 400°F for 30 minutes or until done. Remove from pan, allow to cool, then slice and enjoy!

🎖 Homemade vs. store-bought

Preservatives are added to many foods to keep them from spoiling. Commercially produced bread has preservatives in it to keep it soft and to keep it from molding.

Purpose: To see if homemade bread or store-bought bread lasts longer

Materials: homemade bread, store-bought bread, two plastic zipper bags, "Homemade vs. Store-Bought" worksheet

Procedure:

1. Place a slice of your homemade bread and a slice of store-bought bread on the counter where they will not be disturbed.
2. Place one slice of homemade bread in a plastic zipper bag. Place a slice of store-bought bread in another plastic zipper bag. Place both bags in a warm dark place where they will not be disturbed.
3. Write your predictions of what you expect to see on a copy of the "Homemade vs. Store-Bought" worksheet.
4. Closely observe each slice of bread daily for five days. Write your observations on the worksheet.
5. At the end of five days, answer the questions at the end of the worksheet.

Food Chemistry

Bread through the Centuries

SPECIAL FEATURE

Dinner rolls, crescent rolls, tortillas, sourdough, wheat, white, challah, pita, and bagels are just a few of the many types of bread eaten around the world. Wheat flour is the most commonly used flour for making bread, but it is not the only flour used. Other flours are made from rye, buckwheat, barley, potato, rice, legumes, beans, quinoa, amaranth, and nuts. In many cases, these other flours are mixed with wheat flour because of the high amount of gluten in wheat. But no matter the type of flour you use, you will be eating a food that has been around for thousands of years and has played an important part as a staple of life in nearly every culture.

Bread is not only important for food, it has become part of our vocabulary and is used in many common expressions. It is used symbolically in the Lord's Prayer, "Give us this day our daily bread," meaning our daily needs. We sometimes use the term "bread" to mean money when we refer to work as "our bread and butter." And you might hear someone say that something is the "bread of life," meaning it is very important for survival or happiness. Jesus said that He is the bread of life (John 6:35). All of this shows that bread is a vital part of survival and an integral part of our lives.

The process of baking yeast bread probably started soon after Babel (Genesis 11). It is recorded to have been used in Egypt and later spread to the Greeks and Romans. Bread was considered so important in Rome that it was put on a higher scale than meat. If soldiers were not given their allotment of bread they felt slighted or neglected. The Roman welfare system in the city of Rome was originally set up to hand out grain, but later the government started baking the bread before giving it to the people. Later, in the dark ages, white bread became popular. White bread was preferred by the noblemen and was more expensive than the darker whole wheat bread because of the added expense and work that went into making the flour white.

Bread has also played an important role in the Bible. In the Book of Genesis, when Melchizedek, a high priest of the Most High, came out to bless Abram, he brought with him bread and wine (Genesis 14:18). Later, when the Lord was passing by on his way to destroy Sodom, Abraham offered Him bread (Genesis 18:5). When Jacob made the meal for which Esau traded his birthright, it included bread and lentils (Genesis 25:34). On the night of the Passover, the Israelites were to eat meat with bitter herbs and unleavened bread. God said that each year at Passover, His people were to celebrate by eating unleavened bread (bread baked without yeast) for seven days. This is called the Festival of Unleavened Bread (Exodus 12:8).

Bread is an important part of worship as well. When God gave instructions for the tabernacle, He told them to put the Bread of the Presence on the table so it would be in front of Him all the time (Exodus 25:30). Unleavened bread made with olive oil was also a part of a grain offering to the Lord (Leviticus 2).

130 • Properties of Matter

Bread was also important in the life of Jesus. When Satan tried to tempt Jesus with bread, Jesus responded with, "Man shall not live by bread alone, but by every word that proceeds from the mouth of God" (Matthew 4:4). When Jesus was giving a sermon and the people got hungry, He took five loaves of bread and two fish and fed 5,000 men, plus the women and children with them (Matthew 14:16–21). But the most important reference to bread in the Bible is near the end of Jesus' life: "And as they were eating, Jesus took bread, blessed and broke it, and gave it to the disciples and said, 'Take, eat; this is My body'" (Matthew 26:26). Jesus is the true Bread of life.

Identification of Unknown Substances: Final Project

What is this, anyway?

How can we identify unknown substances?

Now that you have learned many things about matter, you can begin to apply some of that knowledge. One way that you can use what you have learned about the properties of matter is to try to identify unknown substances. How does a scientist, even a young one, identify a substance without knowing what it is? You use the scientific method. First you learn about something, which is what you have been doing in the lessons in this book. Then you ask a question; in this case you want to know what the substance is. Third, you make a hypothesis. Based on what you know about how the substance looks you can make a guess as to what it is. Next, you design a way to test your hypothesis. Using what you have learned about matter, you can try different chemical tests to see how the substance reacts and to test if it is what you think it is. Finally, you see if your results support or contradict your hypothesis.

There are many different ways you can test a substance to see what it might be. First, you want to use your senses. Examine the substance's color, texture, and viscosity. Carefully smell the substance. To do this, hold the substance about six inches from your nose and use your hand to push or wave some of the air from above the sample toward your nose. You do not want to take a big whiff because some substances can burn the inside of your nose.

Never taste an unknown substance! Many substances can be harmful.

These observations can be helpful; however, many substances look alike and may not have much of a smell. So other chemical and physical characteristics must be examined. Some physical and chemical tests can be very complicated and must be done by experienced scientists under controlled conditions. But others are relatively simple and can be done at home.

For example, you can measure the mass, density, boiling point, and freezing point of many substances. You can also do some of the tests mentioned for chemical analysis of food, such as using iodine to test for starch or using brown paper to test for oils. You have also seen that vinegar or lemon juice reacts with baking soda to produce carbon dioxide bubbles. You can also test for acids and bases using the pigment from red cabbage as an indicator. These are all tests that you can do at home.

Some tests that scientists do, you may not want to do at home including flame tests and acid tests. Different substances give off different colors when a sample is placed in a flame. Also, certain metals react differently with various acids. Another way that scientists test unknown substances is with an instrument called a spectroscope. This instrument passes a beam of light through a prism and then the spectrum of light passes through the sample. The color or colors of light that pass through the sample are different for each substance, so this can be used to identify the substance.

Chemical analysis of substances is not only important for identifying unknown substances, but also for monitoring many manufacturing processes. Testing is done to ensure the quality of raw materials. This helps guarantee consistent results when manufacturing food and other items. End products are also tested to ensure quality and consistency. Chemical testing is used to test the purity of metals and dyes. It is used to test the amount of alcohol in liquor. And it is used to test the purity of our water supply. These are only a few ways that chemical analysis is used. Your food, clothing, appliances, building materials, and nearly everything else around you has been tested by a chemical reaction before coming to you. All of this testing assures that the products you buy are safe and reliable. God created matter to be consistent and to react the same way each time it is exposed to other chemicals, and we can use this consistency to ensure safety.

What did we learn?

- What method should be used in identifying unknown substances?
- Why should you avoid tasting unknown substances?
- How can you test the scent of an unknown substance safely?
- What are some physical characteristics of an unknown substance you can test at home?
- What are some chemical characteristics you can test at home?

Taking it further

- Why is it important for food manufacturers to test the ingredients they use and final products they produce?
- Why is it important for water treatment facilities to test the quality of the water?

Identifying unknown substances

Complete each experiment using a copy of the "Identification of Solids" worksheet and a copy of the "Identification of Liquids" worksheet.

Write a paragraph explaining what you learned from each experiment so you can share the results with others.

Design your own experiment

Design your own experiment and have someone else do the tests.

LESSON 34 **Properties of Matter • 133**

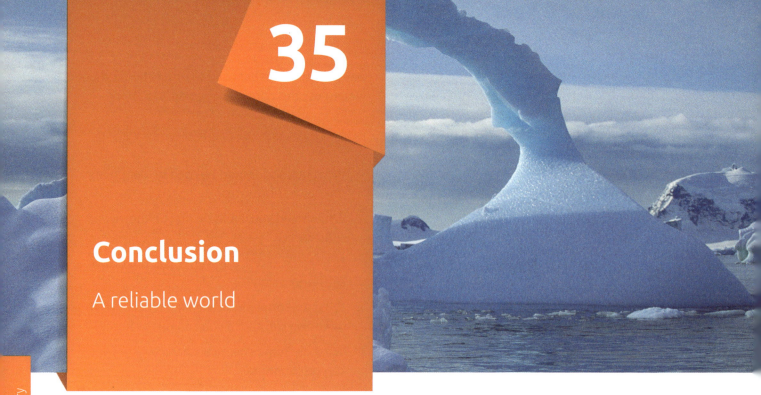

35

Conclusion
A reliable world

God's wonderful creation

Think about all the things you have learned about the matter that makes up this world and the universe. Matter was designed in such a way that particular substances always have the same physical and chemical characteristics. Pure water always boils and freezes at particular temperatures. Water is always made of one oxygen and two hydrogen atoms. Gold has a higher density than silver or bronze. Matter cannot be created or destroyed by any natural means.

All of these laws show that God is a loving Creator. God created water to be the great solvent. He created our bodies to chemically react with the food we eat to provide energy. And God created the perfect recycling system when He made plants that use carbon dioxide and release oxygen and then made animals and humans that use oxygen and release carbon dioxide.

What did we learn?

- What is the best thing you learned about matter?

Taking it further

- What else would you like to know about matter? (Go to the library and learn about it.)

🧪 Reflect on God's creation

Let's do a final experiment to demonstrate God's love for us. One of the most important properties of water is its ability to hold heat. The temperature on Earth is relatively stable and moderate compared to temperatures on other planets. This is due primarily to the fact that the earth has an atmosphere and that over 70% of the surface of the earth is covered with water. Water gains and loses heat much more slowly than air does. This allows the temperatures along the coasts to be milder than inland areas, and allows the temperature on Earth to be more moderate than on any other planet.

Purpose: To demonstrate water's ability to hold heat

Materials: two balloons, candle

Procedure:

1. Fill one balloon with air and tie it closed.
2. Fill a second balloon with water and tie it closed.
3. Carefully light a candle and hold the air-filled balloon over the flame. What happens? It quickly pops. Why? Because the air could not absorb the heat and the balloon quickly melted.
4. Now carefully hold the water-filled balloon over the flame. Be sure the flame is under the part with water and not air. This balloon can absorb the heat for a long time. If you hold it long enough the water will even begin to boil. But do not hold it that long as you could get burned.
5. After 1–2 minutes remove the balloon and blow out the candle. Feel the balloon and you will see that it is slightly warm. The water was able to absorb the heat so quickly that the balloon did not melt or pop.

Conclusion: This demonstrates God's love and care. This property of water was not an accident. God designed it just for us to moderate the temperatures on Earth. Take a few moments and reflect on the world and the matter it is made of, and thank God for His wonderful design.

Letter to God

Write a letter of thanksgiving to God, thanking Him specifically for the matter in the universe and how it affects you.

Properties of Matter — Glossary

Alpha-starch   Starch molecules that are smaller than beta-starch and are easier to digest
Antioxidant   Substance that prevents chemical reaction with oxygen
Atmospheric pressure   Pressure exerted on objects by the molecules in the air

Beta-starch   Starch formed from long rigid chains of glucose
Boiling point   Temperature at which a substance begins to evaporate
Boyle's law   Volume of a gas decreases as the pressure increases if temperature is kept constant
Buoyancy   The ability to float

Carbohydrates   Chemicals that comprise sugars and starches
Charles's law   Volume of a gas increases as temperature increases if pressure is kept constant
Chemical formula   Combination of letters and numbers representing the chemical make-up of a compound
Chemical property   One that describes how a substance reacts with other substances
Classification   Grouping things together according to similar characteristics
Colloid   Suspension with very tiny particles suspended in it
Compound/Molecule   Substance formed when two or more elements chemically combine
Concentrated   A solution with a relatively large amount of solute
Concentration   Quantitative measure of grams of solute per 100 grams of solution
Condense   Change from a gas to a liquid
Controlling variables   Changing only one variable at a time in an experiment

Density   Mass divided by volume
Diatomic molecule   Molecule formed by two of the same kind of atom
Dilute   A solution with a relatively small amount of solute
Displacement method   Measuring volume by calculating how much liquid it displaces

Element/Atom   Substance that cannot be broken down by ordinary chemical means
Emulsifier   Substance that helps to keep particles in suspension
Evaporate   Change from a liquid to a gas
Experiment   Controlled test to see what will happen in a given situation

Fats   Long molecules found in oils and other foods
Fermentation   Chemical process in which yeast reacts to produce carbon dioxide
First law of thermodynamics   Matter and energy cannot be created or destroyed, they can only change form
Flavor enhancer   Substance that makes a flavor stronger
Foam   A liquid with air trapped between the molecules
Food and Drug Administration (FDA)   Government organization overseeing food and drug safety
Freeze   Change from a liquid to a solid

Gas pressure   Pressure exerted by gas molecules inside a closed container
Gas   State in which molecules are far apart from each other and move quickly
Gluten   Substance in wheat that allows dough to stretch
Gram   Unit of measure for mass in the metric system

Henry's law   Solubility of gas increases with pressure
Heterogeneous mixture   Mixture in which the substances are unevenly distributed
Homogeneous mixture   Mixture in which the substances are evenly distributed
Homogenization   The breaking up of fat molecules to suspend them in milk
Hypothesis   Good or educated guess

Immiscible   One substance will not mix with another
Indicator   Substance used to detect the presence of another substance

Kinetic energy   Energy of moving objects

Law of conservation of energy   Energy cannot be created or destroyed, it can only change form

Law of conservation of mass   Matter cannot be created or destroyed, it can only change form
Liquid   State in which molecules are close together and move over each other
Liter   Unit of measure for volume in the metric system

Mass   How much of something there is
Matter   Anything that has mass and takes up space
Melting point   Temperature at which a substance melts
Melt   Change from a solid to a liquid
Meniscus   Curve formed at the top of liquid in a narrow tube
Meter   Unit of measure for length in the metric system
Mixture   Combination of two or more substances that do not chemically combine

Newton   Metric unit of weight

Overflow method   Measuring volume by measuring the water that overflows when an object is placed in a completely full container

Pasteurization   Heating to kill bacteria
Phase change   Changing from one state of matter to another
Physical property   One that can be measured without changing the type of matter
Precipitate   The substance that precipitates from a solution
Precipitation   When a solute comes out of a solution
Preservative   Substance added to prevent food spoilage
Proteins   Chemicals made from amino acids found in meats, nuts, and dairy products

Qualitative observations   Observations that do not involve numbers
Quantitative observations   Observations that involve numerical data

Saturated   Solution in which no more solute will dissolve
Scientific method   Systematic way of observing, testing, and repeating experiments to attain knowledge
Scientist   One who uses observation and a systematic method to study the physical world
Sedimentation   Allowing particles that are suspended to settle out of the suspension
Solid   State in which molecules are tightly packed and only vibrate
Solubility   Measure of how much solute can be dissolved in a given amount of solvent
Solute   Substance that is dissolved
Solution   Homogeneous mixture where one substance is dissolved in another
Solvent   Substance in which the solute is dissolved
Structural integrity   The ability of cream to trap and hold gas molecules
Sublimation   Changing directly from a solid to a gas
Suspension   Mixture in which one or more substances are immiscible

Translucent   Light will pass through

Universal solvent   Solvent in which everything will dissolve

Viscosity   Measurement of how strongly a liquid's molecules are attracted to each other
Volume   How much space matter takes up

Weight   How strongly something is attracted by gravity

Properties of Matter — Challenge Glossary

Amorphous solid   Solid in which the molecules do not line up in repeating patterns

Beaufort scale   Used to describe intensity of wind

Calorie   Amount of energy needed to raise the temperature of 1 gram of water 1 degree Celsius
Calorimeter   Device for measuring the calories in a substance
Carbonates   Minerals containing carbon and oxygen
Centrifuge   Instrument used to separate by spinning

Chromatography   Separation by flowing through a medium, such as paper
Conductivity   Measure of the ability of a substance to conduct electricity
Crystalline solid   Solid in which the molecules line up in repeating patterns
Curds   Solid part of milk used to make cheese

Decantation   Separation by allowing particles to settle then pouring off the liquid
Diffusion   Movement of molecules from an area of higher concentration to an area of lower concentration
Distillation   Separation by evaporation then condensation

Enzyme   Biological substance used to speed up chemical process
Eyepiece   Lens at the top of a microscope

Food calorie   1000 calories, kilocalorie
Fractional distillation   Distillation involving more than one substance with different boiling points
Fujita scale   Used to describe intensity of a tornado

Halides   Minerals containing halogens
Halogens   Elements in column VIIA of the periodic table
Hard water   Water that contains dissolved minerals, especially calcium and magnesium

Limescale   Scaly deposits formed by precipitation of calcium and magnesium

Microscope   Device for viewing very tiny objects
Minerals   Compounds in the earth's crust
Mohs scale   Used to measure hardness of rocks and minerals

Native minerals/Native elements   Pure substances in the earth's crust
Naturalism   Belief that nature is all there is
Nonsoluble   Not able to be dissolved

Objective lens   The lens closest to the object being observed on a microscope
Observational science/Operational science   Science that can be observed and reproduced
Origins science   Science that deals with figuring out past events
Ounce   English unit for weights smaller than a pound, 16 ounces = 1 pound
Oxides   Minerals containing oxygen

Percolate   Movement of water up through the ground
Phosphates   Minerals containing phosphorous and oxygen
Pound   English unit of weight

Reflecting telescope   Telescope which uses a series of lenses and mirrors to magnify an image

Saffir-Simpson scale   Used to describe intensity of a hurricane
Salinity   Measure of the amount of salt dissolved in water
Silicates   Minerals containing silicon and oxygen
Slug   English unit of mass
Sulfides   Minerals containing sulfur

Transpiration   Evaporation of water from plants

Water softener   System used to remove minerals from water
Whey   Watery liquid left when curds are separated from milk

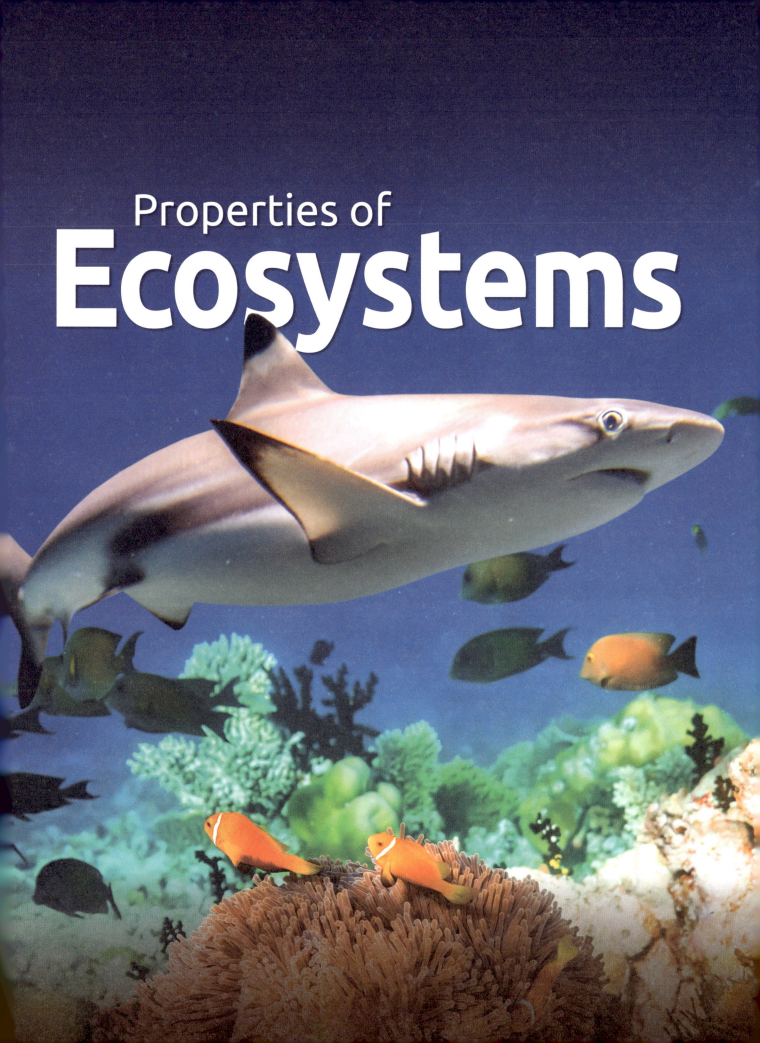

UNIT 1
Introduction to Ecosystems

1. What Is an Ecosystem?
2. Niches
3. Food Chains
4. Scavengers & Decomposers
5. Relationships Among Living Things
6. Oxygen & Water Cycles

◊ **Identify** and **describe** ecosystems and niches.

◊ **Identify** and **describe** food chains and food webs.

◊ **Identify** roles of scavengers and decomposers.

◊ **Explain** the various roles plants and animals play.

◊ **Describe** the oxygen and water cycles.

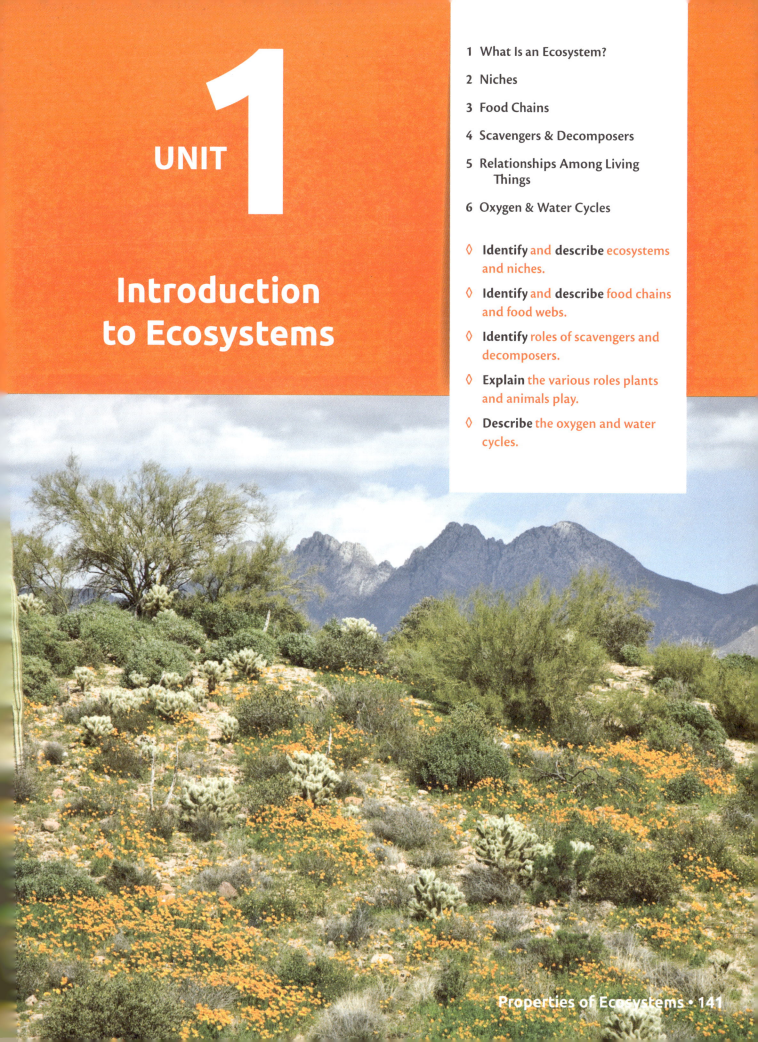

1

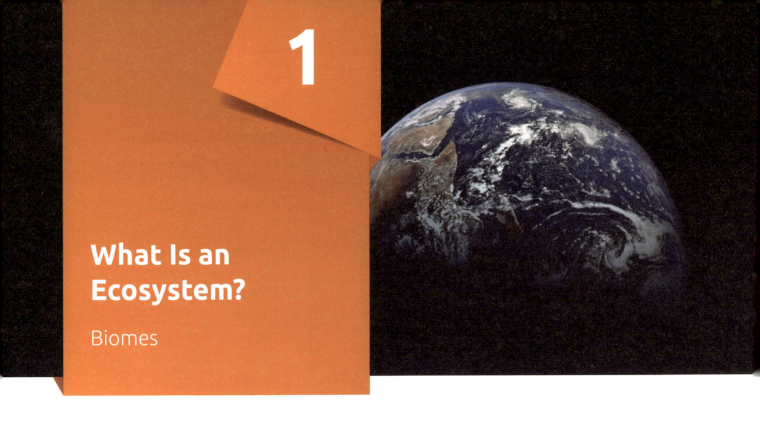

What Is an Ecosystem?

Biomes

How big is an ecosystem?

Words to know:

habitat
ecology
biosphere
biotic
abiotic

ecosystem
biome
flora
fauna
climate

Challenge words:

biogeographic realm   ecozone

Where do you live? You probably live in a house or an apartment with your family. Your home is in a neighborhood with other homes. The people who live around you are your neighbors. Your home and neighborhood make up your **habitat**; it is the environment in which you live. Animals and plants live in neighborhoods, too. The study of plants and animals and the environment in which they live is called **ecology**. In this book you are going to learn about many habitats and how the plants, animals, and other organisms in them interact.

The **biosphere** is the part of the Earth that contains life. The biosphere includes the atmosphere, the surface of the Earth, a small part of the crust of the Earth, and the water that covers most of our planet. As we study ecology, we will be learning about the many different areas within the biosphere of Earth. The Earth is the only known planet that has a biosphere; it is the only planet known to contain life. This life was created by God, and when He created life, He designed the Earth so that things work together to allow life to continue. The Bible does not tell us whether organisms such as bacteria or fungi exist on other celestial bodies. While we would not expect such life to exist, it is not entirely out of the question.

The biosphere contains both biotic and abiotic things. **Biotic** describes things which are alive. Other things around us are **abiotic**, which means they are not alive. What kinds of things are biotic? Plants and animals make up the most visible of biotic organisms. Fungi, bacteria, and single-celled

Fun Fact

The Greek word for habitat is *oikos*. The study of habitats is thus oekologie from which we get the word *ecology*. The word *ecosystem* is a shortened version of ecological system, and the word *biome* is a shortened version of biological home.

organisms are also biotic. What kinds of things are abiotic? Some non-living things include the minerals in the soil, water, chemicals, sunshine, and man-made objects. All of the living and non-living things in a particular area affect each other.

All of the biotic and abiotic items in a particular area make up an **ecosystem** and many ecosystems together make a **biome**. The plants in an ecosystem are called the **flora** of the ecosystem and the animals are called its **fauna**.

There are many different ecosystems in the world. The types of living things in an ecosystem are determined by many factors. The most important factor determining which plants and animals live in a certain area is the **climate**, which is the general or average weather conditions of a certain region, including the amount of sunlight the area receives, the average temperatures, and the amount of moisture available. Because the Earth is tilted with respect to the sun, the amount of sunlight that an area receives depends on where you are located between the poles and the equator. The poles

A coral reef is an example of a marine ecosystem.

receive very little direct sunlight, whereas the equator receives a large amount of direct sunlight.

Ecosystems change as you move from the poles toward the equator. Ecosystems also change as you move from east to west across a continent. The terrain causes changes in the climate so the environments are varied in different locations. As you study the lessons in this book you will learn about the many wonderful ways that God created the life on our planet to interact with its environment.

My backyard habitat

Purpose: To become aware of different elements in your surroundings or habitat

Materials: string, yardstick/meter stick, "My Backyard Habitat" worksheet, magnifying glass

Procedure:

1. Although most animals live in the same habitat their whole lives, humans move about from one habitat to another. Make a list of all of the habitats you spend time in each week.
2. Closely examine the habitat in your backyard. Use string to mark out a square that is 1 yard (1 m) long on each side.
3. While standing up, carefully look at what is inside your square. Write your observations on a copy of the "My Backyard Habitat" worksheet.
4. Now, get down on your knees and use a magnifying glass to closely observe the smaller things in your square. Look for small animals, decaying plants, small twigs, paper, plastic, etc. Record all of your observations on your worksheet.
5. Listen to the sounds that can be heard from your square. Again record your observations.
6. Record the weather conditions on your worksheet.
7. List any ways that you think the animals that you observe use the other things that you have observed in your square.
8. Take photos of your area and what you found. Save your worksheet and your photos and include them in the notebook you are going to be making throughout this study.

Conclusion: Nothing can live in isolation. Even the smallest insect needs food and shelter and uses objects in its environment to provide these things. This in turn affects other animals and plants that are living in the area.

What did we learn?

- What is ecology?
- What is the biosphere?
- Give an example of something that is biotic and something that is abiotic.
- What is flora?
- What is fauna?

Taking it further

- What factor has the greatest effect on the plants and animals that live in a particular ecosystem?
- How does your habitat change throughout the day?
- List some ways that climate affects the habitats of people.

Ecozones

As we study the different types of ecosystems we will see that some animals live only in certain parts of the world even though the climatic conditions in other parts of the world would support those animals. For example, zebras do not live on the great plains of North America even though the environment is very similar to the savannah in which they live in Africa. Another example is the many marsupials that live only in Australia. If conditions are right for these animals to live in other places, why are they only found in a particular area?

There are many different possible explanations, but one likely explanation is called the Ararat migration hypothesis. The Bible tells us that representatives from all of the land animals were saved from the Great Flood on Noah's Ark and that the Ark came to rest on the mountains of Ararat. Thus, all of the animals had to make their way from Ararat to the other parts of the world.

It is believed that after the Flood there was an Ice Age which created large ice sheets around the world. This lowered the water level and would have exposed land bridges between areas of the world separated by water today. This would have allowed animals to migrate over the land into areas that are farther from Ararat, which is in modern Turkey. After several hundred years, the ice sheets began to melt, creating barriers that prevented further migration.

The natural barriers of large bodies of water, such as the oceans, as well as very high mountains or large deserts, keep many animals from migrating over large distances. Today, scientists recognize six major areas of land that are separated by one or more of these large barriers. These areas of land are called **biogeographic realms** or **ecozones**.

The *palearctic realm* is the area containing Eurasia and north Africa. It is isolated from other realms by oceans to the north, west, and east, and the Himalayan mountains and Sahara Desert to the south. Sub-Saharan Africa is part of the *afrotropical realm*, which is surrounded by oceans on the west, south, and east and the Sahara Desert on the north. The *Indo-Malay realm* is the area west and south of the Himalayan mountains and includes India and most of southeast Asia. Australia and the surrounding islands comprise the *Australian realm*. North America makes up the *Nearctic realm* and Central and South America comprise the *Neotropical realm*.

These realms are relatively isolated so the animals that live in one realm cannot easily move to another realm. On a copy of the World Map:

1. Label each of the following barriers:
   a. Atlantic, Pacific, Indian, Arctic, and Antarctic Oceans
   b. Sahara Desert
   c. Himalayan Mountains
2. Color each biogeographic realm a different color.
3. Create a key to label and identify each realm.
4. Save this map to add to the first section of your notebook.

Garden of Eden
The first ecosystem

SPECIAL FEATURE

As Adam walked along with Eve he reached up and grabbed a perfectly ripe fruit from the tree and offered it to Eve. "Hungry?"

"Yes, thank you." Eve took the fruit and bit into it as Adam reached up again and took one for himself.

The two walked on a little further as they thought about the conversation Adam had with God the night before. They both enjoyed their time with God and looked forward to it. God had given them everything they needed. The weather was always perfect, food was always just an arm's reach away, and even when they were separated from each other, God had provided animals of all kinds to keep them company. It was a perfect world—it was the Garden of Eden.

The Garden of Eden was the very first ecosystem. No one knows exactly what the Garden was like or where it was located; but we do know a few things based on what the Bible says. In Genesis 1, as God is making the universe, six times He says that what He made was "good." This meant without any flaws or defects. Genesis 1:31 says, "Then God saw everything that He had made, and indeed it was very good. So the evening and the morning were the sixth day."

All God made worked together perfectly and all was beautiful.

In chapter 2 we see a little more detail about the last day of creation:

> This is the history of the heavens and the Earth when they were created, in the day that the LORD God made the Earth and the heavens, before any plant of the field was in the Earth and before any herb of the field had grown. For the LORD God had not caused it to rain on the Earth, and there was no man to till the ground; but a mist went up from the Earth and watered the whole face of the ground. And the LORD

> God formed man of the dust of the ground, and breathed into his nostrils the breath of life; and man became a living being.

> The LORD God planted a garden eastward in Eden, and there He put the man whom He had formed. And out of the ground the LORD God made every tree grow that is pleasant to the sight and good for food. The tree of life was also in the midst of the garden, and the tree of the knowledge of good and evil. . . .

> Then the LORD God took the man and put him in the garden of Eden to tend and keep it. And the LORD God commanded the man, saying, 'Of every tree of the garden you may freely eat; but of the tree of the knowledge of good and evil you shall not eat, for in the day that you eat of it you shall surely die.'

Properties of Ecosystems

And they were both naked, the man and his wife, and were not ashamed" (Genesis 2: 4–9, 2:15–17, 2:25).

From these passages we can learn a few things about the first ecosystem. God watered the plants by using a mist, which may have been a fog or a very fine rain. We know that man and animals could eat all the fruit that grew on the trees; both man and the animals ate only plants. Since man had not sinned yet there was no death of man or animals in the world. We know that all the trees and plants were very pleasing to look at. The insects that flew or crawled to pollinate the plants did not sting, bite, or bother Adam, Eve, or any of the animals. Man and all the animals lived in harmony with each other and with the plant life around them.

We can conclude from verse 25 that the weather was very mild. Adam and Eve had no need for clothing because they had not sinned. We can surmise from this that the temperatures were mild enough that they were not cold, even at night, without coverings.

The evidence of this mild tropical climate is not only found in the Bible but also in fossil records from around the world. Fossils show tropical plant life in almost every location on the globe. This shows that at one time much of the Earth was warm and moist. Again, our observations of the natural world confirm the history contained in God's Word.

So why and how did the Earth change? Why do we have such extreme conditions now? Did man cause it? As we read in Genesis 3:17–18, after man sinned:

Then to Adam He said, "Because you have heeded the voice of your wife, and have eaten from the tree of which I commanded you, saying, 'You shall not eat of it':

"Cursed is the ground for your sake; in toil you shall eat of it all the days of your life. Both thorns and thistles it shall bring forth for you, and you shall eat the herb of the field."

From this passage we know that thorns and thistles are part of the curse; therefore, before man sinned the plants did not have thorns, and thistles did not grow. We can also see that God took away Adam's easy supply of food. Adam would now have to work for his food. Not only were Adam and Eve punished for their sin, the whole Earth and universe were cursed because of it (see Romans 8:20–22). This means that not only did man lose his ready supply of food and have to work for it, but the curse on the Earth applied to all the animals, plants, and all living things. Now they too would have to struggle to survive.

Many of the things you will learn about in the following lessons did not apply to the original creation. You will learn about competition, food chains, overpopulation, and extinction. These are a result of the curse brought on by man's sin. In the original ecosystem there were no predators and prey because there was no death. However, even though the world we have today is cursed, it is still a magnificent place to live. Even though the original perfection is gone, the mark of the Creator remains and can still declare His glory.

2

Niches

What's your job?

What is a niche?

Words to know:

niche community
population

No organism lives by itself. All plants and animals are interconnected in one way or another. For example, a flower growing in a garden cannot reproduce without a bee or other insect to pollinate its flowers. The bee must have nectar to eat and to make honey. Other animals and humans eat the honey made by the bee. They are all connected to each other. Each plant and each animal in an ecosystem has a particular role to fill or a job to do. The job or role that a particular organism performs is called its **niche**.

An animal's niche is determined by many factors. It is determined by what it eats and what eats it. An animal may eat only particular plants, but may be eaten by several other animals in the area, thus its niche includes eating plants and providing food for other animals. How the animal acts and what it does help to determine its niche as well. The animal may also play a role in spreading disease or in providing waste to help plants to grow. All of these things, and many more, help determine what niche an animal will fill.

Similarly, plants can fill different niches as well. We usually think of plants in terms of the animals that eat them. But plants also provide shelter and homes for many organisms. Animals can use plants in many different ways. Also, plants may compete with other plants. All of these things affect the niche that a particular plant fills.

The **population** of a particular species of plant or animal is determined by the number of individuals of that species in a particular area. All of the different populations living together in an area make

Different bees have different niches within their colony.

LESSON 2 **Properties of Ecosystems • 147**

🧪 An Earthworm's niche

Purpose: To observe an Earthworm's niche by making an environment for it to live in

Materials: jar, dark soil, sand, oats, Earthworms, dark construction paper, tape

Procedure:

1. Fill a jar ¼ of the way with moist, dark soil. Add a layer of lighter colored sand then another layer of soil.
2. Sprinkle 2 tablespoons of oats on top of the soil.
3. Add 10–12 Earthworms to the jar then seal the jar. Punch holes in the lid to allow air into the jar.
4. Wrap a piece of dark construction paper around the jar and tape it in place. Put the jar in a cool location out of direct sunlight.
5. Observe your Earthworm habitat each day for several days by removing the paper and looking at the contents of the jar. What do you see happening inside the jar?
6. Take photos of your Earthworm habitat each day and include them in your notebook.

Conclusion:

After several days, you should observe the different colors of soil mixing together. This is one of the important niches filled by Earthworms; they help to break up and mix soil to make it more fertile.

up a **community**. The size of populations and the particular types of populations in a community also affect the niche that each animal and plant fills within that community.

Niche can refer to two different kinds of roles. As we have described it so far, an animal's niche is the role that it plays in the community—the job that it does in its neighborhood. But niche can also refer to the role that the animal plays within its family or population. For example, a bee's niche within its environment is as a pollinator and as a producer of honey. But within the colony of bees, different bees play different roles. Sterile female bees are the workers that gather the nectar and make the honey. The queen bee is the only one that produces eggs for reproduction. The male bees, called drones, are responsible to fertilize the eggs. Thus different bees have different niches within their colony. ✲

🧠 What did we learn?

- What is a niche?
- Name two factors that determine an animal's niche.
- What is a population?
- What is a community?
- What are two different kinds of niches an animal can have?

🚀 Taking it further

- What different niches do you fill in your family and in your community?
- How does competition for food and other resources affect the niche of a plant or animal?

🧪 Setting up your notebook

Purpose: To build a notebook, recording all of the things you learn about ecosystems. Today, you are going to start your notebook by making dividers for each section of your book.

Materials: 3-ring binder, nine dividers for the notebook

Procedure:

1. Obtain a 3-ring binder.
2. Make nine dividers for your notebook. Label the dividers as follows:
   a. Introduction
   b. Grasslands
   c. Forests
   d. Aquatic ecosystems
   e. Tundra
   f. Deserts
   g. Mountains
   h. Animal behaviors
   i. Ecology
3. Add your photos and worksheet from lessons 1 and 2 to the first section of your notebook.

Conclusion:

As you complete each activity in this book, add your worksheets, reports, photos, etc. to your book. When you have completed the study of ecosystems you will have the book to help you remember what you learned.

🏅 What's my niche?

On a piece of paper, describe the niche occupied by each of the following plants or animals. Be sure to include all the possible uses and different areas where you might find them. Include your list in your notebook.

Tree

Wolf

Mouse

Robin

Grass

LESSON 2 **Properties of Ecosystems** • 149

3

Food Chains

Does it have links?

What is a food chain?

Words to know:

food chain

producer

consumer

herbivore

carnivore

omnivore

food web

Challenge words:

carrying capacity

All organisms in a biome are connected to each other in various ways. The primary connection between these living things is the flow of energy. The flow of energy from one organism to another is called a food chain. Another way to think of a **food chain** is: a series of organisms in the order that they feed on one another.

Nearly all food chains begin with green plants. Green plants change the energy of the sun into food, primarily glucose (a type of sugar), through the process of photosynthesis. These plants are called **producers**. Any organism that does not produce its own food, but instead eats plants or other animals, is called a **consumer**. A first order, or primary, consumer is the animal that eats the plant. The second order, or secondary, consumer is the animal that eats the first order consumer, and so on. Most food chains have three or four levels; only a few food chains have more than five levels.

Animals that eat only plants are primary consumers and are called **herbivores**. All animals were herbivores in God's original creation (Genesis 1:29–30). Most grazing animals such as deer, antelope, cows, and horses are herbivores. Animals that eat only other animals are secondary consumers and are called **carnivores**. Animals did not start eating other animals until after the Fall of man. Some common carnivores include wolves, lions, and snakes. Some animals eat both plants and animals and are called **omnivores**. Black bears

are omnivores. They will eat berries and honey, as well as fish and other animals.

A common food chain might start with acorns from an oak tree which are eaten by a squirrel. The squirrel might then be eaten by an owl. In this food chain, the oak tree is the producer, the squirrel is the primary consumer, and the owl is the secondary consumer. Another food chain might begin with corn which is eaten by a mouse. The mouse is then eaten by a weasel which is eaten by a wolf. This food chain has four levels.

Many animals eat more than one kind of food and many animals have more than one predator; therefore, there can be multiple food chains containing the same plants and animals. The interactions among multiple food chains is called a **food web**. It is called a web because when arrows are drawn showing all of the possible ways that energy flows through the various animals, it resembles a spider's web.

What did we learn?

- What is a food chain?
- What is a producer?
- What is a consumer?
- What is a food web?
- List two herbivores.
- List two carnivores.
- List two omnivores.

Food chains & webs

Draw a picture of a food chain. You can make up your own or use the plants and animals listed below; they are not in the correct order. Draw arrows in the direction that the energy is flowing.

Bush   Fox   Bird   Caterpillar

On your drawing, identify the producer, 1st order consumer, 2nd order consumer, and 3rd order consumer.

Draw a food web. Again, you can make up your own or use the organisms listed below. Make sure the arrows show the flow of energy in the web.

Grass   Insect   Bush   Rabbit
Wolf   Bird   Fox   Frog

Add these drawings to your notebook.

LESSON 3   **Properties of Ecosystems • 151**

🚀 Taking it further

- Is a black bear a first or second order consumer?
- Is man an herbivore, carnivore, or omnivore?
- Explain how a food chain shows energy flow.

Fun Fact

Before the Fall of man, all animals and people were herbivores (Genesis 1:29–30). After the Flood, God allowed people to eat animals as well as plants (Genesis 9:3).

🏅 Carrying capacity

It is important to understand how the various organisms in a particular area fit together in a food chain or a food web. This helps us to understand how many of a particular animal or plant can survive in that area. The number of a particular species in a given area is called the population. For example, if you counted all of the foxes living in a particular square mile area, that would be the fox population of that area.

The population of a particular species depends on many things. What things do you think affect the population? The amount of food available is a very important factor in determining how many of a species can survive. Population is also affected by how much space is available and how many other animals are competing for the same space and food. The population also depends on how many predators are living in the same area. Weather can play a role in population as well. When the weather is mild, the animals and plants survive better than when the weather is unusually harsh.

The maximum population an area can support is called its **carrying capacity**. The carrying capacity of an area may vary from year to year depending on the weather. In years where growing conditions are good, such as years when there is adequate water and sunshine, the plants grow well. This provides more food for the consumers, and the carrying capacity of the area increases. During times of drought, there are fewer plants so the carrying capacity of the area decreases.

As long as the population of a species is lower than the carrying capacity, the population may increase. When there is an adequate food supply the birth rate usually increases and the death rate decreases. This increases the population. However, when the population becomes larger than the carrying capacity, food becomes scarce so the death rate increases and the birth rate decreases. This causes the population to decrease.

We have shown how weather can change the carrying capacity of an area. What other factors can you think of that can affect the carrying capacity of an area? For example, what would happen if a disease suddenly struck a prairie dog colony and killed a significant portion of the prairie dogs? This would decrease the food supply for all of the animals in the food web that eat prairie dogs, causing many of them to die, or to become weak because of lack of food. This would mean that the carrying capacity for the predators in that area was decreased.

Human activity can also greatly change the carrying capacity of an area. People can grow much more food in a given area than would naturally grow there. This means that more cattle can be fed from a given area when humans are growing the food. In general, when talking about populations and carrying capacity, scientists usually refer to conditions that have not been changed by human activity.

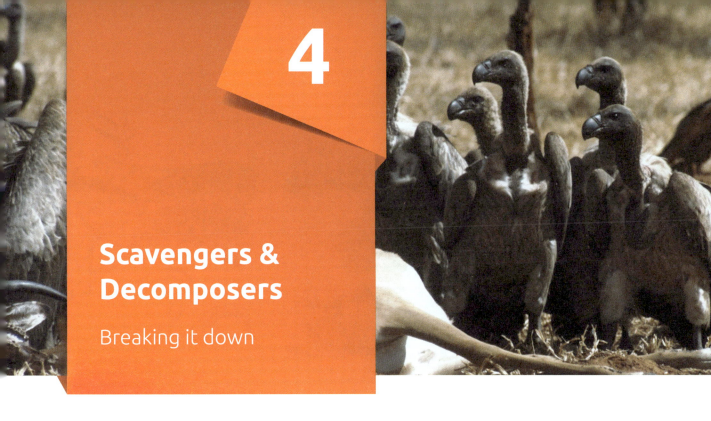

4

Scavengers & Decomposers

Breaking it down

Why are scavengers important?

Words to know:

scavenger
decomposer
decomposition
law of conservation of mass

Food chains and food webs do not end when a plant or animal dies. The dead plant or animal still has a considerable amount of energy tied up in its tissues and cells. Fortunately, God created many different types of organisms that eat dead plants and animals. Some of these organisms eat the dead plants or animals and use the energy for their own survival. These types of organisms are called **scavengers**. Some examples of scavengers include vultures, flies, and Earthworms. Many other animals are scavengers when given the opportunity, but may also hunt or eat live plants and animals as well. Lions are an example of an animal that will hunt when necessary, but will eat dead animals if they are available.

At the very end of every food chain are the **decomposers**. These are organisms that eat dead plant and animal material and break it down into basic elements and molecules such as nitrogen, carbon, and phosphorus. This process frees up those elements so they can be reused to grow new plants and provide food for new animals. The most common decomposers are bacteria and fungi. Bacteria and fungi also decompose animal waste to free up the materials that have been excreted.

The process of **decomposition** is a vital function. Without decomposition, all of the elements necessary for plant and animal life would become tied up in dead plants and animals and new plants and animals could not exist. The **law of conservation of mass** states that matter cannot be created or destroyed by any natural

Fungi like these yellow fairy cups help decompose plant and animal material.

LESSON 4   **Properties of Ecosystems • 153**

decomposers

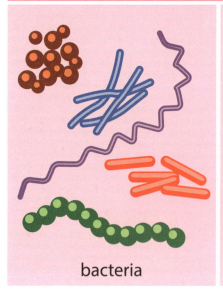
bacteria

fungi

worms

Fun Fact

Although many animals convert only about 10% of the food they eat into body mass, some animals are more efficient. Pigs convert about 20% of what they eat into body mass and turkeys convert as much as 30% of the energy they eat into body mass. This means that it takes less food to grow a pound of pork or turkey than it does to grow a pound of beef. Compare the prices of pork and turkey to the price of beef the next time you are at the store and see which meats are the most and least expensive.

means, it can only change form. There is a limited amount of nitrogen, oxygen, carbon, and all other atoms in the world. This is why God created the world to recycle these materials. Decomposition is one way that these elements are recycled.

What did we learn?

- What are organisms called that eat dead plants and animals?
- Name two different animals that eat dead plants or animals.
- What types of organisms are at the end of every food chain?
- Name two common organisms responsible for decomposition.

Taking it further

- Why is decomposition so important?
- What physical law makes decomposition necessary?

Adding decomposers

Add scavengers and decomposers to your food chain and food web pictures in your notebook.

Population pyramids

In the previous lesson you learned that the population of a species depends on the amount of food available. Take a minute to think about how much grass a deer eats each day. Think about how many mice an owl must eat each night. In a healthy ecosystem would you expect to find more grass plants than deer? Of course, otherwise the deer would starve to death.

There must be significantly more producers in an area than consumers for life to continue. Similarly, there must be a significantly higher number of first order consumers than second order consumers in any particular area. For example, there may be hundreds of mice living in a field, but only one hawk living in the area. This is another way of talking about carrying capacity of an area.

In the same ecosystem the carrying capacity for one species will be very different from the carrying capacity for another species. If we were to draw a picture of the population of each species in a food chain it would resemble a pyramid. The plants would be the wide base at the bottom. The first order consumers would be a smaller number stacked on top of the producers. The second order consumers would be an even smaller number stacked on top of the first order consumers, and so on.

Although the numbers of first order consumers compared to the numbers of second order consumers varies greatly between different species, there is a rule called the 10% rule that is a good estimate of the relative populations that an area can support. On average an animal converts only about 10% of the energy it eats into body mass; the other 90% of the energy is used for searching for food, chewing, maintaining body functions, heating the body, and so forth. So there must be at least 10 times as many first order consumers as second order consumers. Thus if an area can support 500 grasshoppers, it can support about 50 frogs and about 5 snakes.

Even though we said that producers/plants are at the bottom of the population pyramid, there is one group of organisms that is even more numerous than the producers. That group is the decomposers. Since decomposers are so small and must break apart every plant and animal after it dies, there are millions of bacteria for each plant or animal in a given area. Decomposers should actually be the base (largest part) of our pyramid.

Draw a population pyramid for each of the food chains that you drew in lesson 3. Don't forget to put decomposers at the bottom of the pyramid. Include these drawings in your notebook.

LESSON 4 **Properties of Ecosystems** • 155

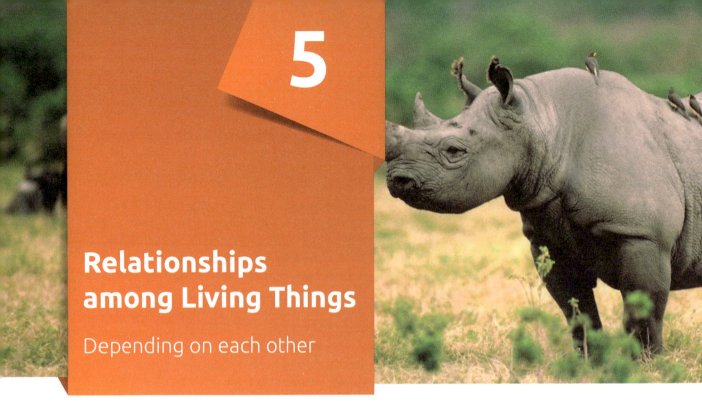

5

Relationships among Living Things

Depending on each other

How are organisms connected?

Words to know:

symbiosis
mutualism
predator
prey
parasitism
commensalism
epiphyte
competition
neutralism

The plants and animals in an ecosystem are connected in a variety of ways. In previous lessons you learned about food chains and food webs, which show the flow of energy from one organism to another. However, there are many other relationships between plants and animals besides the flow of energy. Any relationship between two different species living in close connection is called **symbiosis**.

There are several different types of symbiosis. The relationship most often associated with symbiosis is a relationship in which both organisms benefit. This is called **mutualism**. There are examples of mutualism all around us. You have a certain type of bacteria living in your digestive system, which helps you digest your food. This relationship provides a source of energy for the bacteria and for you, so you both benefit. Another example of mutualism is between the crocodile bird and the crocodile. Bits of food often get stuck between a crocodile's teeth. But he cannot brush and floss to get rid of it. Instead, the crocodile will open its mouth and the crocodile bird will land in the crocodile's mouth and pick out the food stuck between his teeth. Although the crocodile could just snap its mouth shut, the crocodile does not eat the bird. The bird gains a meal and the crocodile gains better dental health. In another example, the oxtail bird eats ticks and fleas off of the back of the rhinoceros. This feeds the bird and gets rid of pests for the rhino, as shown above.

Not all relationships between species are beneficial. Obviously the predator-prey relationship is beneficial for the **predator** (the hunter) but deadly for the **prey** (the hunted). Another type of relationship in which one species benefits and the other is harmed is **parasitism**. A parasite feeds off of another species causing harm, but usually not death, to the host. Common examples of parasites include lice, fleas, and many types of worms. These animals usually suck blood or live in the intestines of the host animal, which harms the host in a

Fun Fact

A very interesting mutualism exists between the yucca plant and certain yucca moths. The yucca moth lays its eggs only on the yucca plant because it is the only plant that the larvae will eat. This benefits the moth. But the yucca moth is the only animal that can pollinate the yucca flower, thus the moth benefits the plant as well. Scientists cannot explain how these two interdependent species could have evolved at exactly the same time. That is because they did not evolve—God created them both.

number of ways. There are parasitic plants as well as parasitic animals. Parasitic plants usually send a shoot into the roots or stems of another plant and steal some of the sap from the host plant.

A third type of symbiosis is **commensalism**. Like a parasite, the guest species benefits in some way from the host; however, the host is not harmed or benefited by the guest. One example of commensalism is the relationship between an epiphyte and a tree. In rainforests as well as other forests, the trees often become large enough to block most of the sunlight from reaching the forest floor. Many species of plants attach themselves to the bark of a tree at a height where the sunlight will reach them. These plants are called **epiphytes**. The attachment benefits the plant, but does not harm the tree. Another example of commensalism is the relationship between the remora and the shark. The remora is a fish that attaches itself to large aquatic animals such as sharks. The remora has a dorsal fin shaped like a suction cup. It attaches itself to the host by this suction cup and hitches a ride. The fish is also believed to eat leftovers that are dropped by the host during feeding. Thus the remora is benefited and the shark is not harmed.

Competition is another relationship between species. When two different species both eat the same food, they will compete for the limited resources in the area. Both species will be hurt by this competition if their populations become too large and there is not enough food for everyone. There can also be competition for space, sunlight, and nutrients between various plant and animal species.

The final relationship between species is neutralism. **Neutralism** is the relationship where neither species benefits and neither is harmed. Many species do not eat each other and do not compete for the same resources so they coexist without much direct effect on each other.

Understanding symbiosis

Complete the "Symbiosis" worksheet.

Optional Activity:

Use a magnifying glass to examine a rock with a lichen on it. Lichens demonstrate a symbiotic relationship between fungi and green algae. The algae perform photosynthesis, which provides food for both organisms, while the fungi provide protection from the weather. This is a type of mutualism.

🧠 What did we learn?

- What is symbiosis?
- What is mutualism?
- What happens to each species in a parasitic relationship?
- Which species benefits in commensalism?
- What is competition among species?
- What is the name of a relationship in which neither species benefits nor is harmed?

🚀 Taking it further

- Why is competition considered harmful for both species?
- Explain how competition could keep the species from becoming too populated.

🏅 Liver flukes

The liver fluke is a flatworm that has a very complex life cycle involving several different relationships with different animals. The most common liver fluke is commonly called the sheep liver fluke. It can infect sheep, cattle, and other animals that eat grass in marshy areas. The fluke is attached to blades of grass and enters the sheep's digestive system with the grass. Inside the sheep the fluke burrows through the intestine and moves to the liver. There it eats liver tissue and matures. This can cause considerable damage to the sheep.

Once it matures, the fluke moves to the bile duct where it produces eggs. The eggs move back into the digestive system with the bile. Eventually the eggs are expelled from the body in the sheep's feces. If there is water nearby, the eggs are carried into the water where the eggs hatch. The infant fluke then enters a particular type of snail. While feeding on the snail it reproduces. The larvae are then excreted from the snail and swim to nearby grass. On the grass the larvae lose their tails and form cysts which attach themselves to the grass and wait to be eaten by sheep.

This complicated lifecycle requires two separate hosts that are completely unrelated except for the fact that they come in close contact periodically when the sheep come near the water where the snails live. There are many other types of flukes that also have similar complicated lifecycles.

Sheep infected with liver flukes can become very sick and often die. So ranchers will treat their flocks with medication to kill the flukes. This not only helps the infected sheep, but helps to prevent other sheep from becoming infected.

Make a diagram showing each step in the fluke's lifecycle. Label the relationships that occur between the fluke and the sheep, the fluke and the snail, and the fluke and the grass with the terms you have learned in this lesson. Add your diagram to your notebook.

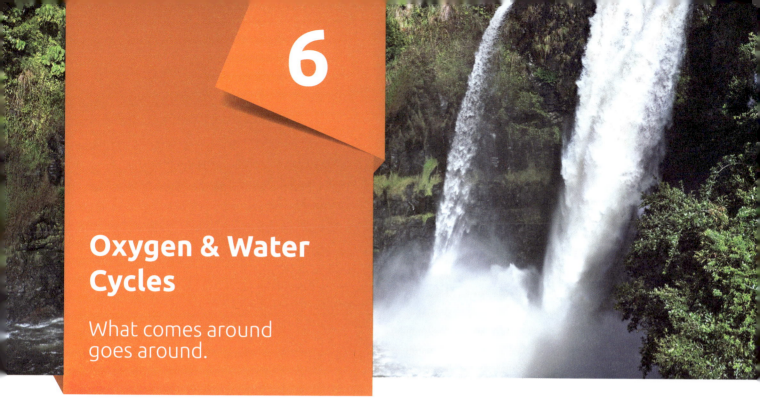

6

Oxygen & Water Cycles

What comes around goes around.

How are materials recycled?

Words to know:

photosynthesis

oxygen cycle

respiration

water cycle

Challenge words:

nitrogen cycle

The law of conservation of mass requires that all elements be conserved or recycled since no new matter can be made by any natural processes. We already talked about how decomposers break down the tissues of dead plants and animals to release the nitrogen, carbon, and other elements in them to be used again in new plants and animals. God has also established several other systems that recycle precious materials.

One of the most amazing demonstrations of the conservation of mass is the relationship between plants and animals. During **photosynthesis** plants utilize water and carbon dioxide, both of which contain oxygen atoms, to capture the energy of the sun as they rearrange these molecules to make sugar and oxygen. Animals then breathe in the oxygen and eat the sugar. During **respiration** the animals use the oxygen to break apart the sugar molecules to release the energy and in the process they release water and carbon dioxide back into the air, which can be reused by plants. This process is called the **oxygen cycle**. It not only demonstrates how the molecules are recycled, but also shows God's amazing plan for providing energy for all living things.

The oxygen cycle occurs in the water as well as in the air. As water flows, oxygen becomes dissolved in the water. Algae and plants growing in the water perform photosynthesis and release oxygen into the air and the water. Animals that live in the water absorb the oxygen from the water and use it to break down food. This creates carbon dioxide, which is released into the water and into the air where the plants and algae can absorb it and use it in photosynthesis.

Another important recycling process is the **water cycle**. Like all other materials, water must be used over and over again. We have already seen that water is recycled in the oxygen cycle as it is absorbed by plants and released by animals. But water is recycled in other ways as well. Water in the world's oceans and lakes evaporates into the air when the sun warms the surface of the water. This water vapor moves through the atmosphere. As it cools it condenses and forms clouds. Eventually, the water becomes heavier than the atmosphere

LESSON 6 **Properties of Ecosystems** • 159

🧪 Demonstrating the water cycle

Purpose: To demonstrate the water cycle in your very own living room

Materials: potting soil, glass jar with lid, grass or other plant, camera or drawing materials

Procedure:

1. Place 2–3 inches of potting soil in the bottom of a glass jar.
2. Dig up a clump of grass, being sure to get as much of the roots as possible. Plant the grass in the jar.
3. Add enough water to moisten the soil, but don't make it soggy.
4. Tightly seal the jar with a lid.
5. Place the jar in a warm location, but not in direct sunlight, and observe it several times a day for one week.
6. Take several pictures of your jar, showing the different stages of the water cycle. Use these pictures to make a water cycle poster to include in your notebook. If you don't have access to a camera, you can draw pictures of what you observe.

Questions:

- What do you observe happening in the jar?
- Does it look the same in the morning as it does in the afternoon?

Conclusion:

You should notice some fogginess or cloudiness in the jar as it warms up. As it cools you may notice water condensing on the sides of the jar. You should also notice that at other times the water is gone. This is the water cycle at work. Water is absorbed by the grass and some of it is used for photosynthesis and some of it is released into the air. Also, as the plant grows it breaks down some of the food it produces the same way that animals break down food. This process also releases water.

As the the inside of the jar warms up, water also evaporates into the air. This can cause the inside of the jar to appear foggy. As the air cools in the jar, the water condenses and falls back into the soil.

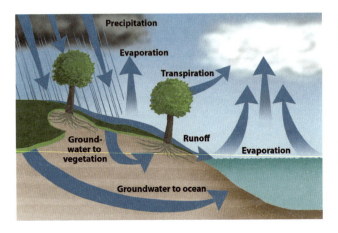

can support and falls back to the Earth as precipitation. Some of this water is used by plants and animals while some of it flows back into the lakes and oceans. Thus, the water is constantly being recycled.

When you think about how well everything works together in our world, you have to stop and thank our Creator, God, who made it all! ✳

🧠 What did we learn?

- How do photosynthesis and respiration demonstrate the oxygen cycle?
- What are the major steps in the water cycle?

🚀 Taking it further

- Water exists in three forms: solid, liquid and gas. What phase is the water in before and after evaporation?
- What phase is the water in before and after condensation?
- What phase is the water in before and after precipitation?

The Nitrogen cycle

Another important cycle that exists in nature is the **nitrogen cycle**. Nitrogen is a very important element needed to form amino acids and proteins, which are the building blocks of all plants and animals. Plants absorb most of their nitrogen from the soil and animals get nitrogen by eating plants. Most of this nitrogen eventually returns to the soil when decomposers break down the tissues of dead plants and animals.

You might think that nitrogen is easy to obtain since the atmosphere is 78% nitrogen. However, this form of nitrogen gas (N_2) is not easily used by plants and animals so it must be converted to more useful forms including nitrates (NO_3^-), nitrites (NO_2^-), and ammonia (NH_3). This transformation occurs in two basic ways. First, when lightning strikes, the heat generated causes the nitrogen in the atmosphere to react with the water in the atmosphere to from nitrates and nitrites. These then fall to the Earth in the rain. The second method for converting nitrogen gas in the atmosphere to usable forms is through a chemical reaction performed by nitrifying bacteria. These bacteria are found in the soil and in the roots of legumes. Legumes include peas, soybeans, lentils, peanuts, beans, and alfalfa. The process of converting nitrogen gas to nitrates and other compounds is called nitrogen fixation or nitrification.

Once the nitrogen is in a useable form, plants absorb the nitrogen through their roots and use it to grow. Animals eat the plants, and the nitrogen compounds are passed on to the animal. Some of the nitrogen compounds are returned to the soil through the animal's waste. The rest of the nitrogen in the plants and animals is returned to the soil when decomposers break down the tissues of the dead organism. Finally, some of the nitrogen compounds in the soil are turned back into nitrogen gas by denitrifying bacteria in the soil, thus completing the cycle for the nitrogen.

The nitrogen cycle works well in most ecosystems. However, in commercial farming, the number of plants growing in a given area is often much higher than in a natural ecosystem, so the nitrogen in the soil is quickly depleted. Farmers must add nitrogen back into the soil. This is done several different ways. Often farmers will rotate their crops, growing corn one year then alfalfa or other legumes the next year to put nitrates back into the soil. Other times, farmers will add nitrates to the soil by spreading animal waste, such as cow manure, on the soil before planting crops. In other instances, chemical fertilizers containing nitrogen are added to the crops. This helps the plants to grow more quickly and to grow larger. But even if man helps the crops by providing nitrogen, that nitrogen was obtained in another location, so the nitrogen is ultimately still being recycled.

Take what you have learned here and write clues for the following words then arrange them in a crossword puzzle. Add your puzzle to your notebook.

- Ammonia
- Nitrite
- Nitrate
- Bacteria
- Legumes
- Lightning

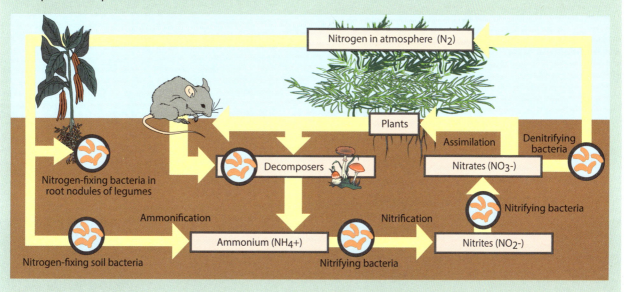

LESSON 6 Properties of Ecosystems

UNIT 2

Grasslands & Forests

7 Biomes Around the World
8 Grasslands
9 Forests
10 Temperate Forests
11 Tropical Rainforests

◊ **Describe** what a biome is.
◊ **Explain** the effects of climate on biomes.
◊ **Describe** characteristics of grasslands.
◊ **Distinguish** between deciduous and coniferous forests.

7

Biomes around the World

Where are they located?

How are biomes distributed?

Words to know:

tropical zone

northern temperate zone

southern temperate zone

northern polar region

southern polar region

Challenge words:

ecotone

succession

pioneer plant

climax ecosystem

An ecosystem is all of the living and non-living things that interact in a given area. From this perspective, a puddle of water could be considered an ecosystem. Similarly the whole Earth could also be considered an ecosystem because every living thing on Earth is connected in some way to every other living thing on Earth. However, when scientists talk about biomes they usually mean a large region with a particular climate and the plants and animals that live in that region.

Areas that are very dry are called deserts. A desert is one type of biome. Another type of biome is the Arctic tundra, which is an area that is consistently cold and has a relatively short growing season. Other biomes include temperate forests and tropical forests. There are also various water biomes such as lakes and oceans. We will be studying each of these biomes as well as many others in the following lessons.

The Earth is tilted with respect to the sun. So as the Earth revolves around the sun the amount of sunlight reaching various parts of the Earth is different. This is one of the most important factors contributing to the different climates around the world. The Earth can be divided into five zones based on the amount of sunlight received. The **tropical zone** is the section of the Earth located between the Tropic of Cancer (23.5° north latitude) and the Tropic of Capricorn (23.5° south latitude). This area centered around the equator receives the most direct sunlight all year round. Tropical biomes are located in this tropical zone.

Between the Tropic of Cancer and the Arctic Circle (66.5° north latitude) is the **northern temperate zone** and between the Tropic of Capricorn and the Antarctic Circle (66.5° south latitude) is the **southern temperate zone**. These zones receive more hours of sunlight in the summer and fewer hours of sunlight in the winter months than the tropic zone does. The temperate forests and grassland biomes are generally found in these zones.

LESSON 7 **Properties of Ecosystems** • 163

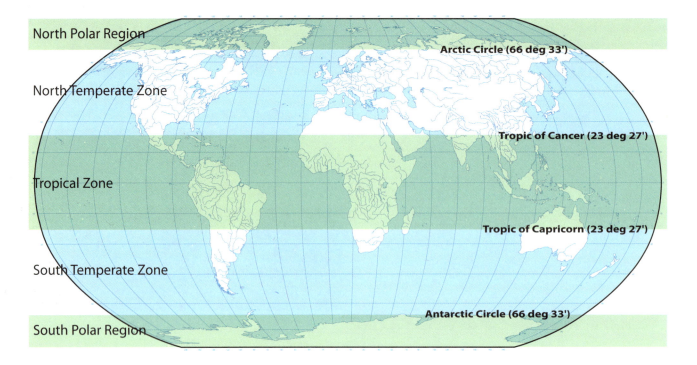

North of the Arctic Circle is the **northern polar region** and south of the Antarctic Circle is the **southern polar region**. The polar regions are usually colder than the temperate regions and although they receive many hours of sunlight in the summer, the sunlight strikes the Earth at a steep angle in these areas and much of the sunlight reflects off of the atmosphere and snow. Thus these areas remain fairly cool even in the summer time. You will find the Arctic tundra located in the polar regions.

As you study each biome, pay attention to where they are located around the world.

Locating biomes

Purpose: To become acquainted with the location of various biomes and to demonstrate how climate affects them

Materials: world atlas or online maps showing temperature and rainfall for the world as well as location of various biomes, copy of "Average Rainfall Map," "Average Temperature Map," and "Biomes Map"

Procedure:

1. Color the "Average Rainfall Map" to indicate the annual average rainfall in each part of the world. Use a different color for each unit of rain, such as light blue for less than 10 inches of rain per year, dark blue for 10–20 inches of rain, etc. Be sure to make a key for your map.

2. Color the "Average Temperature Map" to show the average temperature in the various parts of the world. Create a key to show the temperature range represented by each color. Use blue and green for cooler temperatures and yellow, orange, and red for hotter areas.

3. On the "Biomes Map" use a different color for each biome. Be sure to include deserts, grasslands, mountains, oceans, tropical forests, temperate forests, and polar biomes on your map. Your atlas may show other biomes as well which you may wish to include. Be sure to make a key for your map.

4. Place your maps side by side and compare the temperature and rainfall maps with the location of the various biomes. You should notice that polar tundra occurs in the polar regions where the temperatures are colder than other areas. You should also notice that there is a correlation to rainfall and the location of the deserts.

5. Place these maps in your notebook and use them to locate the various biomes as we study them in the following lessons.

What did we learn?

- Where is the tropical zone located?
- Where is the northern temperate zone located?
- Where is the southern temperate zone located?
- Where are the polar regions located?

Taking it further

- Why are the polar regions generally colder than the tropical regions even though they receive many more hours of sunlight each day during the summer?
- What correlations do you see between the temperature and rainfall maps that you made?

Succession

Although we will be studying several distinct ecosystems, in reality you don't just step from one ecosystem into another. There is generally an area between two different ecosystems which is a transitional area. This transitional area between two ecosystems is called an **ecotone**. It may contain plants and animals that are found in both ecosystems. For example, between a grassland and a forest you may find an area that has more trees than the grassland, but not enough trees to be considered a forest. Some animals are primarily found in the grasslands and others are primarily found in the forest, but in the ecotone between the two you may find animals living together that do not live together in either the grassland or the forest. You may also find animals living in the ecotone that do not inhabit either of the bordering ecosystems.

Many ecosystems are fairly stable. They have the same types of plants and animals living there from year to year. However, some parts of ecosystems change over time. This change from one ecosystem to another is called **succession**. Succession occurs because the conditions in the ecosystem change. This change can be a slow change because of a changing climate, or it can be a sudden change because of a natural disaster such as a forest fire or flood, or it can be a change brought about because of the actions of people, such as the clearing of a field.

An area that is experiencing succession changes over time as different plants become dominant. Some plants can grow in harsh conditions, but as these plants become more numerous, they change the environment, making it more attractive to different kinds of plants. These new plants then choke out the first plants, further changing the growing conditions. As the dominant plant species change, the dominant animal species will change as well.

The easiest way to understand succession is to think about an area of land that has experienced a sudden change. Let's think about a piece of land near the edge of a forest that has been cleared of all plants by a farmer. If the farmer does not plant crops, but instead just leaves the field alone, it will go through a series of changes over the years.

The first plants to move into an area after a change are called **pioneer plants**. These pioneers are usually plants that do not compete well with other plants, but which can reproduce more quickly than other plants. In our cleared field we will likely see mosses and small flowering plants such as dandelions grow up. These plants will attract small insects. As these pioneers become more numerous, they help to keep the water from evaporating and they add humus to the soil. This allows plants that grow more slowly to begin to grow. These plants may include grasses, ferns, and small shrubs. These new plants attract small mammals such as rabbits and mice to the area.

After many years, sun tolerant trees such as cottonwoods and chokecherries begin to grow. These plants provide more shade and cause some of the smaller plants to die out. At the same time, these larger plants attract larger animals including many birds and deer. Over the years, other successions will occur until the land returns to the ecosystem that originally occupied the land. This final, stable ecosystem is called the **climax ecosystem.**

Evolutionists have theorized that the earliest plant and animal species evolved into more complex species over millions of years. Although at first glance succession may seem to support the evolutionary ideas, there are significant differences between what we observe in succession and what supposedly happens in evolution. Evolution says that the simple plants such as moss evolved and changed into more complex plants such as grass and other flowering plants. However, what we actually observe is that moss is replaced by flowering plants as seeds that blow in or are carried in by animals take root and grow. The moss does not change into a different species.

When moss is the dominant plant the dominant organisms are bacteria, but as grasses take over, worms and other small invertebrates move in. Evolutionists claim that bacteria evolved into worms, but we see that the animals are attracted by the available plant life and move in as new plants become available. The worms and other species

already existed in other ecosystems and did not develop from the bacteria. Thus, what we actually observe in succession is very different from the supposed evolutionary processes described by some scientists.

Activity:

Research ecological succession on the Internet or in other sources. Then make a poster showing the same area of land in various stages of succession. Possible ideas include bare rock to forest succession after a volcanic eruption, aquatic succession, or succession after a forest fire or flood. Add this poster to your notebook.

Alexander von Humboldt

1769–1859

SPECIAL FEATURE

Explorer, ecologist, meteorologist, government consultant, mining expert. All of these describe Alexander von Humboldt during his lifetime, but he is most remembered as being the father of ecology and the father of geography. He was born in Berlin, Germany in 1769 to an army major who served under Frederick the Great. When he was nine his father died leaving him and his older bother to be raised by their mother. She was said to be unemotional and showed little love to her boys. But she wanted the best for her children in education, and the wealth of the family made this possible. She had them tutored at home, and wanted both of them to enter either the military or civil service. Both of them had an aptitude for learning and their studies pointed them toward civil service.

Alexander attended several universities in Germany, always adding to his understanding of science, concentrating in geology and biology. When he was 22 he met the scientific illustrator who worked for Captain James Cook on his second voyage. The two hiked around Europe together. Humboldt later went to the Freiberg Academy of Mines and after graduating became a mining inspector.

In 1796 his mother passed away, leaving him with a substantial income. He left his job and started making plans for a scientific expedition. He chose Aimé Bonpland, a botanist, as his assistant. The two men traveled to Madrid, Spain and got special permission and passports from King Charles II to explore South America.

They spent the next five years traveling throughout South America and Mexico. While studying the flora and fauna of the Orinoco River basin in Venezuela, the two men mapped over 1,700 miles of the river, thus giving Humboldt the title of the

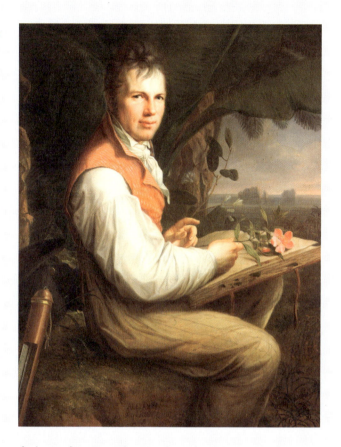

father of geography. During their journey the two attempted to climb two mountains, one of which was Mt. Chimborazo, at the time believed to be the tallest mountain in the world. They were stopped from completing the climb up Mt. Chimborazo at an elevation of around 18,000 feet by a wall-like cliff.

Humboldt's explorations then took him into Mexico. While there, he worked with the Mexican Government on mining reform and gave lectures at the mining academy. He was offered a position in the Mexican cabinet but turned it down.

After a short stay in Cuba, he traveled to Washington D.C. in 1804, where he stayed for six weeks. President Jefferson was very interested in his work and invited Humboldt to spend three weeks with him. During this time Humboldt had several meetings with President Jefferson and the two became good friends.

At the age of 35, Alexander returned to Europe, ending a unique scientific journey. He settled in Paris for a short time, lecturing at the *Institute National*. When he realized that most of his fortune was gone, having spent the money on his travels and reserving the rest for publishing his findings, he returned to Berlin where he accepted a professorship. The king was so impressed with his work that he gave Alexander a stipend without any responsibilities attached to it. This was the first time Germany had done this. (The last time they did this was for Albert Einstein). With this new income, Humboldt again returned to Paris.

Alexander gave many lectures on his discoveries and wrote several books. His lectures were so well attended that new larger assembly halls had to be found. His first published work was a 33-volume collection called *The American Journey*, which detailed his work done in South America. Over half of the volumes detailed the plants and animals that he and Bonpland had observed. He is credited with being the first to really record the interactions between plants and climate and to recognize interactions between plants and animals. This is why he is often cited as the father of ecology.

The second major work was named *Kosmos*, which was an effort to fully describe the whole world from all scientific viewpoints. The last of the five volumes was published after his death.

Later in life, Von Humboldt was invited to Russia by the Tsar, where he explored much of that country. He recommended weather stations be set up around the country and using the data from these stations he determined that the interiors of continents have more extreme weather than along the coast. He also developed the first isotherm map—a map that has lines of equal average temperatures.

So, is Humboldt remembered in the United States? Well, there are eight townships named after him, one bay in California, and three states that have counties that bear his name—not bad for a man who only spent six weeks in the country.

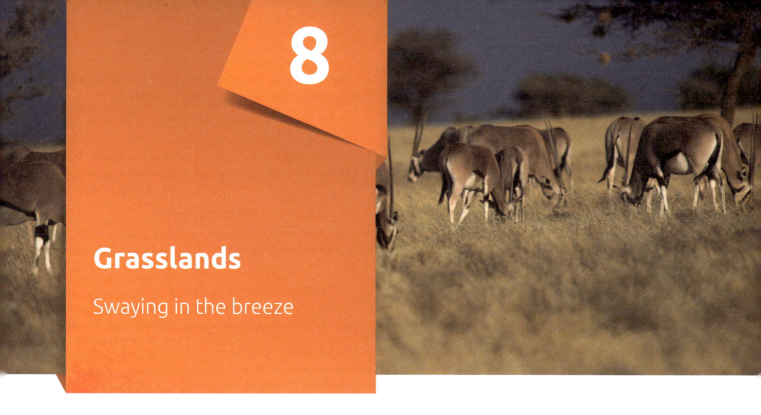

8

Grasslands

Swaying in the breeze

What is a grassland?

Words to know:

grassland
semi-arid
prairie
pampa
savannah
steppe

When you think of zebras, lions, and hyenas what type of land do you picture in your mind? Do you think of a wide open area with tall grass blowing in the wind? This is a description of the African grassland called the savannah.

A **grassland** biome is an area where the main vegetation is grass. Grasslands usually have many different varieties of grass. Some grass grows only a few inches high; other grasses grow 2–3 feet high. And some grass, such as pampas grass, can be over 10 feet tall. Although grass is the dominant plant, grasses are not the only plants to be found in a grassland environment. Wildflowers including sunflowers, coneflowers, milkweed, and sage are plentiful and make the grassland very beautiful in the spring. There are usually only a few trees such as willows and cottonwoods that grow mostly near streams.

Grasslands receive 10–30 inches (25–50 cm) of rain per year. This is generally too dry for most trees to survive. Areas such as grasslands, that are wetter than deserts, but still fairly dry have a **semi-arid** climate. Grasslands usually have distinct wet and dry seasons. Because of this, the plants and animals that survive in grasslands are designed to survive prolonged periods of drought. Many plants go dormant and quit growing when water is scarce, but then begin to grow again when water becomes available. Others grow very long roots to reach water deep underground. Many animals can conserve water; others migrate to wetter areas and return when the rain returns. Grasslands generally have warm summers and cold winters.

Grasslands have periodic fires. Often when the grasses dry out, a lightning strike can start a grass fire. This may seem like it would destroy the ecosystem. Even though trees and shrubs are often killed by the fires, the main growing part of

Fun Fact

There are over 11,000 different varieties of grass that grow in the various grasslands around the world.

LESSON 8   **Properties of Ecosystems • 169**

A black wild horse on the Russian steppe

grass is below ground so the fire burns up the dry leaves, but does not kill the plants. This actually recycles the nutrients in the leaves, adding them back into the soil to be reused.

Grasslands have several different kinds of animals that survive well. Grazing animals are the ones most associated with grasslands. These are the animals that eat the grass. Zebras, antelope, deer, bison, wild horses, and gazelles are a just a few examples of grazing animals. Many different grazing animals can survive in the same area because different animals eat different parts of the grass plant. Some grazers eat just the tops of the grass. Other grazers prefer the middle section of the grass stems. Others will eat the lower portions of grass. This provides food for several different species from the same plant. And since the main growing area of the grass is below or near the ground, grass will grow back even if most of the top of the plant has been eaten.

Grasslands also have many burrowing animals that make their homes in the ground. These animals come out of their burrows to find food, but spend a significant amount of time underground. Burrowing animals that you might find in a grassland include prairie dogs, gophers, and jack rabbits. Burrowing often helps to break up the soil so that water can penetrate the ground and help the plants to grow better.

Finally, there are many carnivores that feed on the grazers and burrowing animals. Carnivores include eagles and hawks, coyotes, foxes, lions, snakes, hyenas, and jackals. Without these carnivores, the grazing animals would soon pass the carrying capacity and begin to die off.

Grasslands are called by many names throughout the world. In North America they are called **prairies**. In South America they are called the **pampas**. In Africa grasslands are called the **savannah**, and in Europe and Asia they are called the **steppe**. Although nearly every continent has large grasslands, the different species of plants and animals vary from region to region. The prairies are dominated by antelope and at one time they were filled with bison. There are many varieties of grass. The trees that grow near rivers in the prairie are usually cottonwood, oak, or willow. The savannah is home to elephants, lions, warthogs, and wildebeests. There are usually only one or two varieties of grass growing in a given area of the savannah. The trees found in the savannah are usually acacia and pine trees.

The steppes have wild horses, and Saiga antelope. The grass that grows in the steppe is usually shorter than the grass that grows in the prairies. In the Pampas in South America you will find Pampas deer, Geoffroy's Cats, and rhea birds, which are related to ostriches. The famous Pampas grass is very tall and often used as ornamental grass in yards. Shorter grass also grows in the Pampas. Soil in grasslands is usually very fertile; this is why much of the grassland around the world has been turned into farmland or grazing land for domesticated animals.

God's design is evident in many aspects of the grasslands. From the ability of grasses to grow even after being eaten over and over to the design of different animals to eat different parts of the plant, we see God's hand in making the grassland a very efficient and beautiful place. ✵

Fun Fact

Tropical grasses can grow as much as 1 inch (2.5 cm) per day.

🧠 What did we learn?

- Name three characteristics of a grassland biome.
- What are four different types of grasslands?
- Where can each of these grasslands be found?

🚀 Taking it further

- Why are there few trees in a grassland?
- How do many plants survive extended periods of drought in the grassland?
- How can grass survive when it is continually being cut down by grazing animals?

🧪 Grasslands worksheet

Whenever you learn about a new biome you will be asked to complete a biome summary worksheet for your notebook. On this worksheet you will have to fill in information about the climate, locations, animals, and plants of that biome. The basic information needed to complete the worksheets will be included in the lessons. However, you can feel free to read more about each biome in other sources and include other information on your worksheet. In addition to the worksheet, you should try to include pictures of the plants and animals that are found in that ecosystem. You can look on the Internet, use clip art, take photos, or draw your own pictures.

Today, complete the "Grasslands" summary worksheet and add it to your notebook. Find pictures of plants and animals from various grasslands and add them to your notebook as well. Try to include plants and animals from prairies, savannahs, pampas, and steppes.

🧪 Different varieties of grass

Purpose: To recognize the various types of grass available in your area

Materials: grasses growing in a natural area, flowering plants field guide, newspaper, heavy books, cardstock or heavy paper, page protectors

Procedure:

1. Visit a natural area where plants are allowed to grow wild. Collect samples of as many different kinds of grass as you can find. Try to include the leaves, flowers, and seeds if possible.
2. Use a field guide to help you identify the various types of grass that you collect.
3. When you return home, press the grass between sheets of newspaper by placing heavy books on top of the paper until the grass is dry.
4. Glue your samples to cardstock or other heavy paper. Label the types of grass and place in a clear page protector. Add your samples to your notebook.
5. If you like, you can include samples of non-wild grass such as the grass growing in your yard, as well as wheat, corn or other crop grasses.

Growing grass

God designed grass plants to provide food for many of the world's animals. Grasses grow from the bottom of the plant, allowing the top to be eaten over and over again. God also designed the grazing animals with special digestive systems that can digest grass. A human could not survive on grass, but cattle, horses, antelopes, and other grazing animals have very special stomachs that can break down grass.

Humans have a single-chambered stomach. Food enters at one end, chemicals are added, and the food is moved around and broken down before leaving the other end of the stomach. However, grazing animals have three- or four-chambered stomachs. The food enters the first chamber where bacteria begin the digestive process. Many grazing animals then regurgitate their food and chew it more. When swallowed the second time, the food then enters the second chamber where digestive chemicals are added. The food continues on into the other chambers of the stomach where it continues to be broken down. Both the design of the grass and the design of the grass eaters demonstrate the genius of their Creator.

Purpose: To demonstrate grass's ability to regrow even when it is continually cut down

Materials: grass plants, scissors, ruler, "Growing Grass" worksheet

Procedure:

1. Obtain four similar grass plants. Place them together in a pot filled with potting soil.
2. Number four craft sticks from 1–4. Place one stick next to each plant.
3. Water the grass as needed to keep the soil moist, but not soggy.
4. Cut off the top ⅓ of plant 1. Cut off the top ⅔ of plant 2. Cut off almost all of plant 3—leaving only about ¼ of an inch of grass remaining above the soil. Do not cut anything off plant 4.
5. Measure the height of each plant and record your measurements in the appropriate column of day 1 on the "Growing Grass" worksheet.
6. Measure and record the height of each plant each day for the next three days (days 2–4).
7. After three days, again cut plants 1, 2, and 3 as you did in step 4.
8. Measure and record the height of each plant each day for three more days (days 5–7).

Questions:

- How did cutting the grass affect its ability to grow?
- Did one plant grow more than the others?
- How does this experiment demonstrate God's provision for grassland animals?

Conclusion:

You should find that cutting the grass does not negatively affect its ability to grow; cutting may actually encourage the plants to grow more.

9

Forests

Filled with trees

What is in a forest?

Words to know:

forest
emergent layer
canopy
understory
shrub layer
herb layer
floor

Challenge words:

outer bark
phloem
cambium
sapwood
xylem
heartwood

Have you ever been in a forest? In a forest, you are surrounded by trees that tower over your head. It may be somewhat dark in the forest because the tall trees block out much of the sunlight. If you are quiet you can hear some of the many animals that make their home there.

An ecosystem that is dominated by trees is called a **forest**. There are several different kinds of forests depending on climate and latitude. Near the equator, where there is abundant rainfall and consistently warm temperatures, we find the tropical rainforests. As we move away from the equator we find deciduous forests that have broadleaf trees that lose their leaves in the winter. Finally, as you move farther north, you find the coniferous forests that are filled with evergreen trees such as pine and fir.

There are many differences between the various kinds of forests, and we will examine many of these differences in the following lessons. But there

LESSON 9 **Properties of Ecosystems** • 173

are also some similarities between the various forests, and we are going to learn about some of the similarities in this lesson. Because many trees grow closely together in a forest, the plants in a forest tend to grow in distinct layers. The top layer is called the **emergent layer**. This is where you find the tops of the tallest trees. These trees rise above the other trees and look like they are poking out of the roof of the forest.

Most of the mature trees form the next layer down, which is called the **canopy**. This is the roof of the forest. Because the trees of a forest usually grow close together, their branches and leaves often block out most of the sunlight, making it relatively dark inside the forest.

The third layer down is the **understory**. This layer consists of two different groups of trees. Some of the trees in the understory are young trees that will continue to grow and eventually become part of the canopy. Other trees in the understory are different species, which may be shorter and are designed (or have adapted) to grow with less sunlight.

Below the understory layer is the **shrub layer**. Here you will find shorter plants including many varieties of shrubs. Again, the shrubs that grow in a forest must be able to survive and thrive with less sunlight than other shrubs that grow in more open areas.

Below the shrub layer is the **herb layer**. This is where you will find small plants such as grass, flowers, ferns, and seedlings. Finally, the lowest layer of

> ## Fun Fact
> Forests cover one third of all land in the world and contain two thirds of the leaf surface area of all land plants.

the forest is the called the **floor**. The plants that live on the floor include lichens, mosses, and fungi. This layer is also called the litter layer because you will find fallen trees and branches, as well as leaves and other decaying material here.

There are many animals that live in forest biomes. And just as the plants grow in distinct layers, most animals spend the majority of their lives in only one or two layers of the forest. Birds may swoop down to the forest floor to catch their prey, but they generally take it back to their nests to eat it; thus they spend most of their time in the canopy or emergent layers. Similarly, monkeys that live in the canopy may occasionally leave the trees, but quickly return, spending most of their lives in the canopy. Other animals spend their entire lives on the forest floor. Therefore, you must study each layer of the forest to fully understand the whole ecosystem.

What did we learn?

- What are the major plants in a forest?
- What are the six layers of a forest?
- Which layer forms the roof of the forest?
- Name three kinds of forests.

Taking it further

- Why is the forest floor relatively dark?
- Why is it important to study each layer of a forest?
- How might new trees find room to grow in a mature forest?

Where would I live?

Complete the "Where Would I Live?" worksheet and identify the layer of the forest in which you are most likely to find each animal.

Tree anatomy

Trees are truly amazing creations of God. They begin as tiny seeds but can grow into giants of the forest. To appreciate how this happens, we need to understand the anatomy of a tree. So let's take a look inside a tree trunk to understand how a tree grows, how it transports all of the water and nutrients, and how it stays strong enough to support all of its branches and leaves.

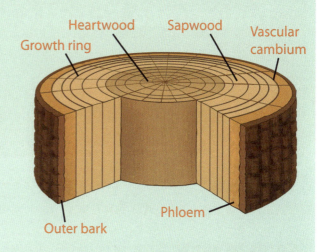

On the outside of a tree is its outer bark. The **outer bark** protects the tree from insects, fire, and other hazards. If we could peel back the bark and look inside the tree trunk we would find that just inside the bark is the **phloem** or inner bark. This part of the tree carries food that is produced in the leaves to the rest of the tree. These cells have a relatively short life and eventually die and become part of the outer bark.

Just inside the inner bark is a layer of cells called the **cambium**. Cambium cells are constantly growing and dividing to produce new cells. The new cells toward the outside of the tree become new phloem cells. The new cambium cells toward the inside of the tree become new wood cells. The continuous production of new cambium cells in the spring and summer causes the bark and wood to become thicker and stronger as the tree gets older.

The cambium produces larger light colored cells in the spring and smaller dark colored cells during the summer. This causes the inside of a tree trunk to have rings. The relative size of the rings indicates the growing conditions for the year. Smaller narrow rings might indicate a period of dry weather, whereas wide rings likely indicate very favorable growing conditions.

As we move inward from the cambium, we find the **sapwood** or **xylem**. The sapwood is the part of the tree that carries the water and nutrients upward from the roots to the rest of the tree. These are the vital ingredients that the leaves need for performing photosynthesis. Just like the phloem cells, the xylem cells only live a short time and eventually die. When they die, they become part of the inner wood of the tree called the **heartwood**. Although heartwood is no longer living and transporting materials, it provides the needed strength for the tree to support the weight of all the branches and leaves.

Activity:

Draw a diagram of the layers of a tree trunk showing the bark, phloem, cambium, sapwood, and heartwood. Add this diagram to the forest section of your notebook.

Temperate Forests

Can you see the forest for the trees?

What is a temperate forest?

Words to know:

deciduous forest

deciduous tree

coniferous forest

coniferous tree

evergreen tree

boreal forest

taiga

Temperate forests are located in the temperate zone between the tropics and the polar regions. They receive more rain than grasslands but less rain than a tropical rainforest. There are two main kinds of temperate forests: deciduous forests and coniferous forests.

Deciduous forests contain many kinds of plants, but the most abundant plants are deciduous trees. **Deciduous trees** are trees with broad leaves that shed their leaves each autumn and grow new leaves in the spring. Deciduous forests can be found in Japan, eastern China, western Europe, Asia, eastern Australia, and eastern North America.

Deciduous forests grow in areas that receive from 30–60 inches (75–150 cm) of rain each year. These areas of the world experience four distinct seasons. The temperature in the deciduous forests varies greatly depending on the time of year. Winters are cold and summers are generally warm and wet. Wintertime temperatures can be as low as -20°F (-30°C) and summer temperatures can be as high as 85°F (30°C), but the average yearly temperature in a deciduous forest is around 50°F (10°C).

Common trees found in deciduous forests include oak, maple, beech, and elm trees. Common shrubs include rhododendron and huckleberry. You will also find a variety of wildflowers such as bluebells and primroses. The forest floor also has many mosses, ferns, and fungi.

The animals that live in a deciduous forest are as diverse as the plants that grow there. You will find many birds from bald eagles to grouse. You will also find opossums, black bears, white-tailed deer, squirrels, and rabbits. In Australia you will also find the duck-billed platypus.

Fun Fact

The largest coniferous trees, the giant redwoods and sequoias, grow in the forests of northern California. These trees contain more biomass than any other living plant. Some of these giant trees are nearly 3,000 years old.

Because the winters are often cold and trees lose their leaves, the food supply in the winter is much less than it is in the summer. Therefore, God has equipped many animals to survive the harsh winter conditions by hibernating, or going into a deep sleep during the winter, thus using less energy and needing less food. Other animals migrate to warmer areas in the winter and return in the spring or summer. So the animals you see in the summer are often different from those you see in the winter.

Because the trees lose their leaves in the fall, deciduous forests are extremely beautiful in the autumn. The leaves of different trees turn a variety of colors as the trees quit producing chlorophyll in preparation for winter. So before the leaves fall, a forest can become a sea of various shades of red, yellow, and orange. Eventually, the leaves dry up and fall off, leaving bare branches to weather the cold winter.

Another type of temperate forest is the coniferous forest. **Coniferous forests** are dominated by **coniferous trees** that bear seeds in cones instead of having flowers and fruit, as most deciduous trees do.

Coniferous trees are **evergreen trees**—they do not lose their leaves in the fall because they have needles or scaly leaves instead of broad leaves. Other names for specific coniferous forest regions in the north include the **boreal forest** and the **taiga**.

Coniferous forests also experience warm summers and cold winters. The winters tend to be colder than in deciduous forests, and the amount of precipitation is less as well. Coniferous forests receive 12–33 inches (30–80 cm) of precipitation each year, much of which comes in the form of snow.

Because of these harsher conditions, the main trees in coniferous forests are evergreen trees that have needle-like leaves instead of broad, flat leaves. Common trees found in coniferous forests include spruce, fir, and pine trees. Although evergreen trees are the most common kinds of trees, there are often deciduous trees such as aspens mixed in with them. Other plants in a coniferous forest include poison ivy, columbines, ferns, and mosses. Lichens also flourish in the coniferous forest biome.

Coniferous forests are the largest type of land biome, covering approximately 17% of the land. They

Tree identification

One of the first things you need to know about a forest is what kind of trees are growing there. You can learn to identify trees by comparing their leaves with the pictures in a tree guide. When you do this you must first ask yourself some questions:

1. Are the leaves broad and flat or are they needle-like?
2. If they are broad, are the edges smooth, toothed, or lobed?
3. If they are broad, do the leaves grow across from each other, or alternating along the twig?
4. If they are needle-like do they grow singly or in groups?
5. If they grow in groups how many needles are in a group?
6. How long are the needles?

These questions will help you find where to look in the field guide to find the type of tree you are looking for.

If you have the opportunity, take a tree guide to a forest and use it to identify the trees that you find growing there. If you do not have a nearby forest, you can practice identifying trees by their leaves by completing the "Tree Identification" worksheet.

Forest worksheets

Complete the "Deciduous Forest" and "Coniferous Forest" summary worksheets and add them to your notebook. Find pictures of plants and animals from various forests and add them to your notebook as well. Be sure to include pictures showing the autumn colors of the deciduous forest.

Fun Fact

The cones of some evergreen trees, such as the lodgepole pine, do not open to release their seeds unless they experience the heat of a forest fire. This prevents the trees from becoming too crowded, but ensures that the forest will be reseeded in the case of a fire.

cover about 50 million acres (20 million hectares) of land. Coniferous forests can be found in North America, Europe, and much of Asia, as well as in many mountain regions around the world.

In spite of the harsh conditions in the winter, the coniferous forest is home to many different kinds of animals. You will find foxes, wolves, Dall sheep, bighorn sheep, caribou, lynx, badgers, owls, and beavers. Many of these animals sleep or hibernate in the winter to conserve energy, just as many of the animals in a deciduous forest do.

Coniferous forests are often dotted with lakes. Many of these lakes were carved out by glaciers that covered much of the northern latitudes during the Ice Age. These lakes and their accompanying streams provide water for the abundant wildlife found in these forests.

What did we learn?

- What are some characteristics of a deciduous forest?
- What are some characteristics of a coniferous forest?
- What is another name for a coniferous forest in the far north?
- What is a deciduous tree?
- What is a coniferous tree?

Taking it further

- What are some ways that plants in temperate forests were designed to withstand the cold winters?
- What are some ways that animals in temperate forests were designed to withstand the cold winters?
- Would you expect plant material that falls to the floor of the coniferous forest to decay quickly or slowly? Why?

Forest Jeopardy

Complete the "Forest Jeopardy" worksheet by writing questions for the answers that are given on the worksheet.

11

Tropical Rainforests

Growing where it's wet

What is a rainforest?

Words to know:
tropical rainforest   arboreal

Forests that grow where there is abundant rainfall, at least 80 inches (200 cm) per year, are called **tropical rainforests**. Tropical rainforests grow near the equator, between the Tropic of Capricorn and the Tropic of Cancer. They are found in Central and South America, Africa, India, Southeast Asia, and western Australia.

Because of the abundant rainfall, rainforests are very humid and have many rivers flowing through them. Because they are near the equator, the temperature in rainforests does not change much. The temperature in a rainforest is usually between 70 and 85°F (20–30°C).

Many plants and animals thrive in this warm, humid environment. Just a few of the animals include spider monkeys, gibbons, tree frogs, lizards, lemurs, boa constrictors, anacondas, bats, capybaras, ibis, piranhas, and toucans. In fact, more than half of the land animals and birds in the world live in the tropical rainforest. Many of the animals are **arboreal**, which means that they live in the trees and seldom come down to the ground. And although there are many different kinds of animals, mammals are more rare in the rainforest than in other biomes. Birds, on the other hand, are more abundant.

The rainforest is also home to many different kinds of plants. One study in a six-acre area of a rainforest found 141 different species of trees. Many tropical plants have become important agricultural products. Some plants you might recognize (but not realize that they were originally found in tropical forests) include pineapples, oranges, bananas, lemons, eggplant, peppers, and cocoa. Today, most of these plants are grown on plantations in tropical areas, but wild plants can still be found in the rainforests.

Another important rainforest plant is chicle, which is used to make chewing gum. Latex, which is used in making many products including rubber and paint, comes from rubber trees that grow in

Fun Fact

Since the rainforest receives so much rain, many of the nutrients in the soil are washed away. The soil is actually more fertile in the grasslands than it is in the rainforests.

LESSON 11   **Properties of Ecosystems • 179**

🧪 Rainforest worksheet

Complete the "Tropical Rainforest" summary worksheet and add it to your notebook. Find pictures of plants and animals from the rainforest to include as well.

🧪 Researching rainforest animals

Purpose: To better understand the animals that live in the rainforest

Materials: research materials

Procedure:

Choose one animal that lives in the tropical rainforest and write a report on it to include in your notebook. Include pictures of your animal as well. Be sure to find the answers to the following questions:

1. What does it eat?
2. Which layer(s) of the forest does it live in?
3. What are its predators?
4. What unusual or interesting habits does it have?
5. What interactions does it have with other animals?

---

the tropical rainforest. Also, balsa and mahogany trees, which are prized for their wood, are tropical forest trees.

The plants in the rainforests are also very important for producing medicines. About one-fourth of all medications are made from plants that grow in the tropical rainforests. The main ingredient in aspirin comes from the rainforest as well as the plants used to make many antibiotics. Quinine, the medication used to treat malaria, also comes from a tropical rainforest plant.

The largest Banyan tree in North America is in Fort Myers, Florida.

Because the trees in the forest block most of the sunlight from reaching the forest floor, there are not many plants growing on the floor. Instead, some plants grow on the sides of the trees where there is more sunlight. These plants are called epiphytes. Epiphytes are plants that use other plants without taking nutrients from them. There are many different epiphytes in the rainforest. Seventy percent of all orchids are epiphytes. Some ferns and cacti are also epiphytes.

One very unusual epiphyte is the banyan tree. The seeds of the banyan are dispersed by birds that eat the fruit of the banyan. When a seed lands on a branch of a tree it may germinate and a new tree begins to grow. The seedling sends out roots that grow down the side of the host tree. As the banyan grows larger, it sends out prop roots which grow out

Fun Fact

Although we often associate the word jungle with the rainforest, a jungle is technically an ecotone, an area between the tropical rainforest and a tropical grassland. There are fewer trees in a jungle which allows for more plant growth near the ground.

of the bottom of its own branches. Eventually, the roots of the banyan trees may completely enclose the host tree. Thus the banyan is sometimes called a strangling fig tree. Banyan trees can continue to spread out as they put out more and more prop roots from their branches. One banyan tree was measured covering ¾ acre of land. Banyans are very long-lived trees; some are more than 300 years old.

What did we learn?

- List some ways in which a tropical rainforest is different from a temperate forest.
- Where are the rainforests located?
- What is an arboreal animal?
- What is an epiphyte?
- Name at least one epiphyte.

Taking it further

- Do you think that dead materials would decay slowly or quickly on the floor of the rainforest? Why?
- If you transplanted trees such as orange, cacao, or papaya trees to a deciduous forest, would you expect them to survive? Why or why not?
- Which animals are you most likely to see if you are taking a walk through the tropical rainforest?

Fun Fact

The long vines that Tarzan swings from are called lianas.

Rainforest products

You are an advertising executive. Your job is to promote products from the rainforest. Choose one or more products discussed in the lesson or choose from the list. Then design an advertising campaign to encourage people to use products from the tropical rainforest. Include your advertising materials in your notebook.

Items from the rainforest:

1. Coffee
2. Brazil nuts
3. Cashews
4. Allspice
5. Cinnamon
6. Cloves
7. Ginger
8. Black pepper
9. Avocado
10. Papaya
11. Guava

LESSON 11   Properties of Ecosystems

UNIT 3

Aquatic Ecosystems

12 The Ocean
13 Coral Reefs
14 Beaches
15 Estuaries
16 Lakes & Ponds
17 Rivers & Streams

◊ **Describe** the characteristics of the ocean, lakes, and rivers.
◊ **Describe** the characteristics of a coral reef.
◊ **Identify** types of beaches.
◊ **Identify** types of estuaries.

12

The Ocean
Marine ecosystem

What lives in the ocean?

Words to know:

phytoplankton

sunlit zone

euphotic zone

zooplankton

twilight zone

disphotic zone

midnight zone

aphotic zone

benthos

nekton

plankton

Challenge words:

bioluminescent

Three quarters of the surface of the world is covered with water, so it should not surprise you that many of the world's ecosystems occur in or near the water. The ocean contains 95% of all surface water and is the world's largest ecosystem. Although we generally divide the ocean into five different oceans (Pacific, Atlantic, Indian, Arctic, and Antarctic), the oceans are really all connected together and form one ocean.

The ocean is perhaps the most important ecosystem in the world. The ocean greatly affects the weather around the world. It also drives the water cycle by providing the large surface from which most evaporation occurs. It contains fish and other animals that feed a large portion of the people in the world and it provides a multitude of jobs for people. Therefore, it is very important to understand the ocean ecosystem.

Because of its vast size, it is difficult to define a single ocean ecosystem. Different areas of the ocean have different conditions and thus have different habitats. However, there are many characteristics that are the same around the world and we will examine some of these characteristics in this lesson.

As you learned earlier, virtually all ecosystems depend on the producers, the organisms that produce food by photosynthesis. In the ocean most of the photosynthesis takes place in microscopic creatures called **phytoplankton**. These creatures are algae and other small organisms that contain

Fun Fact

More than half of all photosynthesis in the world occurs in the ocean by the phytoplankton that live there.

LESSON 12 **Properties of Ecosystems • 183**

🧪 Ocean worksheet

Complete the "Ocean" summary worksheet and add it to your notebook. Find pictures of marine algae, plants, and animals to include as well.

🧪 Currents distribute nutrients

Currents are very important to life in the ocean. Not only do currents move the plankton from place to place, but the currents also move oxygen and nutrients. This helps to keep the ocean an active, living place.

Purpose: To see how currents distribute nutrients in the ocean

Materials: shallow pan, water, food coloring

Procedure:

1. Fill a shallow pan with water and place it on a level surface.
2. Gently blow across the surface of the water. Notice how your breath moves the water, making small waves and currents.
3. Drop a few drops of food coloring at one edge of the water.
4. Again, gently blow across the surface of the water. Watch how the food coloring is distributed throughout the water.

Conclusion:

This is an indication of how currents transport nutrients, oxygen, and plankton around the world.

---

chlorophyll and produce food for other creatures in the ocean.

Since photosynthesis requires sunlight, the life in the ocean changes as you go deeper because the sunlight can only penetrate a limited depth. The upper layer of the ocean is called the **sunlit zone** or **euphotic zone** and consists of the top 660 feet (200 m) of water. This is where you will find most living creatures and nearly all of the plants and phytoplankton. You will find endless varieties of fish, jellyfish, sharks, coral, shrimp, lobsters, and most other sea creatures. You will also find **zooplankton**—microscopic organisms that eat the phytoplankton and then become the food of many other animals in the ocean.

As you go deeper, the sunlight decreases. At depths from 660 to 3,300 feet (200–1,000 m) there is limited sunlight. This part of the ocean is referred to as the **twilight zone** or **disphotic zone**. There are no plants or phytoplankton here because there is not enough light for photosynthesis. The animals that live in the twilight zone must be able to survive in very low light, be able to live in cold water, and be able to stand great pressure on their bodies. You will find octopi, a few fish, such as the coelacanths and viper fish, and a few crabs and eels.

Below 3,300 feet (1,000 m) there is no sunlight even in the daytime. This part of the ocean is called the **midnight zone** or the **aphotic zone**. Only a few animals live in this dark, cold, high pressure environment. Most of the animals that live in the disphotic and aphotic zones are scavengers that eat dead and decaying material that filters down from above, or predators that eat the other animals. Some animals move vertically through the layers, while others spend most of their time in only one layer.

Fun Fact

The average depth of the ocean is 12,451 feet or nearly 2½ miles. This is five times the average height of land formations. Although the ocean is relatively shallow near the shore, it is very deep in other areas.

The huge whale shark feeds on plankton.

The living organisms in the ocean can be divided into three groups. The **benthos** are the plants and animals that live on the ocean floor. In areas that are relatively near the shore, the water is all in the euphotic zone, so many plants and algae attach themselves to the ocean floor. These organisms that live on the ocean floor are considered bottom dwellers. Clams, crabs, worms, and other burrowing animals are part of the benthos as well as starfish, snails, and sponges.

The second group of organisms is the **nekton**. These are animals that are free moving. Fish, whales, shrimp, lobsters, squid and many other sea creatures can swim where they like. Their movement does not depend on the currents of the ocean.

Fun Fact

The only known ecosystem that does not rely on photosynthesis is on the floor of the ocean near hot water vents. There, a certain type of bacteria converts sulfur into food and many animals in that ecosystem eat the bacteria.

Finally, **plankton** are algae (phytoplankton) and animals (zooplankton) that drift with the ocean currents. They live near the surface and cannot overcome ocean currents by swimming. These are the organisms that form the bottom of most food chains in aquatic environments.

All of the plants and animals that live in the ocean must be able to live in a saltwater environment. The ocean is about 3% salt. All together, there are 55 different elements that have been identified in seawater. In 1,000 grams of water there are about 35 grams of minerals. Of the minerals in the seawater, about 75% is salt, although the salt concentration varies from place to place.

In addition to minerals, gases such as oxygen, nitrogen, and carbon dioxide are also dissolved in the water. The concentration of oxygen is highest near the surface of the ocean. Some oxygen dissolves from the air into the water, but most of it is produced by phytoplankton, primarily algae, that grow near the surface of the water.

As you study the ocean, you will gain a greater appreciation for the amazing ecosystems that God has set up around the world.

What did we learn?

- How much of the Earth is covered with water?
- How much of the surface water of the world is in the ocean?
- How many oceans are there?
- What are the three zones that the ocean can be divided into?
- What are the three major groups of living organisms in the ocean?

Taking it further

- What might happen in the ocean if the currents stopped flowing?
- Why do most animals in the ocean live in the euphotic zone?
- Why might the aphotic zone occur at a shallower depth than 660 feet (200 m) in some areas?

Understanding bioluminescence

Although most sea creatures live in the sunlit zone, some of the most interesting creatures in the ocean live in the twilight and midnight zones. Many of these creatures can make their own light, which is handy in an environment with little or no light. Animals which can produce their own light are said to be **bioluminescent**. Research bioluminescent animals and make a report to include in your notebook.

Activity:

Draw a population pyramid for an aquatic food chain. Be sure to start with algae or some other phytoplankton. See if you can make it at least four levels high. Include this drawing in your notebook.

Many jellyfish are bioluminescent.

13

Coral Reefs
Underwater wonderlands

What is coral?

Words to know:

coral reef

atoll

fringing reef

barrier reef

Challenge words:

coral bleaching

One of the most beautiful ecosystems within the ocean itself is the coral reef. A **coral reef** is a limestone formation built from the exoskeletons of millions of tiny invertebrates called corals. Coral reefs are found in warm, clear water near the equator, in the tropical zone. Reefs are almost always found in water that is less than 150 feet (45 m) deep.

Coral polyps are creatures that build the foundation of the coral reef. An individual coral polyp is a tiny tube-like creature with tentacles. It resembles an upside down jellyfish. The polyp builds a protective exoskeleton around itself and spends much of its time inside its protective shell.

Corals use their tentacles to sting and pull in tiny prey, but corals receive most of their energy from a special type of algae that lives inside of them and performs photosynthesis, thus providing food for both the algae and the corals. Corals must live in clear waters in order for the sunlight to reach the algae. If the water becomes murky, the algae cannot perform photosynthesis and the corals can die.

There are several thousand different species of coral. Most corals are named for the shape of the colonies that they build. Corals of the same species will live very closely together. As they grow, their exoskeletons merge together to form colonies. Some colonies look like fans; others look like horns or even brains. As corals die, new coral polyps will build their homes on top of the dead exoskeletons, thus increasing the size of the reef.

In addition to the corals, there are thousands of other animal species that make their homes in the coral reef. Sponges, shrimp, sea stars, and moray eels are just a few of the wide variety of marine animals you might find. There are also over 1,500 species of fish that have been found living near coral reefs. This includes clown fish and butterfly fish. You will also find green turtles, sea anemones, sea urchins, octopi, squid, and clams in a coral reef. A coral reef is a stunning array of shapes and colors because of the wide variety of life there.

Coral reefs generally grow in one of three different configurations. An **atoll** is a circular shaped coral reef that has formed around a sinking inactive volcano. An atoll usually encloses a shallow lagoon.

LESSON 13 **Properties of Ecosystems • 187**

Australia's Great Barrier Reef extends for 1,250 miles (2,000 km) along the northeastern coast of Australia.

Fun Fact

There are more different species of organisms in the coral reef than in any other ecosystem except the rainforest.

A **fringing reef** is a reef that is attached to the mainland and grows out into the open water. A third configuration is a barrier reef. A **barrier reef** forms out in the ocean away from the shore. The reef forms a barrier that helps block waves from reaching the shore, thus giving it the name of barrier reef.

The largest coral reef is the Great Barrier Reef off the northeast coast of Australia. The Great Barrier Reef is the largest biological structure on Earth and is even visible from the moon. It is 1,250 miles (2,000 km) long. The Great Barrier Reef is important in many ways. Not only is it beautiful and home to millions of animals, it protects the coastline of Australia from erosion by blocking much of the wave action. It is also a vital part of the commercial and recreational fishing industries in Australia.

Coral reefs grow fastest in areas with stronger wave action. Corals need calcium and other minerals in order to produce their exoskeletons. Waves bring new nutrients, enabling the corals to grow. Anyone who has ever visited a coral reef can attest to the wonder and beauty that God has created there.

What did we learn?

- Where will you find coral reefs?
- What is a coral reef made from?
- Where do corals get most of their energy?
- What are the three main types of coral reefs?
- What are some of the animals that live in a coral reef besides corals?

Taking it further

- Why are coral reefs found in water that is usually less than 150 feet (45 m) deep?
- Why do corals grow best in swift water?

Coral reef worksheet

Complete the "Coral Reef" summary worksheet and add it to your notebook. Find pictures of plants and animals from the coral reef to include as well.

🧪 Coral model

Coral colonies grow in a wide variety of shapes. Examine pictures of several different types of coral, then use modeling clay to build your own coral colony. Be sure to take a picture of your model and include it in your notebook.

🎖 Coral bleaching

The beautiful colors found in the coral reefs are due in large part to the algae living inside the corals. When two living organisms live together in a mutually beneficial relationship it is called symbiosis or mutualism. The corals provide protection for the algae and the algae provide food for the corals. This symbiotic relationship is vital to coral survival.

In the past 40 years many coral reefs have experienced a phenomenon called **coral bleaching**. Many of the corals have expelled their algae. This causes the corals to lose their main source of food. It also causes the corals to become white or clear in color, thus the name *coral bleaching* because color is removed.

Many times a coral that expels its algae will get new algae and become healthy again. In fact, some scientists believe that coral bleaching may actually be a protective mechanism that allows the coral to get rid of algae that are not suited to the environmental conditions and to gain new algae that are better able to survive.

Other times corals die because they cannot replace the algae and thus they do not get enough nutrients to survive. Many areas of coral around the world have died due to coral bleaching.

No one knows exactly what causes coral bleaching to occur. It is believed that the expulsion of the algae is a reaction to unusual stresses placed on the coral. Scientists have proposed several possible causes. Some people believe it occurs when the water temperature rises too much. Much of the coral bleaching that occurred on a wide basis in the 1980s and 1990s coincided with large El Niño years which may have affected the water temperature around the coral reefs.

A second possible cause for coral bleaching is an increase in ultraviolet radiation. If water is too still, too much ultraviolet radiation can reach the corals and damage the tissues. This is believed to be the cause of some coral bleaching. On the other hand, if water becomes murky it blocks out some of the sunlight. This causes the algae to perform less photosynthesis, which could also result in coral bleaching.

Another possible cause is disease. It may be that a virus or other pathogen is responsible for the way the coral reacts to the algae. Some scientists think that increased acid levels or other pollutants in the water could be responsible for some coral bleaching as well.

Today, scientists do not fully understand the mechanisms or triggers of algae expulsion. More testing and research are needed before we will truly understand the causes of coral bleaching. We know that God has designed many animals to be resilient and able to handle changes to their environment, and the future will show if coral bleaching is detrimental in the long term or just a way to better handle changing conditions.

14

Beaches

Take a walk on the sand

What kinds of beaches are there?

Words to know:

beach
dynamic equilibrium
inter-tidal zone
tide pool

Challenge words:

dune system
primary dune
secondary dune
maritime forest

Where the waters of the ocean meet the land, a shore is born. This shore is often called a **beach**. A beach is an actively changing ecosystem. The water is constantly bringing and depositing new materials. At the same time, the movement of the water is constantly wearing away the land along the shore. Most beaches are in a state of **dynamic equilibrium**. This means that the amount of material being deposited is approximately equal to the amount of material that is being eroded away.

The area of land along a beach that is covered by high tide and uncovered at low tide is called the **inter-tidal zone**. Plants, algae, and animals that live in the inter-tidal zone must be able to deal with constantly changing conditions. They spend part of the day partially or totally submerged in seawater and part of the day exposed to the sun and air. Many of these animals burrow into the sand or dirt to stay moist. Others have shells that protect them. These organisms are designed to conserve water.

The types of plants and animals that live along a beach depend greatly on the material that the beach is made of. Beaches are either rocky or sandy. Rocky beaches provide soil, cracks, and crevices that give plants a place to become firmly attached. These plants provide food and protection for many animals that would not live on a sandy beach. Along a rocky beach you are likely to find starfish, mussels, barnacles, and oysters. You will also observe sea lettuce, swamp periwinkle, and enteromorpha, which is a green algae that many people call seaweed.

Rocky beaches are best known for their **tide pools**. These are areas along the beach that fill with water when the tide is high, and remain filled after the tide goes out. The water in these tide pools is refreshed twice a day with the rising of the tides. These pools provide a refuge from the winds and sun that dry out the rest of the beach.

Some of the plants and animals that you find on rocky beaches are also found on sandy beaches including gulls and other shore birds, but because the shifting sand does not provide a solid foundation,

Ghost crab on the beach

there are also different plants and animals on a sandy beach. You are more likely to find sand dollars, crabs, shrimp, and clams on a sandy beach than on a rocky beach. Other animals you are likely to find at a sandy beach include horseshoe crabs and turtles. And although you will still find sea lettuce, you are not likely to find enteromorpha or periwinkle.

What did we learn?

- What is a beach?
- What are the two main kinds of beaches?
- What is the name of the area of land that is covered at high tide and uncovered at low tide?
- What are some animals you are likely to see in a beach ecosystem?

Taking it further

- Why might you find different plants and animals on a rocky beach from those on a sandy beach?
- How is new sand formed?
- Explain how a beach can be in dynamic equilibrium.

Beach worksheet

Complete the "Beach" summary worksheet and add it to your notebook. Find pictures of plants, algae, and animals from both rocky and sandy beaches to include as well.

Making sand

Sand is an accumulation of sediment. Sand in different parts of the world is comprised of different materials. Some sand is mostly crushed coral. Other sand contains bits of rocks and broken seashells. Other sand is made of volcanic glass—rock that shattered when hot lava was suddenly cooled as it hit cold seawater. This is how the black sand beaches of Hawaii were formed. Much of the white sand around the world is made of quartz.

Purpose: To appreciate the forces required to make new sand

Materials: rocks, seashells, plastic zipper bag, safety goggles, towel, hammer, sand, magnifying glass

Procedure:

1. Place a few small rocks and/or seashells in a plastic zipper bag and close the bag.
2. Wrap the bag in a towel.
3. While wearing safety goggles, use a hammer to break the rocks and shells into small pieces.
4. Carefully remove the smallest pieces of debris from the bag. Compare your sample with a sample of sand. How are they similar? How are they different?
5. Use a magnifying glass to closely examine a sample of sand. Can you see the different shapes and colors of the grains of sand? If you have sand from more than one location you can see how different areas contain different kinds of sand.

Conclusion:

Although you may think of all sand as being the same, when you compare sand from various parts of the world, you will see that there are significant differences.

A dune system

When you visit a sandy beach you can observe ecological succession in progress. Recall that succession is the process where one ecosystem is changing into another ecosystem. At a beach, as you move from the edge of the water toward the land, you can see one ecosystem change into another. This area is called a **dune system**.

The area of land closest to the beach is called the **primary dune**. Primary dunes develop where beach grass begins to grow. This grass anchors the dune by trapping sand and keeping it from eroding away. The grass has an extensive root system that pulls water from the ground and stabilizes the dune. The primary dune area has too much salt, wind, and sun for trees to grow, but the beach grass that grows here is tolerant to salt and can thrive even with varying amounts of salt. The grass provides nesting areas for many shore birds.

As you move inland from the primary dune the landscape changes from one of primarily grass to one of grass and shrubs. This area is called the **secondary dune**. In the secondary dune area you will find bayberry, scrub pine, yarrow, beach grass, and poison ivy. This more diverse plant life provides homes for more diverse animal life as well. Many small animals live in the secondary dunes.

Moving farther inland, you begin to see more trees. The area of shrubs and trees is called a **maritime forest**. This is not your usual forest. The trees in a maritime forest generally are short with twisted limbs. The limbs from these trees provide the interesting drift wood that many people like to collect.

If you have a chance to visit a beach, observe the changing ecosystem as you move inland from the shore. If you cannot visit a beach, look at pictures of beaches. Then draw your own picture of a dune system and include it in your notebook.

15

Estuaries

Where fresh and salty meet

What is an estuary?

Words to know:

estuary

salt marsh

salt meadow

mangrove forest

Challenge words:

watershed

One of the most productive ecosystems in the world is an estuary. An **estuary** is where freshwater flows into saltwater. The estuary is a productive ecosystem because the constant flow of freshwater into the ocean, coupled with the ebb and flow of the tides, stirs up nutrients that can be used by the many plants and algae that grow in the area. This stimulates plant growth, which in turn stimulates animal populations.

Although food is readily available, plants and animals in the estuary must deal with some difficult conditions as well. The salt level in the water is constantly changing. In the summer when there is relatively little rain and higher evaporation, the salt level in the estuary is considerably higher than in the winter when water does not evaporate as quickly and more freshwater is added by rain. Organisms must also deal with strong currents from the constantly changing water flow. This water flow also stirs up mud which can decrease the amount of oxygen available, so plants and animals must be able to cope with these conditions.

There are actually several different kinds of estuary ecosystems depending on which plants are most dominant. A **salt marsh** is a coastal wetland that is flooded and drained by salt water brought in by the tides. Rushes are the predominant plant. Farther inland, where the ground is somewhat drier, salt-tolerant grasses and

There are over 50 different species of mangrove plants worldwide.

LESSON 15 **Properties of Ecosystems • 193**

🧪 Mixing fresh & saltwater

Saltwater and freshwater come together, sometimes forcefully, in an estuary. Saltwater and freshwater do not easily mix. The salt in the ocean makes the saltwater denser than freshwater so the freshwater tends to sit on top of the saltwater and slowly mix with it. You can conduct an experiment to see the effect that salt has on how water mixes.

Purpose: To compare the densities of freshwater and saltwater

Materials: four clear cups, water, salt, green and blue food coloring, eye dropper, marker

Procedure:

1. Fill four clear cups half full of water.
2. Use a marker to label two of the cups "Fresh" and two of the cups "Salty."
3. Add 2 teaspoons of salt to each cup that is marked "Salty." Stir to help the salt dissolve.
4. Add several drops of green food coloring to one cup of salty water. Stir the water. Add several drops of blue food coloring to one cup of freshwater and stir the water.
5. Use an eyedropper to add several drops of green saltwater to the clear cup of freshwater. What happened to the green water? It should sink to the bottom of the cup.
6. Use an eyedropper to add several drops of blue freshwater to the cup of clear saltwater. What happened to the blue water? It should float on top of the saltwater.
7. Allow the cups to sit for several minutes. What happened to the colored water? Eventually the freshwater and saltwater will mix, but it may take several hours.

Conclusion:

Saltwater and freshwater have different densities so they do not readily mix. This causes the density and saltiness of the waters in estuaries to vary. Plants and animals that live in the estuaries are able to deal with changing salt levels

🧪 Estuary worksheet

Complete the "Estuary" summary worksheet and add it to your notebook. Find pictures of plants and animals from the various types of estuaries including mangrove forests, salt marshes, and salt meadows.

Aquatic Ecosystems

small herbs and shrubs help create **salt meadows**, meadows subject to flooding by salt water. And in many areas mangrove trees are the dominant plant, making the estuary a **mangrove forest**.

Mangroves and other estuary plants are very important because they help to filter salt out of the water and help to filter silt and other materials so they do not enter the ocean. This is especially important in areas near coral reefs. The water must remain clear for the coral to thrive, and mangrove trees help to keep the water clear.

In addition to the many plants that live in the estuary, there is a wide variety of animals. There are many mud snails and marine worms that thrive on the mud that is stirred up by the moving water. These animals eat the mud and help to recycle the nutrients. Cockles and other shellfish filter food out of the water. The water in the estuary is full of many varieties of fish including mullet, flounder, and sole. The areas near the water are home to a wide variety of birds such as the bittern, rail, heron, tern, stork, and pelican. Sea lions and other marine animals also live in or near some estuaries.

Some animals in the estuary are seasonal. Many of them migrate to other areas and return during certain times of the year. Some animals, like birds, fly to other parts of the world. Other animals, such as the freshwater eel, swim upstream and later return to the estuary. Some animals swim farther out to sea during certain seasons and later return to the estuary. So the animals you observe at one time of the year could be very different from those you see at a later time. ✳

What did we learn?

- What is an estuary?
- Name three types of estuaries.
- What are some plants you might find in an estuary?
- Name several animals that you might find in an estuary.

Taking it further

- Why is an estuary a very productive ecosystem?
- How do mangrove trees help coral reefs?
- Why is the salt level in the water constantly changing in an estuary?
- Why might you find different estuary animals in the same location at different times of the year?

Watersheds

Things that happen on land can have a huge effect on water ecosystems because the water that flows across a farmer's field or even in your yard eventually flows into a river and finally flows into the ocean. All of the land that has water flowing into a particular river or body of water is called a **watershed**.

Your yard is part of several watersheds. When you water your grass whatever water is not used by the grass flows into the ground and eventually finds its way to an underground river. So your yard is part of the watershed for that underground river. The underground river will eventually reach ground and flow into a river above ground. Your yard will be a part of the watershed for this above ground river as well. Finally, that river may flow into the ocean, making your yard part of the watershed for the ocean, too.

One of the most important watersheds in the United States is the Mississippi River Basin. 41% of the continental United States eventually drains into the Mississippi River, making it the largest watershed in the United States and the third largest watershed in the world. Most of the land from the Rocky Mountains to the Appalachian Mountains is part of this watershed.

The Mississippi River starts at Lake Itasca in Minnesota and flows over 2,300 miles to the Gulf of Mexico. As it flows southward, several major rivers join it including the Missouri, the Ohio, the Arkansas, and the Tennessee Rivers. When the Mississippi River reaches the Gulf of Mexico its freshwater flows out into the salty gulf. The water does not immediately mix with the water in the gulf. In fact, NASA satellites have shown that the freshwater flows through the gulf, around the tip of Florida, and into the Gulf Stream of the Atlantic Ocean before it becomes thoroughly mixed with the saltwater.

Where the Mississippi River flows into the Gulf of Mexico is called the Mississippi River Delta. This is one of the most important estuaries in the world. The River Delta contains approximately 3 million acres (12,000 km²) of coastal wetlands and 40% of the salt marshes in the continental United States.

You can find out about the watershed near your home at the "Surf Your Watershed" website from the US Environmental Protection Agency.

Mississippi River Delta

Fun Fact

The Lewis and Clark Expedition was a project to explore the Louisiana Purchase, which consisted primarily of the lands with water flowing into the Mississippi, i.e. the Mississippi River Basin.

16

Lakes & Ponds

It's fresh

What is the difference between a lake and a pond?

Words to know:

lake

pond

overturn

algae bloom

Lakes and ponds are some of the most beautiful ecosystems in the world. Crystal clear water filled with fish, ducks, and other water birds can inspire and turn your heart to God. Many lakes were carved out by the movement of glaciers during the Ice Age. Other lakes fill the craters of extinct volcanoes.

Lakes are large bodies of freshwater. Although lakes are filled with freshwater instead of saltwater, a deep lake has many of the same characteristics as the ocean. A deep lake has a euphotic or sunlit zone and an aphotic or midnight zone just like the ocean. And the plant and animal life found in each zone is different. Also like in the ocean, the main source of food at the bottom of the food chain is photosynthetic algae. Large lakes can also have beaches like the ocean, but without the distinct tides of the ocean. **Ponds** are lakes that are shallow and do not have an aphotic zone.

Lakes and ponds can be found all around the world. And although the specific organisms vary somewhat, there are many similar organisms found in lakes around the world. Algae, water lilies, and some grasses grow in the water. Along the edge of lakes you are likely to find grasses, rushes, cattails, sedges, sagebrush, and thistles.

The animals that live in lakes and ponds include zooplankton such as rotifers and tiny crustaceans as well as many kinds of frogs, toads, turtles, and insects. However, fish are probably the animals most closely associated with lakes. Trout, perch, walleye, bass, and sturgeon are only a few species of fish that are found in lakes around the world.

Other animals live near lakes and depend on the lake for water. Birds such as cranes, pelicans, egrets, ibis, swans, ducks, and geese are common on or near lakes. Beaver, deer, and moose are also commonly found near lakes.

Since algae is the main source of food for the animals that live in a lake, it is important to understand how God provides the needed nutrients for these tiny organisms to survive. The density of freshwater changes as it cools. As water cools it becomes denser until it reaches 37°F (4°C) at which point is expands as it freezes and becomes less dense. This causes ice to float on the surface of the lake. In the spring, as the ice begins to melt, the cold water on

🧪 Lakes & ponds worksheet

Complete the "Lakes and Ponds" summary worksheet and add it to your notebook. Find pictures of plants and animals that live in or near lakes and ponds to include as well.

🧪 Freezing fresh & saltwater

Purpose: To observe how freshwater and saltwater freeze

Materials: two clear cups, water, salt, thermometer, marker, "Watching Water Freeze" worksheet

Procedure:

1. Fill two clear cups ¾ full of water.
2. Label one cup as "Freshwater" and the other cup as "Saltwater."
3. Add 2 teaspoons of salt to the "Saltwater" cup and stir to dissolve.
4. Measure the temperature of the water in each cup and record the temperatures on the "Watching Water Freeze" worksheet.
5. Place both cups in a freezer.
6. Measure the temperature of the water in each cup every 10 minutes for 60 minutes. Record your measurements and observations on the worksheet.

Conclusion:

As freshwater freezes it expands so ice will form on the surface of the water first. You will not see ice forming on the bottom of the cup. This is a demonstration of God's provision for life in the lakes and ponds.

The salt in the saltwater helps to keep the water molecules apart, making it hard for the water to freeze. It should take much longer for saltwater to begin to freeze, and if your freezer is close to 32°F (0°C) it may not freeze at all. This is an example of God's provision for life in the oceans as well.

the top of the lake is more dense than the warmer water deeper down. When enough ice has melted, this dense, cold water rapidly sinks to the bottom of the lake. This causes the warmer water to rise, stirring up the mud on the bottom of the lake. This action is called **overturn**. Overturn is important because it releases oxygen and many of the nutrients that have become trapped on the bottom of the lake, thus providing nutrients for the algae and other plankton.

Overturns in the spring have been known to cause **algae blooms**, a sudden growth in algae that can turn the surface of a lake or pond green. After a few days or weeks, the algae growth slows down and the other animals that eat the algae enjoy abundant food. A similar overturn usually occurs in the fall as the top layer of water begins to cool. However, this does not usually result in an algae bloom.

The changing density of water is a gift of a loving Creator who designed the water to preserve life in the lakes and ponds. Not only do overturns release nutrients for algae growth, but the fact that ice floats is truly amazing. No other common substance becomes less dense when it freezes. But because frozen water is less dense than liquid water, it floats. This provides an insulating barrier during the winter for the animals that live in the lakes and ponds. If ice did not float, lakes and ponds would freeze solid and the animals would die. ✻

Ice fishing is possible because lakes freeze at the top, allowing the fish to survive underneath.

What did we learn?

- What is a lake?
- What is a pond?
- What are two ways that lakes were formed in the past?
- What is an overturn?
- What is an algae bloom?

Taking it further

- Why is overturn important to lake ecosystems?
- Why does an algae bloom often occur in a lake in the spring?
- In which lake zone would you expect to find most small creatures like rotifers?
- What would happen to fish during the winter if ice did not float?

The Great Lakes

One of the most amazing areas in the world is the Great Lakes region of the northern United States and southern Canada. These lakes hold more than 18% of the world's surface freshwater. The lakes cover 95,000 square miles (246,000 km2) and their watershed covers 288,000 square miles (746,000 km2). The Great Lakes have over 9,000 miles (14,500 km) of shoreline.

These amazing lakes are not only important for the habitat they provide for the fish and other aquatic animals that live there, but there are approximately 35 million people that depend on this ecosystem for water and for their livelihood. Recreation on the lakes is a $15 billion per year industry. Sport fishing is a $4 billion per year industry. In addition, the lakes provide a major traffic route for shipping.

The fish in the Great Lakes have always been abundant. The lakes contain trout, perch, walleye, whitefish, and salmon. The lakes also provide habitat for a wide variety of birds including the bittern, goshawk, meadowlark, and many species of waterfowl.

Because people as well as animals depend on these lakes, constant vigilance is needed to meet the needs of both groups. When people build dams and other structures that stop or slow down the flow of water into the lakes, it has an effect on the plants and animals in that ecosystem. For example, building dams on the rivers flowing into the lake makes it difficult for salmon to return to their birthplaces in the rivers for spawning. This has caused the salmon population to decrease in the Great Lakes. Many groups have been formed to help oversee development around the Great Lakes to ensure that people and animals can both enjoy this beautiful ecosystem.

Human interaction is not the only problem facing the Great Lakes. Many species of animals have been introduced that were not native to the area. Some of these animals do not have natural predators in the area and are crowding out the native species. Some invasive species include sea lampreys, zebra mussels, and rusty crayfish. People are trying to find ways to control the populations of these species and to protect native populations.

There is so much to learn about the Great Lakes that you should do some research on your own. Then take what you have learned and make a Great Lakes page for your notebook. In addition, make a map of the Great Lakes to include in your notebook and complete the "Great Lakes Fact Sheet," too. There is room on the fact sheet to add some facts of your own as well.

Fun Fact

The Great Lakes are so large that they create their own weather systems. One weather phenomenon caused by these large bodies of water is called *lake effect snow*, which brings snow to many nearby cities including Chicago.

Satellite image of the Great Lakes

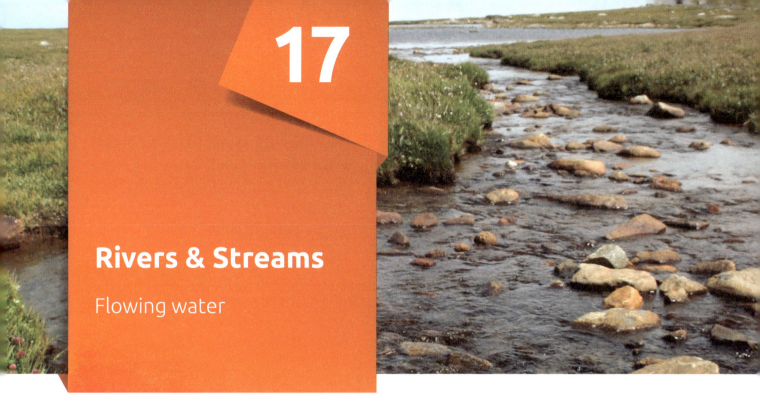

17

Rivers & Streams

Flowing water

Why can't the same animals that live in lakes live in rivers?

Words to know:

river
stream
riparian zone
tributary

Do you enjoy the gentle trickle of a small stream or the majesty of a great waterfall? Then you enjoy the beauty of a river ecosystem. **Rivers** and **streams** are ecosystems that contain flowing freshwater. Although many of the same animals that live in lakes also live in rivers and streams, the current of the river makes it difficult for some plants and animals to survive here. The current also allows more oxygen to become dissolved in the water, allowing different species to thrive.

Although lakes and the ocean get most of their food from the photosynthesis performed by algae, there is very little algae in most rivers and streams. Much of the energy in the river ecosystem comes from the breaking down of leaves and other plant materials that fall into the water. The trees and other plants that grow along the banks of a river contribute to the food supply of the river.

Just as with lakes, in a river there are many animals that live in the water, and many other animals that live on the banks or regularly visit the river. Some of the animals that live in rivers include snails, salamanders, and many varieties of fish including trout and salmon. Other animals that live near and depend on the water include snakes, herons, eagles, insects, turtles, raccoons, muskrats, storks, ducks, otters, deer, and bears.

Although there are few plants in the water, some algae can survive in the moving waters. There is abundant plant life along the banks in what is called the **riparian zone**. These plants include grasses, pussy willows, willow trees, alders, elkslip, and water hyacinths. The plants in the riparian zone provide habitat for the many animals in the area. Riparian plants also help to filter the river runoff.

Moving water can be a powerful force. As the water flows downhill it picks up bits of rock, sand, silt, and other debris. The faster the water is flowing, the more debris (and larger debris) it can carry. As a river reaches a flat area it slows down and begins to drop the sediment it was carrying. Slow-moving rivers tend to flow in lazy curves. As the water goes around a curve it deposits more sediment on the inside of the curve because the water is flowing more slowly there and picks up sediment on the outside of the curve, since the water is flowing more quickly there. This causes the curves to get bigger over time. The plants and animals that thrive in and near fast moving water are often different from those that thrive near slow moving water.

Large rivers seldom start out large at their source. Water flows downhill because of gravity. When rain falls at higher elevations, the water flows into small streams. Streams flow and eventually join together to form small rivers and then small rivers flow into larger rivers. A smaller stream or river that flows into a larger river is called a **tributary**. Some rivers have only a few tributaries, but most rivers have many tributaries.

Fun Fact

The Nile River is the longest river in the world. The Missouri/Mississippi River is the longest river in the United States. The Amazon River has the largest river basin in the world and thus drains water from more land than any other river.

What did we learn?

- What is a river?
- Where does most of the energy for a river ecosystem come from?
- Name some plants you might find in a river ecosystem.
- What is a tributary?
- What is the riparian zone?

Taking it further

- Why do fewer plants grow in the water of a river than in a lake or ocean?
- Would you expect a river to be larger at a higher elevation or a lower elevation?
- Do rivers move faster over steep ground or in relatively flat areas?
- Would you expect water to cause more erosion in a steep area or in a relatively flat area?

Rivers & streams worksheet

Complete the "Rivers and Streams" summary worksheet and add it to your notebook. Find pictures of plants and animals that live in or around rivers to include in your notebook.

Rivers of the world

On a copy of the "Rivers of the World" worksheet, draw in and label the major rivers of the world on the world map.

Below is a list of rivers to include.
- Amazon
- Congo
- Nile
- Mississippi
- Yangtze
- Orinoco
- Yukon
- Volga
- Rio De La Plata
- Hwang Ho/Yellow River

River facts

There are so many interesting rivers in the world that it is worth doing a little investigation and learning more about some of the major rivers. Learn what you can about each of the rivers listed in the activity above, complete the "River Facts Sheet," and include it in your notebook.

SPECIAL FEATURE

The Amazon River

Dark jungles, fierce warriors, and some of the deadliest animals on Earth; these are some of the images many of us have of the Amazon River—the biggest river in the world. But what do we really know about the Amazon?

The first record of this river's exploration was by a Spanish adventurer named Francisco de Orellana, in 1541. Orellana told of being chased by fierce women warriors during his exploration, so the king of Spain named the river the Amazon after the mythological female Amazon warriors.

For centuries people have known that the Amazon is the biggest river in the world, but its true length has only recently been measured. Using the most modern measuring techniques, some people now claim that the Amazon may also be the longest river. Its length is very close to that of the Nile. The Amazon is approximately 4,000 miles (6,400 km) in length while the Nile is approximately 4,160 miles (6,695 km), only a little longer. Measuring the length of a river is not an exact science so different studies sometimes come up with slightly different numbers.

The Amazon supplies 20% of the freshwater going into the ocean. That is more water than the next top 10 rivers of the world combined! The flow rate of the Amazon is 10 times greater than that of the Mississippi River. The river is so large it moves 106 million cubic feet of suspended sediments into the ocean each day. That would be equal to about 1 million pickup trucks full of dirt and debris dumping their loads into the ocean each day. The Amazon's flow at its mouth is 53 million gallons per second (200,000 m³/sec). The Nile's flow is only around 925,000 gallons per second (3,500 m³/sec), or less than 2% of the Amazon's flow. The Amazon is so large that an ocean liner can sail about half way up it, or 2,000 miles inland.

During the dry season the river is considerably smaller than during the rainy season. During the dry season the width of the Amazon River varies between 1 and 6 miles, but when the rains begin to fall, the river swells to over 30 miles wide. Not only does its width increase but it becomes much deeper, too. Its depth may rise by as much as 50 feet in many areas during the rainy season. The water floods the forest, keeping many of the trees partially to mostly submerged for five months out of the year.

The Amazon River Basin is the largest watershed in the world. Water flows into the river from an area about the size of the continental United States. The Amazon drains nearly half of Brazil, as well as land in Peru, Ecuador, Bolivia, and Venezuela. There are over 1,000 tributaries that drain into the Amazon.

So what else makes the Amazon unique? Like all rivers, the Amazon has its own ecosystem, but because of its size, the Amazon River ecosystem is different from any other river ecosystem in the world. The Amazon is sometimes called the River Sea because of its vast size. So you might expect to see some similarities between the Amazon River ecosystem and an ocean ecosystem, and you do.

Properties of Ecosystems • 201

Many of the animals that you find in the ocean have freshwater equivalents in the Amazon River, and the interactions between these animals are similar to what you find in parts of the ocean.

One of the most famous animals that lives in the Amazon River is the piranha. But did you know that there are 20 different species of piranhas in the Amazon? Unlike the image of a ferocious attacker, many piranhas eat plant material. Others do eat meat, mostly fish, and only the red-bellied piranhas swarm in larger schools or packs and can speedily strip flesh off of an animal.

The Amazon River is also home to freshwater dolphins. The Amazon River dolphin, or boto dolphin, is considered the most intelligent of the freshwater dolphins. It has the ability to turn its head from side to side as it swims, which aids greatly when hunting for fish to eat. Boto dolphins use echolocation, or sound waves similar to sonar, to hunt. This ability also gives them the ability to swim through the submerged trees during flood season, which is where many of the fish live during this time.

The tucuxi dolphin looks similar to a bottlenose dolphin found in the ocean. However, it is much smaller than either the bottlenose dolphin or the boto dolphin. The adult tucuxi dolphins are around 110 lbs. (50 kg), whereas the boto can be up to 440 lbs. (200 kg).

Dolphins are not the only large animals in the river. The large size of the Amazon River allows some fish to grow without the restrictions of smaller rivers and lakes. Catfish weighing over 200 lbs. (90 kg) have been captured. Also, the largest freshwater fish in the river is the arapaima. These fish can weigh as much as 440 lbs. and be as long as 15 feet (4.5 m). They are one of the most popular food fish in South America.

Large animals also live on the surface of the water. The Amazon River is home to the anaconda snake, sometimes called the water boa. It lives in water and swamps in and near the Amazon. The female green anaconda can grow up to 32 feet (10 m) long and weigh up to 550 lbs. (250 kg), but averages around 20 feet (6m) long. They can eat fish and small mammals but can also eat crocodiles and deer, swallowing them whole.

There are thousands of small animals that live in and around the Amazon River, too. There are over 1,000 types of frogs that live there. Turtles also live in the water and lay their eggs on the banks of the river during the dry season, providing food for many of the animals that live along the river's shore, in addition to producing more turtles.

The Amazon is unique in both its size and diversity with many more interesting animals than we could possibly cover here. So if you are interested, you can do more research by getting books from your local library or searching the Internet.

UNIT 4

Extreme Ecosystems

18 Tundra
19 Deserts
20 Oases
21 Mountains
22 Chaparral
23 Caves

◊ **Identify** characteristics of the tundra, deserts, mountains, chaparral, oases, and caves.

◊ **Explain** the importance of oases to people.

◊ **Understand** the importance of fire to the chaparral.

◊ **Describe** how some plants and animals were designed to survive with little water.

Properties of Ecosystems • 203

18

Tundra

Is it frozen?

What is tundra?

Words to know:

tundra
Arctic tundra
Antarctic tundra
Alpine tundra
ephemeral
permafrost

Challenge words:

papillae

The northernmost lands of Canada, Alaska, Scandinavia, Greenland, and Russia are often considered a frozen wasteland. These northern areas are very cold and covered with ice and snow much of the year. But don't let the forbidding appearance fool you. There are many plants and animals that survive in this cold land.

Areas that are treeless, with long cold winters and cool summers are called **tundra**. Tundra comes from the Finnish word *tunturi*, which means treeless plain. Most tundra is located above the Arctic Circle and is often called **Arctic tundra**. A small amount of **Antarctic tundra** exists on Antarctica, and tundra conditions exist near the tops of many mountains and are called **Alpine tundra**.

In the Arctic tundra, the winters are very long. In the winter there are many weeks with little or no sunlight and the temperatures can be very cold with average temperatures as low as -30°F (-34°C). However, the summers are very different. The average temperatures are 37–54°F (3–12°C). The sun stays in the sky for many weeks without setting. This provides a growing season of 50–60 days.

Although most plants cannot survive in these harsh conditions, God has designed many plants that can survive. Many of these plants are considered **ephemerals**, which are plants with an accelerated life cycle. Many of these plants can go through their complete life cycle from germination to fruit and seed production in less than two months. There are over 400 varieties of flowers as well as sedges, rushes, cinquefoil, heather, and a few small

Fun Fact

In Point Barrow, Alaska, the most northern point in the United States, the sun rises around May 11 and does not set again until around August 1, providing 83 days of continuous sunlight. On the other hand, the sun sets around November 18 and does not rise again until around January 23, resulting in about 65 days of continual darkness.

Extreme Ecosystems

Alpine tundra

shrubs that live in the Arctic tundra. Also mosses and lichens flourish in the tundra.

Not only do most plants have an accelerated life cycle, but they are specially designed for the tundra in other ways as well. Most plants in the tundra are small and low to the ground. This helps them to withstand the high winds that are common in the winter. Many of the plants have a fuzzy appearance. These small hairs help to protect the plant from wind and help to insulate the plants as well. Plants generally grow together in clumps, providing further protection from wind and cold.

The tundra does not receive much precipitation, usually only 6–10 inches (15–25 cm) per year, mostly in the form of snow. This does not mean that the tundra is filled with dry land. The land in the tundra is actually filled with ponds and bogs in the summertime. This happens because the soil does not defrost more than a few inches on the surface. This perpetually frozen layer is called **permafrost**. The permafrost stops melting ice and snow from draining, so the surface of the land actually stays very wet during the summer, providing the needed moisture for plant growth.

Various animals also live in the Arctic tundra. Many of these are large mammals such as polar bears, caribou or reindeer, and moose. Smaller animals also abound including Arctic foxes, lemmings, Arctic ground squirrels, and swarms of mosquitoes. The tundra swan, Canada goose, and rock ptarmigan are a few of the birds you will find in the tundra, especially in the summer months. The Antarctic tundra is home to many types of penguins.

Like the plants, many of the animals in the tundra also have an accelerated life cycle. For example, the northern robin feeds its young 21 hours a day and the baby birds mature in only eight days. Southern robins require 13 days to mature. God has equipped these animals to survive the short summers.

God has also equipped animals on the tundra to survive the long cold winters. Many animals hibernate during the winter in dens that are insulated from the extreme cold. Other animals migrate to warmer climates in the winter and only return to the tundra in the summer. Although the tundra is a difficult ecosystem for most plants and animals, it is far from barren. It supports a wide variety of plant and animal life.

What did we learn?

- Where is most tundra located?
- What is permafrost?
- What kind of plants grow in the tundra?
- What are some animals you might find in the tundra?
- How much precipitation does the tundra receive?

Taking it further

- Why do many animals in the tundra have white fur or feathers?
- Why do many animals and plants have an accelerated life cycle in the tundra?
- Why do you think the temperatures are so cool in the summer when there is often 24 hours of sunshine?

Fun Fact
The Arctic tundra has the lowest species count on earth.

🧪 Tundra worksheet

Complete the "Tundra" summary worksheet and add it to your notebook. Find pictures of plants and animals that live in the tundra to include in your notebook.

🧪 Animals changing colors

Many animals in the tundra change color to match the environment. Some are brown or grey in the summer and turn white as winter approaches and the land becomes covered with snow. The Arctic fox, Arctic hare, and ptarmigan are all animals that are white in the wintertime. This provides them with camouflage from their predators. Baby seals are also white, providing them a measure of protection while on land. Polar bears, who spend most of their lives on ice floes, are white all the time.

Purpose: To appreciate how white fur or feathers provide protection to tundra animals

Materials: small box, white cotton balls, white tissue paper or white quilt batting, photos of Arctic animals with white fur or feathers

Procedure:

1. Fill a small box with white material to represent snow. You can use cotton balls, strips of tissue paper, pieces of quilt batting, or other white materials.
2. Place photos of Arctic animals with white fur or feathers such as foxes, polar bears, baby seals, and ptarmigans among the white material in the box.
3. Step back a few feet and see how well the animals blend in with their surroundings. Their coloring helps to protect them from their predators.

🧪 Two layers of fur

Many mammals in the tundra have two layers of fur. The first layer is very short and close to the body. This provides a layer of insulation to help trap heat next to the skin. The second layer is longer and provides a second layer of insulation to further trap heat and to keep out chilling wind. In the summer, many of these animals lose their longer fur since it is not needed during the warmer days. They then grow a new coat of fur as winter approaches.

Purpose: To appreciate how two layers of fur help to keep an animal warm

Materials: large bowl of ice, two pairs of gloves (one pair must fit inside the other, for example one could be cotton gardening gloves and the other could be leather work gloves)

Procedure:

1. Fill a large bowl with ice cubes.
2. Place a thin glove on one hand and then place a larger thicker glove over the thin glove on the same hand. You now have one hand that is uncovered while the other hand is covered with two layers of protection.
3. Hold both hands over the ice. Can you feel the cold from the ice with each hand?
4. Push both hands into the ice for a few seconds. How cold do each of your hands feel?

Conclusion:

You should find that two layers of protection keep your hand quite comfortable, while no protection leaves your hand very cold. God's design of two layers of fur, helps many animals survive the cold winters in the tundra.

Polar bears

When you think of ice and snow you probably think of polar bears. These cuddly-looking giants have captured the hearts of many people. However, these bears are not cuddly pets; they are the largest predators on land. An adult male weighs 660–1300 pounds (300–600 kg) and a female weighs about half as much as a male.

Polar bears spend a large part of the time hunting for their favorite food, which is the ringed seal. They will dive from the sea ice into the frigid waters to catch their prey or they will wait by a seal's air hole until it surfaces and then attack. Polar bears swim from one piece of ice to another to follow the movement of the seals. Polar bears have an extremely good sense of smell and can detect a seal more than a mile away. Although polar bears prefer seals, they eat many other animals as well. They eat fish, beluga whales, and have even been known to eat caribou.

Polar bears are specially designed for swimming as well as for travel across the frozen ice of the tundra. Their bodies are streamlined for efficient swimming. Also, they have a thick layer of blubber which makes them more buoyant in the water and helps to insulate them from the cold temperatures. Their toes are somewhat webbed, helping them to be very good swimmers. Polar bears can swim for more than 60 miles (97 km) at a time.

On land the polar bear's large paws help to spread out its weight so it can more easily move across the snow. Also, its soft pads are covered with tiny *papillae* or small bumps, which help to give the bear traction on the ice.

Although a polar bear looks white, its fur is actually translucent. Light reflects off of the hollow hair shaft, making it appear white. As a polar bear

ages it may appear more yellow or even greenish. This happens when algae begins to grow inside the hair shafts. Its fur is oily and does not mat so the bear can easily shake off the water after it emerges from the sea.

A female will have her first babies when she is around 4 or 5 years old. She will go on land and dig a snow cave. There she will give birth to one or two tiny babies. A baby polar bear weighs only about 1 pound (2.5 kg) at birth, but it grows very quickly. At the end of winter the mother and her cubs emerge from their winter den. Babies nurse for about 2 to 2½ years.

In 2008 polar bears were declared a threatened species by the United States government. Although they are protected from hunting by all countries in which they live, there is some concern that global warming may be reducing the sea ice on which the bears hunt. The extent of global warming and the long term effects are not agreed on by all scientists. More research is needed to determine the effect on polar bears.

SPECIAL FEATURE

Robert Peary
1865–1920

In search of a land without an east or west

Has there ever been anything you've wanted to do your whole life? Is it something dangerous that no one has ever done before? A young boy by the name of Robert Peary had a dream like that. He wanted to be the first person to go to the North Pole. And he eventually made his dream come true.

Robert was born on May 6, 1865 in Cresson, Pennsylvania. When Robert was two years old, both of his parents became sick; his mother recovered but his father did not. After the death of his father, Robert and his mother moved to Maine. There, Robert spent much of his time hiking in the woods and exploring new areas. He preferred the solitude of the woods to being around people. This time spent in the woods enabled him to become self-reliant, which was an important aspect of his character.

After high school, Robert went on to earn his civil engineering degree and later joined the U.S. Navy. Before leaving for his first assignment, Robert met an African-American store clerk by the name of Matthew Henson, in Washington D.C. After talking with Henson, Robert learned that they shared the same passion for exploration. Peary hired Henson to be his personal valet, and Henson accompanied Peary wherever the Navy sent him. The two became lifelong partners and friends. Matthew Henson was with Peary on every expedition he went on from that point.

Lt. Peary's first assignment was to Nicaragua where he worked on the Inter-oceanic Ship Canal Project, whose goal was to explore the possibility of putting a canal across Nicaragua. In 1886 Peary and Henson went on their first expedition to the interior of Greenland. Here, Peary and Henson learned about dog sledding, hunting, fur clothing, and building an ice shelter from the Inuit people. These skills would prove invaluable in the future.

In 1888 Robert Peary married Josephine Diebitch and the couple, along with Henson, made two trips to McCormick Bay, half way between the Arctic Circle and the North Pole. After these two trips Josephine stayed home while Peary and Henson made several more trips to Greenland during the 1890s. There they spent much of their time exploring. During this time Robert Peary discovered that Greenland was an island and he also discovered three very large iron meteorites in Cape York.

After discovering that Greenland was an island, Peary decided the best place to launch an expedition to the North Pole was Ellesmere Island. From there he made several attempts to reach the North Pole with his final and successful attempt starting in 1908. He left New York with a group of 23 men on the *Roosevelt* and sailed to Ellesmere Island, Canada. The party wintered there as they prepared for the trip and conditioned themselves to the cold.

On March 1, 1909 they set off for the North Pole. Over the next month several small groups, which had been with them to help move supplies, turned back and returned to the base camp. This was done in order to set up points along the way where supplies would be stored for Peary's return trip. On April 7, 1909 Robert Peary, Matthew Henson, and four Inuits reached what they and the world thought was the North Pole. It has been learned in recent years that his calculations were off slightly because of the shifting ice and he may have been off by 30 to 60 miles. Nevertheless, the world hailed Peary as a hero and he is credited with the first successful trip to the North Pole.

Two years later Admiral Peary retired from the Navy and took up writing. He produced three books about his explorations. The hard life of the north had prematurely turned Peary into an old man, and he passed away in 1920 at the age of 63. He is buried at Arlington National Cemetery.

His lifelong friend Matthew Henson was mostly ignored by the public until 1937 when he was admitted as a member of the Explorers Club in New York. Henson died in 1955 at the age of 89. In 1988 Henson's remains were moved from New York to a site close to Robert and Josephine Peary in Arlington National Cemetery. In 1945 Congress awarded Henson a silver medal for outstanding service to the U.S. Government. And in 2000 the National Geographic Society awarded the Hubbard Medal to Henson posthumously for distinction in exploration, discovery, and research—identical to the one awarded to Peary in 1906.

19

Deserts

Sand and more sand

What different kinds of deserts are there?

Words to know:

desert

cold desert

hot desert

succulent

stomata

transpiration

dromedary camel

Bactrian camel

nocturnal

estivate

mirage

When you think of a desert, do you pic-ture waves of heat rising from a sea of sand? That is what most people think of as a desert. And many deserts are very hot, dry, and sandy. But some deserts can be very cool. A **desert** is a biome that receives less than 10 inches (25 cm) of rain per year. Most deserts receive less than 6 inches (15 cm) of rain per year.

Deserts are generally divided into two groups: hot deserts and cold deserts. **Cold deserts** are dry areas where the daytime temperature drops below freezing during part of the year. The Atacama Desert in Chile, the Gobi Desert in China and Mongolia, and the Great Basin of the western United States are all cold deserts.

Hot deserts are dry areas that do not experience freezing temperatures. The average temperature in a hot desert is 68–77°F (20–25°C), but don't let these numbers fool you. Hot deserts can be very hot during the day in the summertime, often reaching temperatures of 109–120°F (43–49°C). Because deserts are so dry, the temperature drops very quickly when the sun sets; so even though the daytime temperature may be very hot, the desert can become very cool at night. Hot deserts can be found around the world. Some of the largest hot deserts include the Arabian Desert which covers most of the Middle East, the Mojave Desert of the southwestern United States, the Sahara Desert in Africa, and the Australian Desert in Australia.

Deserts often lose more moisture to evaporation than they gain from precipitation. Thus, the environment is very dry. You might expect that there would be very little life in the desert. Although it is true that many plants and animals

Fun Fact

The hottest temperature ever recorded in the Sahara Desert was 136°F (58°C).

210 • Properties of Ecosystems   LESSON 19

Fun Fact

The saguaro cactus, the one most often associated with deserts, only grows in the Sonoran desert in Arizona, California, and Mexico. The saguaro can grow up to 50 feet tall, weigh several tons, and live for 200 years.

Saguaro cactus

cannot survive in such a harsh environment, there are actually many plants and animals that are specially designed to live in the desert. Both plants and animals that live in the desert can conserve water, storing it for use during the long dry periods.

Many plants that live in the desert are **succulents**. These plants have fleshy stems that can absorb and store large amounts of water when it rains so it will be available to the plant later. Many types of cacti and aloe vera plants are examples of desert succulents. Other plants you are likely to find in the desert include the sagebrush, mesquite, Joshua tree, creosote bush, and desert trumpet.

When green plants perform photosynthesis, they usually suck up more water from the ground than they use and release that water into the air through tiny holes in their leaves called **stomata**. This process is called **transpiration**. But desert plants cannot afford to release excess water into the air. So God has designed them to conserve water. Succulents often have needles instead of leaves. These needles help prevent the plant from losing water through transpiration, thus conserving the water that is stored. Other plants have leaves but the leaves have very few stomata. This design also helps to conserve water.

Despite harsh conditions, many animals live in the desert. You are likely to see the spadefoot toad, lizards, kangaroo rats, voles, badgers, scorpions, burrowing owls, ostriches, roadrunners, and vultures. There are also many desert snakes including rattlesnakes, coral snakes, and sidewinders. In many deserts you may also find camels. Camels with one hump are called **dromedary camels** and are native to hot deserts in Africa and the Middle East. Two-humped camels are called **Bactrian camels** and are native to the Gobi Desert. Because the Gobi is a cold desert, the Bactrian camels grow a coat of long hair over their short hair for the winter and shed the long hair in the summer.

Like plants, the animals that live in the desert have many ways to cope with the dry climate and extreme temperatures. First, most animals are **nocturnal**, meaning they sleep during the hot day and are active at night when the temperatures are more bearable. Also, many animals can conserve water. Camels can store large amounts of water. Other animals **estivate**, which means they go into a deep sleep during the summer when it is hottest and awake in the fall or winter when temperatures cool down.

Not only do deserts receive only a small amount of precipitation, the water they do receive does not come in regular intervals. Deserts often go months or even years between rainstorms. When it does rain, there are often flash floods. The ground has become very dry and hard and water does not easily penetrate it. Thus when a large amount of rain falls all at once, it often rushes through canyons and much of it does not sink into the ground.

Rain brings a dramatic change to the desert. Where there was mostly dry brown foliage, the rain produces a rainbow of colors from all of the blooming plants. When water is available, the plants and animals get to work. Just as in the tundra where the growing season is short, many of the plants in the desert are also ephemerals—plants with accelerated life cycles. Many desert animals also have

The desert tortoise estivates in summer.

LESSON 19   **Properties of Ecosystems • 211**

🧪 Desert worksheet

Complete the "Desert" summary worksheet and add it to your notebook. Find pictures of plants and animals that live in the desert to include in your notebook.

🧪 Storing water

Purpose: To demonstrate how plants can conserve water

Materials: thin plastic bag, such as a produce bag

Procedure:

1. Fold a plastic bag accordion style along the length of the bag with each fold about 1 inch wide.
2. Notice the diameter of the bag. This represents a cactus when it has been dry for a long period of time.
3. Carefully fill the bag with air, holding the air in the bag with your hand. Notice what happens to the size of the bag as it fills up. The diameter expands. This is what happens to the stems of cacti as they fill with water.
4. Slowly release the air from the bag. Gently refold the creases to return the bag to its original size. This is what happens to the stems as the plants use up the water.

Conclusion:

The stems of many desert plants are folded like the bag was in this experiment. When water is available, the plant fills up and the folds are pushed out, making room for a large amount of water. When water is not available, the water in the plant is used up and the stems shrink back down.

---

accelerated life cycles. The spadefoot toad can grow from egg to tadpole in only nine days!

In between the rains, the desert dries out. Much of the water evaporates instead of soaking into the soil. As the water evaporates it leaves behind any minerals that were dissolved in it. Salt is the most common mineral that is dissolved in water, and when water evaporates the salt is left behind. Many deserts have large salt flats where salt has been deposited for hundreds of years.

As the ground dries out, wind often whips the sand and soil around, slowly building massive dunes in some deserts. A dune can be started as prevailing winds blow sand against a clump of plants creating a small hill. The wind carries sand uphill until gravity pulls it back down. This creates a smooth slope on the windward side and steep slope on the other side of the plants. If wind continues to blow in the same direction, a dune will slowly form.

In addition to building and moving dunes, wind can cause problems for living things in the desert. The wind can pick up massive amounts of sand or dirt, creating giant dust clouds and sand storms. The sand can bury plants and cover the homes of animals. The sand also wears away rocks and other structures in its path. People and animals have learned to seek shelter from the desert storms.

The heat in the desert can cause an interesting phenomenon. As the sun heats the Earth, the air near the ground becomes hot and rises. This hot air can bend the light from the sun, resulting in a reflection of the sky onto the ground. This makes it look like there

A mirage in the Sahara Desert

is water on the ground where there really is no water at all. This phenomenon is called a **mirage**. Thirsty travelers have been known to be tricked by mirages.

Even though the desert is a harsh environment, the plants and animals that live there are diverse and beautiful. We cannot forget that even in the desert evidence of God's design is everywhere.

What did we learn?

- What is a desert ecosystem?
- How is a cold desert different from a hot desert?
- What are some plants you would expect to find in the desert?
- What are some animals you would expect to find in the desert?
- What is the difference between a Bactrian camel and a dromedary camel?

Taking it further

- In what ways are plants well suited for the desert environment?
- In what ways are animals well suited for the desert environment?
- Why does rain often cause flash flooding in the desert?
- What are some dangers you may face in the desert?
- Why do salt flats often form in the desert?
- Would you expect to find more salt flats in a cold desert or a hot desert?

The Sahara Desert

The world's largest hot desert is the Sahara Desert in northern Africa. The Sahara is about the same size as the continental United States and larger than the continent of Australia. The Sahara Desert covers approximately 3,500,000 square miles (9,000,000 sq. km). It comprises parts of eleven different African countries and contains 25% of all of the sand in the world.

Because the Sahara Desert is so large, its climate and geography vary greatly from one area to another. Although all of the Sahara Desert is dry, the northern part of the desert is significantly hotter and dryer than the southern part. This desert contains miles and miles of sand dunes, but it also contains mountains, plateaus, and valleys. The highest point in the Sahara is Mt. Koussi in Chad, at an elevation of 11,204 feet (3,415 m). The lowest point in the Sahara is the Qattara Depression at an elevation of 436 feet below sea level (-133 m).

Within the borders of the Sahara Desert are several rivers. The primary river is the Nile River, flowing north through the Sahara into the Mediterranean Sea. There are also many salty lakes and one fresh lake—Lake Chad. So although the Sahara Desert is hot and dry, there are many areas of water throughout this amazing desert.

Although most of the Sahara desert is hot and dry today, we know that the climate in this area was significantly different in the past. Hundreds of petroglyphs exist showing many animals including crocodiles and elephants in areas that today do not have enough water to support that kind of life. Archaeologists have also discovered the remains of several towns in the desert where there is very little water today. Fossils of hippos and crocodiles have been found in the desert as well. Another clue comes from satellite views of the Sahara Desert, which show large river channels that are now dried up.

There are several proposed reasons for the dramatic change in climate, but the most likely explanation is that after the Flood, weather patterns and air flow were significantly different from today. The weather patterns after the Flood likely carried storms to the area that is now the Sahara Desert, bringing rainfall and allowing many animals and people to live in that area. After the oceans began to cool and ash from volcanoes began to clear, the weather changed to be more like what we see today, and the Sahara Desert no longer receives the rain it once did. In fact, today we see the desert actually growing. The southern border of the Sahara is moving south by as much as 30 miles (48 km) each year.

20

Oases
A refreshing spot

Extreme Ecosystems

Why are oases important to people in the desert?

Words to know:

oasis

Swaying palm trees and cool shade might seem like a dream when you are in the middle of a desert; however, there are places in the desert that make that dream come true. An **oasis** is a special ecosystem located in a desert where there is abundant water. Usually this water comes from an underground spring. When the water table is high, wind may erode away enough sand to allow the water to bubble up to the surface, revealing a spring. Oases often occur along geographic fault lines, which allow the water table to be closer to the surface. Most deserts have several oases.

At an oasis, the water from the spring brings life. Many plants and animals that are naturally found in the desert are also found in the oasis; however, many other plants and animals are found at an oasis that are not normally found in the desert. In addition to cacti, lizards, snakes, and rodents, at an oasis you will find palm trees, shrubs, and grasses. These larger plants provide habitat for other animals such as bats, warblers, orioles, and robins. The water also provides habitat for fish and other aquatic life that do not live anywhere else in the desert.

In an oasis, the temperature can be as much as 10 to 20 degrees cooler than in the surrounding desert, so not only do you find shade from the sun, but the air temperature is actually cooler in an oasis. The air cools down in an oasis because of the increased evaporation taking place. The trees and other plants in an oasis perform photosynthesis. When they do this, they release water into the air through their leaves by transpiration. This water evaporates, removing energy from the air and thus cooling the air.

Travel through the desert can be very dangerous because of the heat and lack of water. Over hundreds of years, people have learned of and passed on the locations of oases so that they could travel safely across the desert. Trade routes sprang up from one oasis to another, and whoever controlled the oases often controlled the trade routes through the desert.

Some oases are very small with only a few trees and other plants growing around the spring. Other oases are much bigger with a large area that is watered by multiple springs. The Kharga oasis in the Sahara Desert is about 10 miles (16 km) long and from 12–50 miles (19–80 km) wide. This oasis has many springs and has been attracting people for hundreds of years. It contains a Persian monument dated from the 6th century BC as well as the ruins of several Roman forts. Today, the city of Kharga has about 60,000 residents and is about 100 miles (150 km) wide. It is a popular tourist attraction in southern Egypt. Las Vegas, Nevada began as an oasis in the desert. As the town outgrew the existing water supply, additional water has been brought in, but originally the town was an oasis in the desert.

As in the cases of Kharga and Las Vegas, many desert oases have been expanded to create artificial oasis-type ecosystems in the deserts around the world. One of the most important man-made oases is the Imperial Valley in California. Although this area of southeastern California receives an average of only 3 inches (8 cm) of rain per year, it has become a very important agricultural area largely because of the All-American Canal which brings water from the Colorado River up to 80 miles (128 km) away. This water has turned the valley into a giant oasis. Similar canals are being used to turn parts of the desert in Israel into oases as well. Many of these canals are open to the air and much of the water evaporates before it ever reaches the farmland. However, some canals are being covered with plastic to help prevent unwanted evaporation.

The oasis is a unique ecosystem bringing refreshment and life to the hot dry desert.

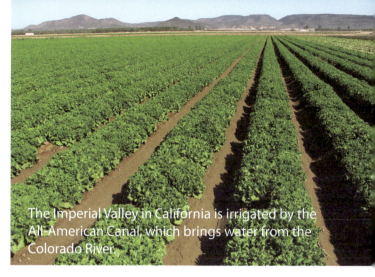

The Imperial Valley in California is irrigated by the All-American Canal, which brings water from the Colorado River.

What did we learn?

- What is an oasis?
- What kinds of plants grow in an oasis?
- What kinds of animals live in an oasis that don't usually live in a desert?

Taking it further

- Why is it often cooler in an oasis than in a desert?
- Why are oases important for trade routes?
- How might a man-made oasis change the ecosystem in a desert?

Transpiration

Purpose: To observe transpiration

Materials: several plant leaves, plastic zipper bag

Procedure:

1. Place several plant leaves in a plastic zipper bag and seal the bag.
2. Place the bag in a sunny location for about 1 hour.
3. After 1 hour in the sun observe the bag. What do you see?

Conclusion:

You should see water condensing on the inside of the bag. Where did this water come from? The water was released from the leaves as they performed photosynthesis. In the dry desert air, the water from transpiration does not condense; instead, it evaporates and cools the air.

Oasis worksheet

Complete the "Oasis" summary worksheet and add it to the desert section of your notebook. Find pictures of an oasis to include in your notebook.

LESSON 20 **Properties of Ecosystems • 215**

Products of the desert

People often think of the hot dry desert as a useless and lifeless place. You have already learned that there are many plants and animals in the desert, so you know it is not lifeless. The desert is also not useless. There are many things that make the desert a valuable place as well. Many important products come from deserts.

One of the most important products found in many deserts is petroleum. Significant amounts of oil have been found under the deserts in the Middle East, America, Australia, and Africa. The existence of petroleum in these desert areas is evidence of a very different climate in the past as well as evidence of the global Flood. Petroleum can be formed from marine organisms that have been buried and compressed. It is quite possible that much of the oil that exists today under the deserts was formed as a result of the Genesis Flood.

Gold and diamonds are also found in several deserts. Deserts in Australia, Namibia, and South Africa contain some of the largest deposits of these precious materials. In fact, South Africa contains the world's largest diamond mines. The deserts in Australia are also a significant source of uranium, nickel, and aluminum. Deserts in Chile supply copper as well as sodium nitrate, which is used in making fertilizer.

One of the most abundant resources available in any desert is solar energy. Because of the lack of cloud cover, many deserts experience more than 300 days of sunshine per year. Many people are working to find ways to harness the sun's energy to help meet the energy needs of people around the world.

Do a little investigation of your own. Find out how the products of the desert are being used. Then, create a poster or report about desert products to include in your notebook.

Fun Fact

Several new solar power stations, each claiming to be the biggest solar energy station yet, are in the process of being built in the deserts of California, Nevada, and Arizona. These solar power stations are expected to save consumers millions of dollars and reduce carbon emissions by hundreds of tons each year.

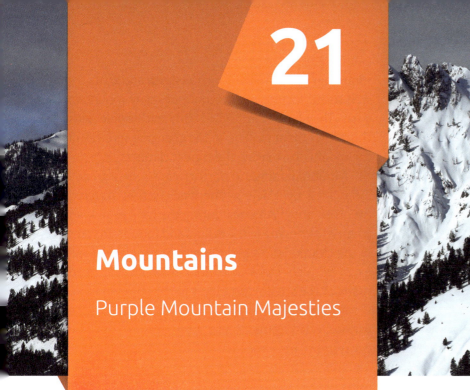

21

Mountains

Purple Mountain Majesties

Why do mountains have different ecosystems at different altitudes?

Words to know:

timberline   snow line

Mountains are some of the most amazing geological formations in the world. They can tower thousands of feet above the surrounding landscape and be so beautiful they take your breath away. Mountains are found in every part of the world, even under the ocean. When an underwater mountain sticks up above the surface of the ocean we call it an island.

It is impossible to describe a single mountain ecosystem. Because most mountains are tall, the amount of light, wind, and precipitation changes as you gain altitude. The temperature also decreases with altitude. This creates varying ecosystems at different altitudes on a mountain.

Mountain ranges in temperate zones usually start with grasslands on the lowest parts of the mountains. As you go up, you will usually find deciduous forest ecosystems. These forests often contain beech, oak, basswood, maple and ash trees. Rhododendron and azalea plants also thrive in deciduous mountain forests.

At higher altitudes the deciduous trees give way to evergreen trees. At these higher altitudes the temperatures in the winter are often too cold and the climate is often too dry for many deciduous trees. Instead you find pine, spruce, fir, and juniper trees. Aspen trees are one of the few deciduous trees that do well at higher altitudes. These evergreen forests also contain cacti, and many wildflowers.

As you continue going to higher altitudes, you reach a point where it is too cold and there is not enough air pressure for trees to grow. This point is called **timberline**. Above timberline, you will find the alpine meadows. Here you find shrubs, grasses and many mountain flowers such as columbine, larkspur, and glacier lily.

Above the alpine meadow the temperatures are very cold and the growing season is very short. This is where you find the alpine tundra. Only very small hardy plants can grow in the alpine tundra. Although the alpine tundra is similar in many ways

Extreme Ecosystems

Fun Fact

Volcanoes are mountains. If the volcano has been inactive for a long time it may not seem different from other kinds of mountains. However, an active volcano has a constantly changing ecosystem.

LESSON 21   **Properties of Ecosystems • 217**

to the Arctic tundra, it does not necessarily have the permafrost layer that the Arctic tundra has. Also, the animals found in the alpine tundra are different than in the Arctic tundra. You will not find polar bears or seals, but you will still find foxes, Arctic hares, and ground squirrels. You may also see small birds, insects, and pikas.

If you go even higher on a mountain, you eventually reach an altitude at which the snow no longer completely melts, even in the summertime. This is called the **snow line**. Because the snow does not completely melt above the snow line, there are very few plants and only a few animals that visit the very highest peaks of these mountains.

The heights of mountains vary greatly from one location to another. So you will not find all the ecosystems listed above on every mountain. Many mountains, such as the Appalachian Mountains are primarily deciduous forest. Many areas of the Rocky Mountains start at 5,000 feet above sea level, so they begin with evergreen forests. Only a small percentage of mountains are tall enough to have a permanent cover of snow, so many mountains do not have a snow line. Nevertheless, you usually find more than one ecosystem on a mountain.

The animals that live in the mountains often move between ecosystems by moving up and down the mountain. However, many animals live primarily in one or two ecosystems. Animals you are likely to find in mountain ecosystems include bears, timber wolves, mountain lions, porcupines, chipmunks, hummingbirds, bluebirds, and eagles. Above timberline you often find pikas, marmots, big horned sheep, and mountain goats. Animals such as sheep and goats are well suited for the rocky terrain of many mountain areas. They have split hooves with soft flexible pads. This allows them to cling to the changing surfaces of the mountains, making them great climbers.

Mountain ranges in northern areas were greatly influenced by glaciers. The glaciers that covered much of North America during the Ice Age helped to carve out many of the valleys and mountain lakes in the United States and Canada. Some glaciers still exist on high mountaintops. These glaciers feed many of the streams and lakes in the summertime.

Mountains in tropical zones also experience various ecosystems as you gain altitude, but the ecosystems are somewhat different than in temperate zones. Often tropical mountains have rainforests near the bottom of the mountain. Then as you go up, the rainforest gives way to bamboo forest. Above the bamboo forest is the heath, which is an ecosystem composed of a variety of shrubs. At higher altitudes you find areas of small plants and flowers and above that you find alpine tundra.

What did we learn?

- What ecosystems are you likely to encounter on mountains in temperate zones?
- What ecosystems are you likely to encounter on mountains in tropical zones?
- What is timberline?
- What is snow line?

Taking it further

- Why do the ecosystems change as you gain altitude on a mountain?
- Why don't you find every ecosystem on every mountain?
- What other ecosystems are you likely to find on mountains that were not listed in this lesson?
- How have glaciers influenced the shapes of mountains?
- Why is there less oxygen as you gain altitude?

🧪 Mountain worksheet

Complete the "Mountain" summary worksheet and add it to your notebook. Find pictures of plants and animals that live in various ecosystems on mountains to include in your notebook.

🧪 Modeling a mountain

Purpose: To demonstrate the diverse ecosystems found on a mountainside

Materials: art supplies, newspaper, paint, twigs, leaves, grass, small flowers, cotton balls

Procedure:

1. Form newspaper into the shape of a mountain.
2. Paint your mountain with greens and browns to represent grass and rocks on the side of the mountain. Allow the paint to dry.
3. Use whatever materials you have to glue to the sides of your mountain to represent the different ecosystems you might encounter as you go up the mountain. Use your imagination. Here are some ideas to help get you started:
   a. Glue grass near the bottom to represent grassland.
   b. Glue leaves and twigs higher up to represent forests.
   c. Glue small flowers higher up to represent alpine meadows or tundra.
   d. Stretch out and glue cotton balls near the top to represent snow.
   e. Use any other materials you have available.
4. Take a picture of your mountain and include it in your notebook with an explanation of each ecosystem that is represented.

Conclusion:

Mountains are rich with diversity of plant and animal life. You can enjoy several different ecosystems in one day just by visiting the mountains.

🎖 The Himalayas

The highest mountain range in the world is the Himalayan Range, often just called the Himalayas. This mountain range is actually three parallel ranges that stretch 1,500 miles (2,400 km) between the Indian subcontinent and the Tibetan Plateau of Asia. This mountain range is home to 100 peaks that are over 23,000 feet (7,200 m) high.

As you would expect, with the Himalayas covering such a large area of land, the mountains have many different ecosystems. These ecosystems change not only with altitude, but also from west to east. The amount of rainfall increases as you move from west to east, so the ecosystems change as well. The snow line also changes from east to west and from north to south. In some areas the snow line is as low as 14,100 feet (4,300 km). In other areas the snow line is as high as 19,700 feet (6,000 km).

At the lowest altitudes are the Lowlands. This is an area of forest found at the base of most of the mountains. As you gain altitude you enter the Teri belt, which is an area of mostly grasslands. These are some of the highest grasslands in the world. Above this are the Midlands. In the west the Midlands consist of temperate deciduous forests. In the east they consist mostly of pine forests. As you go up you encounter alpine meadows and alpine tundra. Finally, since so many of the mountains in the Himalayas are so tall, most of them have glaciers at the top. In fact, there are over 15,000 glaciers in the Himalayas.

Obviously with so many ecosystems, the Himalayas are home to large numbers of different plant and animal species. Some of the animals that are unique to the Himalayas include the snow leopard, clouded leopard, Bengal tiger, and red panda. You also find many common mountain animals such as deer, goats, sheep, wolves, and marmots.

The highest mountain in the world, Mount Everest, is found in the Himalayas. It towers up at 29,035 feet (8,848 m) with K2 not far behind at 28,251 feet (8,611 m). The weather on the tops of these high mountains is very unpredictable and very dangerous. Only a small number of people have ever successfully scaled these heights. Sir Edmond Hillary and Tenzing Norgay were the first people to reach the summit of Mount Everest on May 23, 1953. K2 was first successfully scaled in 1954. At these extreme heights, there is very little oxygen and climbers must carry oxygen tanks with them. No plants and few animals are found at these altitudes.

The Himalayas are very important to the climate and well being of the Indian subcontinent. First, the Himalayas provide a barrier for the monsoon moisture that sweeps inland from the Indian Ocean. This causes much of that moisture to fall in the mountains. This water then drains into three major river systems that carry the water to the 1.3 billion people that live on the Indian subcontinent. These rivers not only provide water, but also wash down tons of silt which enriches the soil of the plains and valleys fed by the rivers.

There is much more to learn about the Himalayas. Choose an area that interests you and find out more about it. Then take what you have learned and make a Himalayas page for your notebook.

Mount Everest

22

Chaparral

The Mediterranean climate

Where does the word *chaparral* come from?

Words to know:

chaparral

Challenge words:

fire cues

The Spanish word for scrub oak is *chaparral*, and the **chaparral** is an ecosystem that is unique to hot, dry mountain slopes that are often filled with shrubs. The chaparral is found in semi-arid climates in California, around the Mediterranean Sea, and in Australia. It is sometimes called the Mediterranean ecosystem or maquis (mah-KEE). In a chaparral, the summer days are warm and sunny and the winters are mild and rainy.

As the name would imply, the chaparral is dominated by shrubs. These plants usually grow in very dense thickets. Sometimes these thickets are so thick that large mammals and even people cannot penetrate them. Common plants in the chaparral include scrub oak, live oak, yucca, buckbrush, and lotus. In many areas of chaparral, citrus fruit trees and grapes are cultivated by farmers.

Because most large mammals cannot move about in the chaparral, the wildlife is mostly smaller animals. In the chaparral you will find woodrats, brush rabbits, gray foxes, coyotes, bobcats, quail, blue jays, wrens, and sparrows. Where the foliage is less dense you may find some deer as well. In the chaparral in Australia you commonly find eucalyptus trees and koalas. In areas along the Mediterranean Sea you also find many goats and other climbing animals.

Summers are hot and dry in the chaparral. Summertime temperatures often reach 104°F (40°C) with very little rainfall. In the winter the daytime temperatures average 50°F (10°C) with 15–40 inches (38–100 cm) of rainfall per year.

Because summers are hot and dry, and because the shrubs are so dense, wildfires are very common in the chaparral. Lightning strikes are the most common cause of these fires. Low humidity and high winds help the fires to spread and become very large.

Fun Fact

Chaps are special coverings worn by many cowboys to protect their legs. The name chaps comes from the word chaparral. Cowboys needed protection from the thick shrubbery when riding through the chaparral.

LESSON 22 **Properties of Ecosystems • 221**

🧪 Chaparral worksheet

Complete the "Chaparral" summary worksheet and add it to the mountains section of your notebook. Find pictures of plants and animals that live in the chaparral to include in your notebook.

🧪 Fire in the chaparral

Find pictures of the chaparral before a fire, shortly after a fire, and several years after a fire. Use these pictures to make a page about the effects of fire in the chaparral for your notebook.

Although this may seem like a devastating event for an ecosystem, God has designed the plants in the chaparral to recover after a wildfire just as they do in the grasslands. New seeds will quickly sprout and areas that were burned will soon be growing and thriving again. Some species that grow in the chaparral cannot germinate unless they experience a fire. In fact, in areas where there has not been a fire for more than 10–15 years, some species begin to die out from overcrowding and inability to germinate.

Wildfires, particularly in southern California, cause many problems for the people that live in the chaparral. Although the fires may be good for the ecosystem, they are not good for people's houses. There is a constant battle each summer against these fires in an effort to protect people's homes and businesses.

🧠 What did we learn?

- What is a chaparral ecosystem?
- What are two other names for a chaparral?
- Name some plants you might find in the chaparral.
- Name some animals you might find in the chaparral.
- What animal might you find in the Australian chaparral that you would not find in the American chaparral?

🚀 Taking it further

- What conditions make fire likely in the chaparral?
- How are plants in the chaparral specially designed for fire?
- Should people try to put out fires that naturally occur in the chaparral?

A wildfire in southern California

Fire germination

Several species of plants that live in the chaparral only germinate when their seeds experience a fire. Some of these plants include scrub oak, ceanthus, toyon, manzanita, and holly-leafed cherry. Although it is known that many of these seeds will not germinate unless they experience a fire, it is not completely clear why or how a fire triggers germination.

Many experiments have been done to try to understand the role that fire plays in the germination of many of these seeds. It is believed that there are several different types of **fire cues**, or conditions, which could cause a seed to germinate. The most obvious cue is heat. When a fire sweeps through an area the ground can experience temperatures of several hundred degrees. It is believed that some seeds need high temperature to trigger germination.

A second fire cue is smoke. Scientists believe that some seeds need to experience smoke from a fire before they will germinate. Tests have been conducted that put aerosols into the air. Other tests have dissolved smoke in water. Both airborne smoke and dissolved smoke have been shown to trigger germination in some seeds.

Other fire cues include charred wood, oxidation due to burning, and adding of acids to the soil during a fire. All of these cues have been shown to be effective in triggering germination in some species. Although there are many different cues, it is clear that God designed many of these species to survive and repopulate the area after a fire.

A manzanita bush

23

Caves

Are they just holes in the ground?

What will you find in a cave?

Words to know:

cave
trogloxene
troglophile
troglobite
guano
chemosynthesis

Challenge words:

echolocation

Have you ever been inside a real cave? Did you have a flashlight or other source of light with you? You probably needed it because a cave is a dark place. A **cave** is a cavern inside a mountain or underground. A true cave has no light except at the entrance to the cave.

Because there is no light in a cave there are no green plants in a cave. There may be a few small plants or some algae growing at the entrance, but once you actually get inside a cave there are no plants. This greatly limits the food supply in a cave, so a cave is considered a low energy ecosystem. In general, it is relatively difficult to explore most caves, so very little is known about their ecosystems.

We know that there are three major groups of animals in a cave ecosystem. The first group is called **trogloxenes**. These are the animals that visit the cave but do not spend most of their lives in the cave. Raccoons and woodrats are examples of animals that may be found in a cave at a given period of time, but do not spend most of their time there. The most famous of the trogloxenes is the bat. Bats spend the day roosting inside the cave, but they leave every night in search of food and return at dawn. Many bats only spend certain seasons in caves, and some bats roost in trees or man-made structures instead of in caves.

The second group of animals is the **troglophiles**. These are animals that like to live in the cave, often near the entrance. Similar animals may be found living outside of caves. These animals often leave the cave in search of food and then return to the cave. Salamanders, millipedes, and snails are examples of animals that spend most of their time in the cave but periodically leave and then return. Some scientists group bats with the troglophiles instead of with the trogloxenes.

Troglobites are animals that live in the cave their entire lives. They live in complete darkness and many of them have very little or no pigment. Some of them are completely blind. Since there is no light

224 • Properties of Ecosystems   LESSON 23

in a cave, these animals have no need for eyesight. However, many of these animals have a heightened sense of hearing or other senses that help them to survive in the darkness. Crayfish, shrimp, crickets, fish, and scorpions are animals you may find in the darkest parts of a cave.

Since there are no plants inside a cave, you may be wondering what the animals eat. Obviously, the animals that live near the entrance can go out of the cave to get food. But the animals that spend their entire lives inside the cave must also have something to eat. Some organic material—parts of dead plants or animals—gets washed into the cave when it rains or gets blown in by the wind, providing food for some of the animals. Some animals are predators and eat other animals that live in the cave. But the greatest source of nutrients for the animals in the cave is bat **guano**, or bat droppings.

The bats get their food from outside the cave, but spend a significant portion of their lives roosting in the caves, often leaving a thick layer of guano on the floor of the cave. Bacteria and fungi in the cave eat the guano and excrete nutrients that are eaten by other animals such as the crickets. The crickets then become food for the larger animals such as crayfish and scorpions.

Another source of food has been found in one cave in Romania. Scientists have found a certain kind of bacteria living in this cave that can use sulfur, rather than sunlight, to produce carbon-based molecules, thus providing food for the other animals. This process is called **chemosynthesis** and is uncommon. Chemosynthesis also takes place near the thermal vents on the bottom of the ocean where there is no light available.

The amount of life in caves is very low compared to other ecosystems. There are three main limiting factors that keep life from flourishing. First, there is a limited food supply. This limits how many animals can survive. Second, the temperature must be fairly constant for the animals to survive. This is related to the food supply. If the temperature rises inside the cave, the cold-blooded animals' body temperatures increase. With a higher body temperature and higher metabolism the animals will now need more food, which may not be available. Finally, humidity is also

a limiting factor in the amount of life found in a cave. When a cave dries out, animals such as salamanders dry out, too. This causes many of them to die.

Although cave life is somewhat fragile and is certainly unusual, the fact that life exists there at all is testimony to God's creative power.

What did we learn?

- What is a cave?
- What kinds of plants will you find in a cave ecosystem?
- What are the three categories of animals in a cave ecosystem?
- Explain the different habits of each category of cave animal.
- What is the main source of nutrients in a cave ecosystem?

Taking it further

- Why is a cave considered a low energy ecosystem?
- Why can a rise in temperature inside a cave threaten the ecosystem?
- Which sense is least useful in a cave?
- Which senses are most useful in a cave?

Cave worksheet

Complete the "Cave" summary worksheet and add it to the mountain section of your notebook. Find interesting pictures of cave formations as well as animals to include in your notebook.

Plants in caves?

Purpose: To understand why plants don't live in caves

Materials: houseplant, box

Procedure:

1. Place a box over the top of a houseplant. Be sure that the box completely covers the plant and does not allow any light inside.
2. Each day, lift the box and water the plant just enough to make the soil moist without making it soggy. It would be best to do this at night when there is no sunlight.
3. Observe the plant each day for several days. How does it look after 1 week with no sunlight?

Conclusion:

Green plants require sunlight to perform photosynthesis and grow. Without sunlight, plants will die. Inside a cave there is no sunlight so no green plants can grow there.

Bats

The most important animal in most cave ecosystems is the bat. Bats are very interesting animals. They are the only mammals that truly fly. The wings of a bat are actually thin membranes of skin stretched between very long, slender fingers. There are also membranes connecting the hind feet with the tail for a complete wingspan.

Most bats eat insects; however, some bats eat fruit, fish, or even blood. Bats that eat insects usually have poor eyesight, but have good hearing. They also are able to hunt using **echolocation**. Echolocation is similar to sonar used on submarines. The bat sends out a high frequency sound wave. The sound bounces off of things around it and the bat can detect the reflected waves. This helps the bat create a mental picture of its surroundings. Echolocation works so well that bats can quickly navigate through completely dark caves and snatch insects out of the air while flying.

The most common insect-eating bat in North America is the brown bat. These bats roost in trees or buildings during the summer, but spend their winters in caves. Another insect-eating bat is the Brazilian free-tailed bat, which winters in Mexico and spends the summers in the southern United States. These bats collect by the thousands and roost in the famous Carlsbad Caverns in New Mexico.

The smallest insect-eating bat is the Kitti's hog-nosed bat. It is only 1 inch (2.5cm) long and lives in caves in the rainforests.

The largest bat is the fruit bat, often called the flying fox because of its long pointed face. Its body is up to 16 inches (40 cm) long and it can have a wing span up to 6 feet (1.8 m). Fruit-eating bats do not use echolocation to find their food. Instead, they have very keen eyesight and can see the fruit. They also listen for the sounds emitted by other bats to help them locate food. Most fruit bats roost in the trees of swamps and forests instead of in caves.

Fruit bats present an interesting problem for evolutionists. Many scientists looking at the very sharp teeth of the bat would assume that it had sharp teeth because it was a carnivore. However,

fruit bats eat only fruit. They use their sharp teeth to penetrate the fruit and extract the juice, not to eat meat. This shows that God's design does not always match man's expectations.

The fisherman bat eats fish. It flies low over a river or lake and snatches fish from the water with its hind feet. It may eat the fish while it is flying or take it back to its roost to eat. Like most bats, the fisherman bat is nocturnal and generally hunts at night. Other bats eat frogs and other small animals.

The most frightening bat may be the vampire bat. Vampire bats seldom feed on people, but because they get their nourishment from the blood of other animals, people often fear them. Vampire bats usually eat the blood of large mammals such as cattle or horses. The bat lands near the animal and uses its wings to move across the ground. Then the bat makes an incision with its teeth. The bat then uses its tongue to lap up to two teaspoons of blood from the animal. This does not usually harm the animal; however, since bats can carry diseases such as rabies, the incision runs the risk of spreading diseases.

While some bats migrate to warmer climates in the winter, many bats hibernate. They will find a cave and sleep while hanging upside down. Bats can hang upside down for extended periods of time because God designed them to be able to do this. There are tendons attached from the bat's feet to its upper body, so when its body pulls down, the tendons naturally cause the toes to squeeze shut without the bat having to exert any effort. Also, its circulatory system is designed to keep the blood from rushing to the bat's head when it is hanging upside down. The bat is most at rest when hanging upside down.

Bats are amazing creatures and are vital to the health of a cave ecosystem. Use what you have learned about bats and make a page about bats to include in your notebook.

Fun Fact

A brown bat can eat as many as 600 mosquitoes in one hour.

UNIT 5
Animal Behaviors

24 Seasonal Behaviors
25 Animal Defenses
26 Adaptation
27 Balance of Nature

◊ **Understand** the characteristics of hibernation and estivation.
◊ **Describe** how and why animals migrate.
◊ **Identify** animal defenses and adaptations.
◊ **Describe** methods of maintaining the balance of nature.

228 • Properties of Ecosystems

24

Seasonal Behaviors

It happens every year.

How do the seasons affect animals?

Words to know:

hibernation   migration

Challenge words:

animal courtship

Because of the tilt of the Earth with respect to the sun, most parts of the Earth experience different seasons. Summers are warmer than winters in nearly every part of the Earth. The change in seasons triggers some very interesting behavior in many animals. Scientists believe the change in the number of hours of daylight is the primary factor in many of these behaviors.

In areas with very cold winters, many animals hibernate. **Hibernation** is a state in which the animal goes into a very deep sleep. The animal's heart rate and breathing slow down significantly and its body temperature drops. A hibernating animal does not wake up to eat; it survives the hibernation by using up fat stored in its body. A hibernating animal's metabolism may be so slow that it may actually appear dead to the casual observer. This slow state allows the animal to spend the winter in a burrow protected from the harsh environment and sleep until the weather becomes more favorable. When the daylight hours start lengthening, the animal wakes up and begins searching for food.

Bats, squirrels, woodchucks, and different insects are some of the many animals that hibernate. Some people think that bears hibernate as well. Many bears do spend much of the winter in a deep sleep; however, the bear's metabolism does not significantly slow down and it can awaken and search for food in the middle of winter. So bears do not experience true hibernation. In deserts, many animals sleep through the summer to avoid the hot dry weather. This is called estivation, and is a similar state to hibernation, only it occurs in the summer.

Another way that many animals deal with changing seasons and unfavorable weather is by **migration**, the moving from one area to another for a particular season. Although we usually think of bird migrations,

Fun Fact

The Arctic tern has the longest migration of any animal. It flies 25,000 miles (40,000 km) each way when it migrates. It flies from pole to pole.

LESSON 24   **Properties of Ecosystems • 229**

🧪 Monarch butterflies

The migration of the monarch butterfly is very fascinating. The butterflies start out in the northern United States and in Canada. In the fall, as the days become shorter, the butterflies begin their long journey south. Monarchs that are east of the Rocky Mountains fly to southern Mexico while those west of the Rocky Mountains fly to southern California. They stop to eat along the way, but also use fat stored in their abdomens for additional energy. When they reach their destination, the butterflies hibernate for the winter.

When the days get longer in the spring, the butterflies begin heading north again. Along the way, they mate and lay eggs. The butterflies that spent the winter in the south usually die before returning to their original homes. Their offspring hatch and turn into butterflies within a few weeks. These new butterflies continue the trip north. When they reach the northern point of their journey, this second generation mates and lays eggs and then dies in the north. Their offspring hatch in late summer and, after changing into butterflies, begin the journey south that their grandparents made the year before.

Scientists are not sure how these grandchildren, and sometimes great grandchildren, know where to go during their migration. But they have determined that the butterflies return to the same areas, and sometimes even the same trees that were inhabited by the previous generations. God has given them the instinctive ability to return to their winter roosting grounds, having never been there before.

Do some research on the migration of monarch butterflies and make a map showing the routes of each generation. Include this map in your notebook in the animal behaviors section.

---

other animals such as sea turtles, whales, and butterflies also migrate. Again, migration is usually triggered by the changing number of daylight hours. Animals generally migrate to find better food supplies, milder weather, and often for reproduction.

Most animals that migrate move from north to south and back again. The animals move toward the equator in the wintertime and back toward the poles in the summertime. However, in some areas, such as in parts of Europe, animals migrate from east to west. Some animals move from the mainland to the British Isles during the winter because Great Britain usually has milder winters than most of Europe. Then the animals return to the mainland in the summer.

Migrating is a very complex behavior. Not only do the animals know when to leave an area, but they often travel thousands of miles during each trip without getting lost. A particular bird often returns to the same nesting area each year after traveling thousands of miles. Scientists are not sure how these animals navigate. They believe that some animals use the stars as a reference. This certainly seems to be the case with sea turtles as well as with some birds. Other birds seem to use landmarks to help guide them, and some are believed to use the Earth's magnetic field lines. Fish and other aquatic animals often use smell to help them find their way.

Birds generally fly to their summer and winter homes. Flying requires a great deal of energy. Most birds spend time building up stores of fat before migrating so they will have the required energy. Many birds stop periodically and eat, so they do not have to have enough body fat for the entire journey, but the Golden Plover flies 2,000 miles (3,200 km) non-stop over water and must have enough energy for the entire flight. God has amazingly designed this bird to make the flight without dying.

Birds generally migrate in flocks and often fly in a V-shaped formation. This is because the V formation is a very efficient way to fly. This formation breaks up the air currents and reduces friction. The lead bird has the most difficult time, so migrating

Salmon migrate to the exact place they were born where they reproduce and then die.

birds take turns being the leader. In North America there are four distinct migratory flyways that different species of birds use. They are the Pacific, central, Mississippi, and Atlantic flyways. Within each flyway, particular species of birds seem to have their own corridors in which they fly.

Although most migration is seasonal, some migrations happen only once. Pacific salmon are born in rivers. They swim to the ocean where they spend most of their lives. Then, when it is time for them to mate, they swim back up stream to the exact place where they were born. There they reproduce and then die. Their migration is a once-in-a-lifetime experience. Scientists believe that the fish use their sensitive sense of smell to follow the right path to their spawning grounds in the river where they were born.

When you look at the complex patterns of migration and the abilities of animals to hibernate and estivate, you see the hand of a wonderful Creator. God has designed these animals to survive the changing seasons.

What did we learn?

- What is hibernation?
- What is estivation?
- What is migration?
- List three different kinds of animals that migrate.
- What is the most likely trigger for seasonal behaviors?

Taking it further

- How can animals know where they are supposed to go when they migrate if they have never been there before?
- How do animals navigate while migrating?
- Why might a group of animals move from one location to another, other than for their annual migration?
- If you see a monarch butterfly in the fall and then see another one in the spring, how likely is it that you are seeing the same butterfly?

Animal courtship

Springtime brings many changes in the wildlife in most areas. The migrating birds return and begin their songs. The plants begin growing again after their winter rest. And many animals begin courtship. **Animal courtship** is a way for animals to attract mates. Some species mate for life and only go through their courtship process once. Other species choose new mates each year. The courtship rituals are as varied as the animals themselves, but they all serve the same purpose. The male demonstrates his superiority over other males so the female will be willing to choose him, and his competition will back off.

Birds have some of the most creative and interesting courtship rituals. The Bird of Paradise does an elaborate dance as part of its courtship ritual. The male peacock spreads it brightly colored

LESSON 24 **Properties of Ecosystems** • 231

Many animals other than birds also have courtship rituals. The hissing cockroach hisses and rubs the potential mate's antennae. The pink dolphin, which lives in the Amazon River, presents stones to a potential mate. Elk make loud bugling noises to attract a mate. In addition, male elk may fight each other using their massive antlers to prove their strength and ability to mate with the females they have attracted.

tail feathers in a brilliant display to impress the ladies. The red crowned crane not only prances around with a stiff neck, but it also tosses sticks and grass around and sings a duet with its prospective mate. The male sage grouse has inflatable neck sacs which he

uses to amplify the plopping noises he makes as he courts the females. These sacs make the noises loud enough to be heard up to three miles away. The male bowerbird builds an elaborate structure and struts around in front of it to attract a female.

Many courtship rituals involve sounds and elaborate shows, but other animals are more subtle. They produce chemicals called pheromones which attract the opposite sex of the same species. These chemical scents may be on the animal itself or may be spread by the animal onto plants in its area. This smell advertises to anyone in the area that the animal is looking for a mate.

Many courtships take place in the spring, with babies being born later that year. However, some courtships take place in the fall or other times of the year depending on the species. Most birds mate in the spring, but elk mate in the fall. Regardless of the time of year, the changing seasons are still the trigger for the courtship behavior. This behavior helps to ensure that the animals find mates and are able to reproduce.

25

Animal Defenses

A matter of protection

How do animals protect themselves?

Words to know:

camouflage

Challenge words:

dormant

Before the Fall of man in the Garden of Eden, animals did not prey on one another. All animals ate plants and there was no death. But after man sinned, God cursed the Earth. Part of the Curse includes the fact that many animals now survive by eating other animals. However, God did not make the prey totally defenseless. Animals have many ways that they can defend themselves against their attackers. Most of these defenses can be grouped into three types: flight, trickery, or fight.

The first instinct of most animals when they feel threatened is to run away. Some animals are much faster than their predators and can use that speed as their main defense. Deer and other grazing animals are usually fast runners and can outrun many of their predators. Other animals can dive into burrows or quickly hide to get away from predators. Some animals such as antelope and prairie dogs post guards to watch for trouble. When they sense danger, they signal an alarm so the herd or colony can retreat to safety. Prairie dogs are well known for their barking alarm that is passed from one part of the colony to another to warn of approaching danger.

Trickery is also widely used by animals as defense techniques. One of the most effective defenses is **camouflage**. Many animals can blend in with their surroundings, making it difficult for predators to find them. Chameleons can change the color of their skin to match their surroundings. Many insects are shaped to resemble sticks, leaves, or other things in their environment, making them difficult to spot. Tigers, leopards, and other animals are multi-colored, which breaks up their shape and helps them blend in. Other animals, such as the Arctic fox and Arctic hare turn white in the winter when the ground is covered with snow, and others have fur or feathers that help them match the colors of their surroundings.

In addition to camouflage, many animals try to trick their enemies into leaving them alone by their behavior. Some animals try to intimidate their attackers into leaving them alone. The male mountain gorilla will stand up on its hind legs and beat its chest. It also growls and bares its teeth. This aggressive behavior often encourages enemies to back

LESSON 25 **Properties of Ecosystems • 233**

🧪 Animal defenses

Purpose: To create a card game that will help you appreciate the many ways that animals can defend themselves

Materials: card stock or tag board, drawing materials, pictures of animals

Procedure:

1. Cut several sheets of card stock into identical pieces to make a deck of cards.
2. Make pairs of cards. One card should have a picture or name of an animal. The other card should have a description of its defense. Be sure that the defense is specific enough that it only applies to one particular animal in your deck. Following are several ideas for card pairs. Use these and as many of your own ideas as you like.
3. When you have all of your cards made, you can play a matching game by mixing up the cards then spreading them on the table face down. Take turns turning over two cards to try to make a match. When you make a match you get to keep the cards and take another turn. If you do not make a match, turn the cards back over, and it becomes the next person's turn.
4. Take a picture of your cards when they are spread out face up and include the picture in your notebook.

Ideas for cards

Prairie dog—Bark to alert colony

Tiger—Stripes for camouflage

Arctic fox—White fur for camouflage

Chameleon—Changes color

Gorilla—Beats its chest for intimidation

Puffer fish—Fills its body with air

Rattlesnake—Rattles on its tail

Poison dart frog—Body covered with poison

Stick bug—Looks like a stick

Octopus—Shoots ink

---

off. Many other animals will bark, growl, or yelp to warn predators to stay away. The puffer fish can fill itself with air, making it seem much bigger than it actually is and making it difficult for its enemy to get a good grip on it.

Another way that animals try to confuse their enemies is by squirting out ink. Octopi and squids can shoot out a cloud of ink to hide their movements; allowing them to make a quick get away. The regal horned lizard can actually shoot blood out of its eyes to frighten away its enemies.

Most animals prefer to either get away from their enemies or to frighten their enemies away. But even when these defenses do not work, most animals have some way to protect themselves. Many have horns, antlers, claws, or teeth with which to fight. The porcupine has hundreds of sharp quills which discourage anyone from taking a bite. Some animals, such as the electric eel, can give a nasty electrical shock to a predator. Others coat themselves in nasty tasting chemicals and some even have poisonous bites with which they defend themselves.

Although the animals were not originally created to need these defense mechanisms, they were designed by God with the ability to develop these defenses. Predators also have developed ways to enable them to capture their prey so they do not starve. This struggle is part of the Curse, but still shows God's provision for the continuance of life for both the predator and the prey. ✷

The porcupine has hundreds of sharp quills which discourage anyone from taking a bite.

What did we learn?

- What are three main ways that animals try to defend themselves?
- List three ways that animals can trick their enemies into leaving them alone.
- How do some eels protect themselves?

Taking it further

- Why do you think animals prefer to run away or frighten off enemies rather than fight?
- Why do many animals prefer trickery to running away?
- How might a defense also serve as an attack method?

Plant defenses

Animals are not the only organisms that are able to defend themselves. Many plants also have defenses. Most plants are easy food for the animals that rely on them. We have already seen how grass was designed by God to grow quickly even when it is being eaten regularly. However, because of the Curse, many plants now live in harsh conditions. Since the Fall they too have to struggle for survival. Many of these plants have defense mechanisms that help them to survive.

Many desert plants have sharp needles, which discourage animals from eating them. Other plants have thorns, which also discourage animals from eating them. Some plants put out chemicals that prevent other plants from growing too closely to them. This prevents other plants from competing with them for water, sunlight, and nutrients in the soil.

Other plant defenses are not defenses against animals or other plants; instead, they are defenses against the harsh weather. The most common defense against the cold is for plants to shed their leaves in the fall and go **dormant** during the winter. Other plants go dormant when there is not enough water. Dormancy is a plant defense.

Just like with animals, plants did not need these defenses in God's original creation. In the Garden of Eden growing conditions were just right, and there were no thorns or thistles. But after the Fall, even the plants were cursed, and now have to struggle to survive.

LESSON 25 **Properties of Ecosystems** • 235

Animal Behaviors

26

Adaptation

Fitting in

How do animals change?

Words to know:
adaptation   natural selection
survival of the fittest

Challenge words:
adaptive radiation

Have you ever had something unexpected happen and you had to adapt to the situation? Maybe a storm suddenly came up and you needed an umbrella; or perhaps you had unexpected visitors and you had to change your plans for the day. However, this is not the way we speak of adaptation in nature. Plants and animals have been designed with special adaptations. **Adaptations** are physical traits or behaviors that allow an organism to be better suited to a given environment.

Adaptation is a word that is greatly misused and misunderstood. Evolutionary scientists say that an organism has adapted to its environment by developing new traits over time. These new traits make it better suited for where it lives. Waterfowl have webbed feet making them good swimmers. Birds' feathers have a hook and barb design making them easy to repair and good for flying. Perching birds have toes that go both directions on their feet, making them good for grasping branches. Creation scientists believe these are design features created by God. These are not new traits.

Scientists also refer to differences between two similar species as adaptations. Some squirrels are grey and others are brown, depending on which kinds of trees they are likely to live in. Polar bears are white and have slightly webbed feet, which makes

them better able to survive in the Arctic than their cousins, the brown bears, that live in the Rocky Mountains. Pink dolphins are able to live in freshwater, and bottlenose dolphins live in salty water. These differences between similar species can be used to show that animals have characteristics that are related to their niches and their environment. However, these differing traits did not require new information. The original squirrel kind that came off of the Ark had the ability to produce different colors of fur, just as different humans can have different colored hair. The original bears carried the information in their DNA to produce brown bears and white bears. These characteristics are not due to new information being added to the DNA.

Evolutionists claim that adaptations, such as webbed feet in ducks, are the result of changes in an organism over time as they evolved from one kind of organism into another (originally coming from a single-celled ancestor). But the Bible says that God created distinct kinds of animals. Ducks were always ducks and were designed with webbed feet from the very beginning. Webbed feet are not adaptations. The foot structure may change slightly over time, but it did not come from a claw.

Adaptations represent real changes that we see between different species of the same kind of animal. This is the result of the process of adaptation. So, how do animals change to adapt to their environment? Chance mutations cause a creature to have slightly different characteristics from the others of its kind. If this characteristic is beneficial, that creature survives better than the others and is likely to pass this characteristic on to the next generation, making its offspring more likely to survive and reproduce as well. This is referred to as **survival of the fittest**, or **natural selection**. Evolutionists claim that these mutations add new characteristics and take long periods of time, perhaps millions of years, to become well established. Eventually, enough mutations cause changes and a lizard has evolved into a duck.

However, the Bible tells us that all animals were created by God just 6,000 years ago as distinct kinds. For example, God would have created the cat kind, from which all the different cats we see today have come. Cats did not evolve over time from some other kind of animal. Each original kind was created with great variety in its DNA. After the Fall, mutations and other genetic mechanisms altered the DNA, adding more variety. After the Flood, animals left the Ark reproducing and spreading out across the Earth. Their offspring had great genetic variety.

As the animals moved to different parts of the Earth, the environments were different, and those with advantageous characteristics for a particular environment were the ones to survive and have offspring. This is natural selection; however, this does not require new characteristics to be added through genetic mutation. Instead, it accepts the fact that God designed the animals with the ability to produce many different characteristics. The variety is the result of the original DNA being recombined in new ways as well as being altered by mutations. Natural selection simply acts on the traits that are already present—it does not make new traits in the population. These changes in populations do not require millions of years to happen, but can take place in only a few generations as existing characteristics are modified.

The characteristics that make organisms suited for the environments in which they live are designed characteristics or modifications of the traits that already existed. They are not the result of evolution. ✳

🧠 What did we learn?

- What is adaptation?
- Are all helpful characteristics a result of a change in the organism?
- What process causes different species to develop among the same kind of animal or plant?

🚀 Taking it further

- How does natural selection work?
- Does natural selection require millions of years to develop distinct populations?
- Does natural selection require genetic mutation?

Design worksheet

Complete the "How Was I Designed?" worksheet and include it in your notebook. For each plant or animal that is listed, describe one or more design features it has that make it adapted or well-suited to its particular environment.

Darwin's finches

One of the most famous examples of adaptation is found among the Galapagos Island finches, sometimes called Darwin's finches. These birds were observed by Charles Darwin and the other scientists aboard the HMS Beagle in the 1830s. At first Darwin thought there were several different kinds of birds, but after much examination it was determined that there were actually 13 different species of finches that had likely developed from one original type of finch. This process of developing different species from a common ancestor is called **adaptive radiation**.

The biggest difference between the various species of finches is in the size and shape of their beaks. Some of the finches have long, narrow beaks. Others have short, wide beaks. Others have hooked beaks. The dominant type of beak in the community seems to depend on the food that is most readily available on the different islands. Some birds eat leaves, others eat fruits and buds. Some eat insects, while others eat grubs or seeds. The beak that each species has is suited for the kind of food that is available in that area.

It is believed that a single kind of finch arrived in the Galapagos Islands. Their offspring spread out among the islands. The original pair had the genetic ability to produce offspring with many different shapes and sizes of beaks. In each area, the offspring that had the best kind of beak for the available food supplies were more successful in surviving and reproducing, so eventually all of the offspring on a particular island had the preferred kind of beak for that particular area.

Darwin and other evolutionists claim that this natural selection required millions of years to develop. Evolutionists claim that the differences in beaks have come about because of genetic mutations and therefore long periods of time are required for enough mutations to take place. However, modern research in the Galapagos Islands has shown that long periods of time are not required.

Researchers, like Peter and Rosemary Grant, have been studying the finches in the Galapagos since 1973. They have recorded the climatic changes that have affected the food available in different areas. They have shown that the changes in climate can cause changes in available food supplies, and have also shown that the species that are dominant can change very quickly, within only a few generations. Millions of years are not required for the dominant species to change.

Although Darwin's finches are a great example of adaptation and natural selection, they do not demonstrate evolution. They show that a creative God made finches with a wide variety of possible characteristics and that particular characteristics become dominant in certain environments, but they do not show that a finch changed into anything other than a finch. The finches also do not demonstrate any possible mechanism for one organism to change into different kinds of organisms. So although you may see claims that Darwin's finches prove evolution, you can be sure that they do not. They actually support what the Bible has said all along.

27

Balance of Nature

Keeping it working

How do populations stay in balance?

Words to know:
balance of nature
predator-prey feedback loop
territoriality

Challenge words:
GMO—genetically modified organism

Have you ever seen a gymnast perform on a balance beam? She has to keep in the middle without going too far to one side or the other, otherwise she will fall off. She has learned to keep her balance. Nature has several ways of staying in balance, too. The **balance of nature** refers to the condition of an ecosystem in which the producers and consumers are in a state of equilibrium—their populations are not changing significantly.

There are several ways in which we see this state of equilibrium. First, we see balance in the oxygen cycle. The amount of oxygen produced by plants is equal to the amount of oxygen used by the animals, and the amount of carbon dioxide produced by animals is equal to the amount of carbon dioxide consumed by plants. Similarly, the water and nitrogen cycles are examples of balance in nature.

We also see that in most ecosystems the number of plants is adequate for the number of primary consumers and the number of primary consumers is adequate for the number of secondary consumers, and so forth. This balance is accomplished in several different ways.

The primary way that balance is achieved is through competition for food supplies. When the food supply is low, there is more competition for a limited amount of food. The animals that eat that food must struggle to get it. Some will fail and will starve or will not be strong enough to reproduce. This will result in an increased death rate and a decreased birth rate until there is an adequate food supply for the remaining animals. Similarly, if the amount of food increases, there is less competition for the food so the death rate will go down and the birth rate will increase until equilibrium is achieved.

This is especially obvious in the predator-prey relationship. Let's look at an example. If there are too many coyotes in one area, there will not be enough rabbits to feed them all. The slower coyotes will not get enough food and some of them will die. Others will not be able to attract a mate and will not reproduce. This will decrease the number of coyotes. As the number of coyotes decreases, more of the

LESSON 27 **Properties of Ecosystems • 239**

rabbits will live long enough to reproduce, which will increase the number of rabbits. This process is a **predator-prey feedback loop**. When the number of prey increases the number of predators increase, but as the number of predators increase the number of prey decreases until equilibrium is reached.

This balance in nature is very important to maintaining healthy ecosystems. Many birds eat insects. If the bird population decreases, the insect population could get out of control and cause damage to food crops. We see this balance in every food chain and food web. The woodpeckers keep the wood beetle under control. There is a balance between the black-footed ferret and prairie dogs and between snowy owls and Canadian snowshoe hares. Everywhere you look, you see the balance of nature.

Equilibrium can be achieved through the predator-prey relationship, and this is a significant method by which the balance in nature is maintained today. However, this is not the only way that balance is achieved and it is not the original way that God designed nature to stay in balance. In the original creation before the Fall of man, animals did not eat each other, so God did not originally intend balance to be maintained through predator-prey relationships.

Territoriality is another way that balance can be maintained. In general, animals will respect one another's territory. You see this happening all the time, but you may not recognize what is going on. Each spring, male birds stake out a territory based on how much food is available in a given area. A bird will choose enough land to support himself, his mate, and his offspring. He will go from one edge of his territory to another singing loudly to let other males know that the area is taken. Other males will move on to unoccupied territory.

If a male cannot find an unoccupied area, he will choose not to mate that year. If all the area is taken, he will not mate until an existing pair of birds leaves or dies, making room for him and a new mate. This is likely part of God's original plan for population control.

Territoriality is not limited to birds. Many other types of animals also stake out territories and only breed if they can get a large enough area to support themselves and their family. Sea lions, wolves, wild cats, lemurs, and insects are just a few of the animals that defend a given territory. Birds often defend their territories through songs or elaborate shows that encourage encroaching males to back off. Many other animals, such as dogs and cats, leave scents to mark their territories. Some of these scents serve a dual purpose—to attract a mate and to define the territory.

There are other social activities that affect the populations in an area as well. Flocking has been shown to affect the number of eggs that females lay. When birds flock together, if the flock is unusually large, the females lay fewer eggs than normal. If the flock is unusually small, the females lay more eggs than normal. This helps to keep the population of the species within normal ranges.

In general, nature will find a way to balance itself, but man can greatly upset the balance. In

🧪 Population diagrams

Choose one of the food chains you have studied. Using the animals in that food chain, draw a diagram showing what would happen to the population of each animal in the food chain if the number of primary consumers suddenly decreased. Draw a second diagram showing what would happen to the population of each animal if the number of primary consumers greatly increased. Include these diagrams in your notebook.

Growing your own populations

(Optional Activity)

Purpose: To observe how an ecosystem develops its own balance of nature

Materials: cooking pot, grass, distilled water, jar, microscope, eyedropper, slides, slide cover slips, pH testing paper, "Growing an Ecosystem" worksheet

Procedure:

1. Place a handful of grass in a cooking pot with two cups of distilled water. Bring the water to a boil and boil for 3–4 minutes. This should kill all microscopic life in your ecosystem so you can establish a new ecosystem.
2. Pour the water and grass into a glass jar and place the open jar in a sunny location.
3. Each day, for seven days, test the pH of the water and record the pH level on your "Growing an Ecosystem" worksheet.
4. Each day, use an eyedropper to take a drop of water from the jar and place it on a slide. Cover the sample with a cover slip and observe the sample with a microscope. Record your observations on your worksheet. You should see different organisms appear over time.

Conclusion:

It takes about a week for a stable ecosystem to develop in a "pond" such as your grassy water in a jar. It takes about two days for decomposers such as bacteria to appear. By day 4 you should see some algae or other producers. There should be some primary consumers, such as paramecium by day 5 or 6. Finally, you may see some carnivores, such as rotifers by day 7. If the pH of the water remains in the 6 to 8 range, your ecosystem should maintain a balance between the producers and consumers.

the early 1900s many people believed that the predators, primarily mountain lions and coyotes, were killing too many deer around the north rim of Grand Canyon. So from 1907 to 1939 hunters killed hundreds of mountain lions and over 7,000 coyotes. This had a huge impact on the deer population in that area. In 1907 there were about 4,000 deer. By 1918 there were over 40,000 and by 1923 there were nearly 100,000 deer. This huge increase in deer took a heavy toll on the plant life in the area. They were eating anything that grew. But there just was not enough food to feed that many deer. In the winters from 1923 to 1925 approximately 60,000 deer starved to death. By 1939 only about 10,000 deer were left and the plant life was still greatly damaged.

The predator-prey relationship had kept the deer population in line with the amount of food that was available. Man did not need to help control the populations; the balance of nature was already taking care of it. But when man interfered, populations got out of balance. Eventually they came back in balance, but many animals starved to death in the process. People need to be very careful about artificially changing the populations of plants or animals in stable ecosystems.

What did we learn?

- What is meant by the balance of nature?
- Name two ways that the balance of nature is maintained in an ecosystem.
- What are two ways that animals use to stake out their territory?
- What happens if a male cannot find a territory to defend?

Taking it further

- What would be the likely effect on the ecosystem if a prairie dog colony was devastated by the plague?
- What would happen if animals did not respect each others' territories?
- How does the oxygen cycle demonstrate the balance of nature?
- Which methods of population control may have been present originally, and which have developed since the Fall?

Artificial population control

God designed nature to control populations and establish balance. But many times people have changed an ecosystem for various reasons and then tried to restore balance or control the changes. People have been successful in some situations, but more often than not, people have created more problems than they have solved. Let's look at a few ways that man has tried to control the balance of nature.

Many early instances of man's interference in the balance of nature were unintentional. One of the most famous examples led to the extinction of the dodo bird (shown at right). This bird lived only on the island of Mauritius in the Indian Ocean. When people arrived on the island in the 1500s, they brought animals such as dogs, pigs, and macaques with them. These animals did not naturally live on the island and were not a part of the dodo bird food chain. Many of these animals, especially the pigs and macaques, began eating the dodo's eggs. Because there were no natural predators for the pigs and macaques, eventually these animals killed off all of the dodos and the bird is now extinct.

Man greatly upset the balance of nature and the dodo was not able to recover.

Other examples of man's interference are more purposeful, such as the killing of the mountain lions and coyotes near Grand Canyon. People thought they were helping the deer, but in the long run they upset the balance and caused suffering. Another example is the use of pesticides or insecticides in farming. Insects can cause severe damage to food crops. So farmers are always looking for ways to eliminate pests and preserve their crops. Some insecticides have been more successful than others in eliminating pests without harming the environment.

DDT is a chemical that was originally used during World War II to kill mosquitoes in an effort to eliminate the spread of malaria and typhus. This was very successful in many areas and these diseases were greatly reduced. After the war, DDT was also used as a pesticide in farming and it was successful in eliminating many insects; however, it had many unintended consequences. DDT got into the food supply and in the water and did not easily break down. Thus when a mouse ate corn that was sprayed with DDT, the chemical stayed in its body. Although it did not kill the mouse, the DDT was spread to the hawk or eagle that ate the mouse. If the bird ate enough mice with DDT it would not be able to successfully reproduce. Thus, by controlling the insect population with chemicals, man was inadvertently killing off hawks and eagles.

DDT was eventually banned and is no longer used as a pesticide for agriculture. However, because of its success in reducing malaria, it is being used in limited areas in Africa to help protect the people there from the disease-carrying insects. This practice is praised by some and condemned by others.

People are always researching new ways to eliminate pests without upsetting the balance of nature. One of the newer methods is to use genetically modified plants, often referred to as **GMOs** or **genetically modified organisms**. Scientists have learned how to change the DNA of some plants to make them distasteful to certain insects. This causes the insects to leave the plants alone without upsetting the food chain. Other genetically modified plants are able to grow in areas that the normal plants cannot, such as cold or very dry areas.

God has given man the task of being caretakers of His creation. We are to take care of the plants and animals that are here without hurting the environment. We must be careful to make changes that do not upset the balance of nature.

Eugene P. Odum

Father of Modern Ecology 1913–2002

SPECIAL FEATURE

In 1940 most scientists thought that the study of ecology was nothing more than observing and cataloging different plant and animal species. That was the type of ecology that was started by Alexander von Humboldt in the 1800s. Ecology was considered a small part of biology. But this changed when Eugene P. Odum joined the University of Georgia as a professor of zoology and proposed that every biologist study ecology. His idea of ecology was much more than observing different species; it involved understanding interconnections within an ecosystem. Because of his revolutionary ideas, Odum is considered the father of modern ecology.

Eugene Pleasants Odum was born on September 17, 1913 in Chapel Hill, North Carolina. Odum credits his father for helping him to view the world from a holistic viewpoint; this means to look at the whole as more than just the sum of its parts—that the parts interact and work together better than they would independently.

Eugene's love of nature started at a young age, and as a teenager he wrote a weekly column about bird life for a local newspaper. He went on to receive his undergraduate and master's degrees in zoology from the University of North Carolina, and his doctorate from the University of Illinois.

Eugene married an artist named Martha Ann Huff in 1939. They spent their first year of marriage in Rensselaerville, NY where Eugene was the resident naturalist for the Edmund Niles Hyuck Preserve. He spent his time cataloging and studying the plants and animals there, while Martha spent her time painting landscapes.

In 1940 Odum began his work at the University of Georgia where he taught until his death in 2002. Together with his younger brother, Odum wrote *Fundamentals of Ecology*, an influential textbook, which was published in 1953. This was the only textbook on ecology available for the next 10 years. This book became the stimulus for the study of how plants and animals affect each other within ecosystems.

Eugene and his brother continued their research the following summer on the coral reefs in the Eniwetak Atoll in the Marshall Islands. They discovered the symbiosis between coral and algae. Their studies showed that not only does everything in nature work together, but man has an impact on ecosystems as well.

Odum's ideas of interconnectedness and balance in nature were widely publicized in his works. The impact of Eugene Odum's life is reflected in these words by the president of the University of Georgia: "We often speak about creating new knowledge through research at the University of Georgia. Eugene Odum did exactly that. No one has been able to think about the environment in the same way since he began writing about the complexities and dependencies of the relationships among organisms."

Properties of Ecosystems • 243

UNIT 6

Ecology & Conservation

28 Man's Impact on the Environment

29 Endangered Species

30 Pollution

31 Acid Rain

32 Global Warming

33 What Can You Do?

34 Reviewing Ecosystems— Final Project

35 Conclusion

◊ **Describe** man's impact on the environment.

◊ **Understand** the truth about global warming.

◊ **Explain** what people can do to be good stewards of the environment.

28

Man's Impact on the Environment

Where do we fit in?

How do people fit into ecology?

Ecosystems are designed to be balanced.
You just learned how populations of plants and animals are kept in balance. But man does not fit into any ecosystem. People are unique—they are different from the animals. They make their own "ecosystems." Remember back to lesson 1 when you described the different habitats you experience throughout your week? People create homes, farms, and cities that become their habitats. But this does not mean that people do not affect the plants' and animals' habitats. People can actually have a great impact on the ecosystems of the world.

Can you name some ways that people affect different ecosystems? From the time Adam and Eve were banished from the Garden of Eden, people have had to turn wild land into a home and a place to grow food. So one of the first ways that man affects ecosystems is by clearing land for homes and farms. Clearing the land and constructing buildings forces many animals to leave and find new homes. It also reduces the number of native plants, so many animals must leave to find new food sources. If people move into a forest they must chop down trees to make room for homes and crops. If people move into a grassland they have to plow under or remove the grass to make room for crops. This reduces native habitat and impacts the plants and animals in the area.

Does this mean that people should not clear the land and build on it? Not at all! God gave the Earth to us for our use. Farmers not only clear the land, but use pesticides and herbicides to control the insect and plant populations on their farms. They use water to irrigate their crops. Often, they add fertilizers to the soil to increase the productivity of their crops. All of these activities have an impact on the native plants and animals, but are often necessary to ensure adequate food supplies for people.

In the past few decades, farming practices have changed to become friendlier to the environment. Many crops are planted along the contours of the land to prevent soil erosion. Crops are rotated so that different crops are grown each year. This allows the soil to naturally build back the nutrients that have been taken out and reduces the amount of fertilizers needed. Drip irrigation is used in many areas to help reduce evaporation and thus reduce the amount of water needed to grow the crops. Wind breaks have also been planted to help prevent soil erosion.

Ranching is another way that man affects the environment. Ranchers introduce non-native animals such as cattle or sheep to an area. This causes

Coal-burning power plants can add chemicals to the air that can contribute to acid rain.

more competition for food and water and often drives out native animals. However, many animals have learned to live with the ranch animals. On some ranches the ranchers supply food for the animals, so they are not competing with the original animals for food; they still compete for space, however. Ranchers may also hunt predators to keep them from killing their animals. This impacts the whole food chain in that area.

Industry is another way that people impact ecosystems. Buildings take up space. Some industries use up resources such as water. Also, some industries put chemicals into the air that affect the plants and animals. Coal-burning power plants can add sulfur dioxide and nitrogen oxides to the air that can contribute to acid rain. Automobiles also add pollutants to the air that can affect people as well as animals. The logging industry cuts down large numbers of trees, reducing the habitat for many animals.

People also affect ecosystems with many of their recreational activities. Boating and other water activities affect the plants and animals that live in the water. Hiking can disturb the plants and animals in an area. Hunting has a definite effect on the populations of the animals being hunted and on the plants and animals in the food chain. Skiing affects mountain ecosystems. Trees are removed to make way for ski slopes, which changes the habitat for the animals that live there.

As you can see, nearly everything that people do has an impact on the ecosystems of the world and often plants and animals are hurt by reduction of their habitats or by harmful chemicals added to the air and water. Does that mean that people are bad and should do everything they can to avoid any contact with nature? Some people would have you believe that, but that is not what God intended. God created the world for man's enjoyment and then gave man the job of overseeing the use of it. We are to take care of the land and use it for God's glory. It is man's job to use the resources of the world wisely to benefit man and to glorify God. This does not mean that we have the right to abuse nature and to waste its resources. Instead, we are to recognize that the Earth belongs to God and that we are to take good care of it for Him.

What did we learn?

- What are some ways that farmers impact ecosystems?
- What are some ways that farmers and ranchers have changed their practices to be more friendly to the environment?
- What are some ways that industry impacts ecosystems?

Taking it further

- What are some ways that people can minimize their impact on nature?
- How can hunting licenses positively affect man's impact on ecosystems?

Recording my impact

Keep track of your activities throughout the day and record how they may impact the ecosystems around the world. Record your activities on the "How I Impact Nature" worksheet. Think about where the things you use come from, how they are made, and how those processes might affect different ecosystems. Try to think of positive impacts as well as negative impacts. What are you doing to try to minimize negative impacts?

A biblical view of ecology

Should Christians be involved in ecology? Is there a Christian view of conservation? These are questions that many Christians ask. And the Bible gives us a very clear answer. In Genesis 1:28 God told Adam and Eve, "Be fruitful and multiply; fill the Earth and subdue it; have dominion over the fish of the sea, over the birds of the air, and over every living thing that moves on the Earth." Thus man was given a job to do. He is to subdue and have dominion over the plants and animals.

In order to obey this command, Christians must recognize three things. First, man did not evolve and is not just a higher life form or another animal. Man was created in God's image. Therefore, man has a different role in ecology than plants or animals. Man is to take an active role in taking care of nature.

Second, man was commanded to be fruitful and multiply. This means man is to populate the Earth. Many evolutionists claim that the huge human population is more than the Earth can bear and that people should not have children. However, this was not God's plan. God does not view man as an enemy of nature.

Third, God entrusted people to be stewards of the planet. Some people think that having dominion over nature is an excuse to exploit and use nature for our own selfish purposes. But that is not what God intended. God intends for man to administrate and care for the plants and animals. Man must be responsible for his actions and do his best to use what God has given him for his benefit, but in such a way that nature is not exploited.

Most of the ecological problems that we have today, such as pollution, damaged ecosystems, acid rain, etc. are results of human indifference, shortsightedness, selfishness, and greed. These are all moral issues. People have turned their backs on God and have used His creation for their own selfish purposes. Many people feel that government regulations and public awareness campaigns will solve our ecological problems. While these methods have had some success, the ultimate problem is that man does not acknowledge his role as God's caretaker. To truly solve ecological problems will require people turning their hearts to God.

Although God expects us to take care of nature, many people have taken this idea to an extreme. They have become so concerned about the plants and animals that they have begun to worship nature—worshiping that which has been created, rather than worshiping the God who created it (Romans 1:20 25). This is also not what God intended. We must recognize that God is the only one worthy of our worship and that we are working on His behalf when we take care of the planet. If we keep our focus on God, we can be good caretakers of our planet.

29

Endangered Species

Are they disappearing?

What makes an animal an endangered species?

Words to know:

extinct

endangered

habitat reduction

captive breeding

By studying the fossil record, it is obvious that many species of plants and animals have become **extinct**, meaning there are no longer any of that species alive today. Many of those organisms have become extinct because of changes in climate, changing food sources, disease, and other natural causes. Many creation scientists believe that most of the dinosaurs became extinct because they could not deal well with the changes in climate after the Flood. Other plants and animals, such as the dodo bird, have become extinct because of the actions of people.

Today there are approximately 1,000 species of animals worldwide that are in danger of becoming extinct. These animals are classified as **endangered** species. In the United States alone there are approximately 700 plant species and 500 animal species that are classified as endangered. Some of these species are endangered because of the actions of people.

The main reason plants and animals become endangered is because of **habitat reduction**. As people move into new areas or clear new land for farming, many animals and plants are killed or displaced. Many animals and plants survive well in other areas and so they are relatively unaffected by man's expansion. However, some species only live in a limited area. If that area is significantly changed by man, that species may not survive. One example is the clearing of bamboo forests in China. Pandas eat bamboo and have traditionally moved from one area to another to find adequate food supplies. But as the forests are cleared, the Pandas are now limited

🧪 Researching endangered species

Many of the animals you are familiar with are endangered species. Choose one or more of the animals below and write a report on it. Include this report in your notebook.

Blue Whale	Humpback Whale	
Sperm Whale	Gray Wolf	
Pacific Salmon	Giant Panda	
Grizzly Bear	Siberian Tiger	
California Condor	Black Rhinoceros	
Snow Leopard	Hawaiian Monk Seal	

to only a few isolated areas. When the food supplies are not adequate, the Pandas have nowhere to go. This is contributing to their declining numbers.

In addition to habitat reduction, people have introduced non-native or exotic species to various ecosystems. These new plant or animal species are often more aggressive than the native species and can kill off or greatly reduce the native populations. This is what happened to the dodo as pigs and other animals moved in with man and ate the dodo's eggs. Another example is the zebra mussel. This tiny, shelled creature, about the size of a fingernail, is native to Russia. But in 1988 a population of zebra mussels was discovered in the Great Lakes in the United States. It is believed that these creatures were introduced to the lakes in the ballast water of ships that had been to Europe. The zebra mussels do not have many natural predators in the Great Lakes and have spread quickly to other waterways. They eat the same food as zooplankton and as they spread they are starving out many of the organisms that form the bottom of the food chains in these lakes. This affects the whole food chain. The zebra mussels are pushing many species closer to the endangered category.

Over hunting and exploitation of various plants and animals have also resulted in some species becoming endangered. Several species of whales were hunted nearly to extinction in the 1800s before most governments made it illegal to hunt them. The gray wolf has also been extensively hunted by ranchers and was once an endangered species in certain areas. Black rhinoceroses are hunted for their horns and elephants are hunted for their tusks, greatly reducing their populations. The American Bison was nearly hunted to extinction in the 1800s as well.

Pollution is also a large factor in declining populations of some species of plants and animals. When factories dump chemicals into waterways, the chemicals can have devastating effects on the wildlife in the water. Pollution released into the air can also greatly harm plants and animals. Many governments around the world have passed laws to require companies to eliminate or greatly reduce the pollution that they release into the environment. This has helped to greatly improve the situation in many industrialized countries.

Once an animal or plant becomes endangered, it may take great efforts to change the situation and help it back toward a stable population. Many organizations have started acquiring or restoring habitats for endangered species. They have also reintroduced species to areas where they once lived in hopes that they will again thrive there. Some animals are also being bred in captivity to increase their numbers. These plans have had some success. Of all the animals listed as endangered in 1973, 68% of birds and 64% of mammals were declared to be

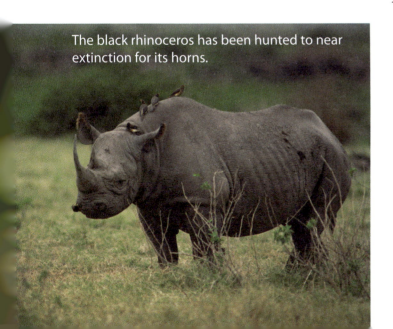

The black rhinoceros has been hunted to near extinction for its horns.

The California Condor almost went extinct.

improving and/or stable by 1994.

One of the most successful **captive breeding** programs is the breeding of the California Condor. This bird was near extinction due to hunting, lead poisoning, and poor birth rates. In 1987 all 22 of the known remaining condors were captured and taken to the San Diego Wildlife Park and the Los Angeles Zoo. There, an extensive breeding program was adopted in hopes of increasing the population. Scientists knew that if a female lost an egg, she would lay a second egg. So when the first egg was laid, the scientists took the egg away and incubated it artificially so the female would lay a second egg. This greatly improved the rate at which the population increased.

By 1991 scientists began reintroducing the California Condor into the wild. Some of these birds have reproduced in the wild, while many have not. Captive breeding continues and new birds are being released. There are now more than 440 California Condors living, about half of which are living in the wild. The breeding program for the California Condor has been the most expensive captive breeding program ever, costing over $35 million.

With the great costs associated with saving some of these endangered species, we have to ask

Fun Fact

There are approximately 1.5 million identified plant and animal species in the world, and scientists believe there may be 20 times that many species that have not yet been identified. Of these, only about 1,000 animal species are endangered.

why should we worry about saving these plants and animals? First, God made us stewards of the Earth and we need to do the best we can to take care of it. In addition, many plants and animals have great value to people. Many plants and animals are used to make medicines, food, and other agricultural products. There are also many commercial uses for plants and animals and we need to ensure that the populations are stable. For example, logging of forests can be very beneficial as long as new trees are planted and the trees are not cut faster than they can be replaced. It is important that we do not push species to extinction. However, a good steward must take care of all resources so the balance between saving endangered species and spending people's resources such as time and money must continue to be carefully considered.

What did we learn?

- Name two possible natural causes of extinction of a species.
- Name three possible man-made causes of extinction of a species.
- Name three things people are doing to help endangered species.

Taking it further

- Why might people overhunt a particular animal?
- Can people use the land without harming endangered species?

Wildlife management

Most people would agree that we need to do things to protect the environment and to save endangered species. However, there are no easy solutions and there are many competing ideas on how best to do this. Most nations around the world have some sort of program for providing wildlife refuges where animals cannot be hunted and habitat cannot be damaged by people. In fact, 3% of the total land areas of the world, about 2 million square miles (5 million sq. km), have been designated as protected wildlife areas.

The United Nations is one organization involved in wildlife conservation. The Food and Agriculture Organization of the United Nations (FAO) works with partner organizations around the world to not only protect wildlife, but to ensure adequate food supplies for people. Their goal is to allow people to reap the benefits of the ecosystems today, while using them in such a way that they will be sustained for future generations as well.

In the United States there are several government organizations that help protect wildlife. The U.S. Fish and Wildlife Service is one of the largest organizations. Its goal is to "protect U.S. fish, wildlife, and plants and their habitats for the continuing benefit of the American people." They do this in many ways including enforcing wildlife laws, protecting endangered species, restoring habitat, and managing over 520 wildlife refuges. Other government agencies include the National Park Service, which manages many wildlife areas, and the Environmental Protection Agency.

In addition to government agencies, there are numerous private organizations dedicated to preserving wildlife and protecting endangered species. Some of these organizations raise money; others raise awareness. However, many of them promote propaganda and do little more than cause panic. If you are interested in joining an organization be sure to understand its goals and methods. Some environmental groups have unbiblical goals or use questionable methods to achieve their goals.

Overall, people are much more aware of the impact of humans on the environment than they were only a century ago. There are many people trying to develop comprehensive plans for using the environment while sustaining it for future generations. Although many of these people are not Christians, they are still fulfilling the biblical mandate to subdue the Earth. Christians can help lead this fight in a biblical way.

Theodore Roosevelt
1858–1919

SPECIAL FEATURE

"Keep it for your children and your children's children, and for all who come after you, as one of the great sights which every American, if he can travel at all, must see."

—Theodore Roosevelt on Grand Canyon

Who is the man that set aside more land for public protection than all the presidents before him? A homeschooler, outdoorsman, lawman, a commander of the Rough Riders, the youngest president in history—he was all of these and more. Theodore Roosevelt was born to a wealthy family in 1858, in New York. He suffered from severe asthma and had poor eye sight. For these reasons his family homeschooled him. His father encouraged him to exercise and build up his body, which he did and in the process he became a lover of the outdoors and of nature.

Roosevelt married Alice Hathaway Lee when he graduated from Harvard in 1880, at the age of 22. The following year he was elected to the New York state assembly. In 1884 his wife gave birth to a baby girl and two days later, on Valentine's Day, both his wife and his mother died. This caused him great pain. He left for the badlands of the Dakotas where he mostly isolated himself. He took up ranching and law enforcement. At this time, this part of the country was still a very rough area and he would on occasion hunt down notorious outlaws along the Little Missouri River.

A year later, after a blizzard wiped out his herd of cattle, he headed back to New York, where he purchased a home in Oyster Bay. He kept this home until his death. The following year he ran for mayor of New York. He came in a distant third. After the election he left New York for London, where he married his childhood sweetheart Edith Kermit Carow. The couple had five children together.

In 1888 Teddy campaigned for Benjamin Harrison who was running for President. Harrison won and appointed Roosevelt to the U.S. Civil Service Commission, a post he held until 1895 when he left it to become President of the New York Board of Police commissioners. As police commissioner he made many changes, putting an end to much of the corruption in the police force. He also required the police officers to pass a physical fitness test. He saw to it that phones were installed in each police station (phones were very new at this time). And to check up on the police officers, he would sometimes walk the officer's beat late at night or in the early morning. He was also the first person to add women and Jews to the department payroll.

After working for the police force for two years, President William McKinley appointed Theodore to be Assistant Secretary of the Navy. Roosevelt loved this job and his work prepared the Navy for the coming conflict with Spain. In 1898 he resigned, and with the aid of Colonel Leonard Wood, organized the First U.S. National Cavalry, from a questionable crew of cowboys, Indians, and outlaws from the west, as well as some Ivy League boys from New York. This group later became known as the Rough Riders. They were famous for their ride up both Kettle Hill and San Juan Hill in Cuba in 1898. (In 2001 President Clinton awarded Theodore Roosevelt the Medal of Honor posthumously for this act.)

Upon his return to New York, Roosevelt ran for governor and this time he was elected. Just as when he was police commissioner, Theodore made a concerted effort to get rid of corruption in the New York government. He was too successful in his work and it is said that the political leaders in New York suggested him as a running mate for William McKinley to get him out of the way. The Vice Presidency was considered for many years as an end to political careers.

McKinley was elected with Roosevelt as his running mate. Less than a year after Roosevelt become the second youngest Vice President, he became the youngest President when McKinley was shot and died eight days later on September 14, 1901.

Roosevelt was an extremely active president. He took Cabinet members on fast-paced hikes in the gardens around D.C., horsed around with his children in the white house and on the white house lawn, kept up the boxing that he started as a boy, and read voraciously. His children were also well loved by the country. His oldest daughter became the toast of D.C. and when friends asked him to rein her in he would answer, "I can be President of the United States, or I can control Alice. I cannot possibly do both."

In 1904 he won the election for President in a landslide victory. He continued his work to make many changes to the country. He worked hard to create what he called the "Square Deal" between business and labor. This helped to balance the power between companies and labor.

So why is a president of the United States in a book about ecology? It's because of what he did for our parks while he was president. He withdrew 235 million acres of public timberland from sale and set it aside as national forest. He created 16 national monuments, 51 wildlife refuges, and five new national parks. He set aside 800,000 acres in Arizona as Grand Canyon National Monument, which later became a national park. He also set land aside for Crater Lake in Oregon and designated the Anasazi ruins of Mesa Verde, Colorado as a National Park.

In addition to all the land he set aside for future generations to enjoy, Roosevelt also started conservation groups by inviting governors, university presidents, businessmen, and scientists to the White House to establish policies to preserve the nation's resources. As a result of his work, 41 states established conservation commissions. His work to help the West did not stop there; he also set up dam projects to irrigate farmland in 16 semi-arid states.

After serving as President for seven years, he decided not to run again but deferred to Taft. He later regretted his decision and ran again under the Reform Party's Bull Moose ticket but was unsuccessful. After WWI he was planning to run again but died in 1919 at the age of 61.

30

Pollution

What happened to clean air?

What is pollution, and what can we do about it?

Words to know:

pollution
biodegradable

Challenge words:

ultraviolet radiation
chlorofluorocarbons/CFCs
hydrochlorofluorocarbons/HCFCs
hydrofluorocarbons/HFCs

Pollution is a problem that can affect any ecosystem. **Pollution** is the presence of any contaminant that harms the ecosystem. It could be chemicals in the water or soot in the air. Pollution can harm the plant and/or animal life in many ways.

Some pollution occurs naturally. A sandstorm in the desert can pollute an area by covering up plants with sand or filling up water holes. An erupting volcano spews tons of ash and other debris into the air. This can block sunlight and make breathing difficult. The eruption of Mount St. Helens in 1980 in Washington state shot ash 16 miles into the atmosphere where it was carried across the United States. That eruption also put tons of mud into the nearby rivers and lakes as well as destroyed thousands of trees. Wildfires are another natural source of pollution.

Although there are many natural sources of pollution, much pollution is caused by the activities of people. There are many ways that people put pollution into the environment. Pollution can come from power plants, factories, automobiles, fireplaces, controlled burns, trash, and more.

Water pollution takes a toll on all the life connected to the water. It can hurt the plants and animals that live in the water as well as the plants that grow near the water and the animals that drink the water. There are many different forms of water pollution. Chemicals often find their way into water sources from farms and factories. Also, rain washes oils and other chemicals from the streets into the storm sewers and eventually into rivers and streams.

In many undeveloped areas of the world, sewage is dumped directly into the water. This can hurt the plants and animals that live in the water as well as breed diseases that harm people. It is of vital importance that sewage be treated before the water is returned to streams and rivers.

Air pollution is another problem for many ecosystems. Carbon monoxide is a poisonous gas that is a result of incomplete burning of fuels. Automobiles and power plants give off carbon monoxide. In

Factories put chemicals, detergents, and oils into rivers.

Every week Americans produce millions of tons of garbage that goes into landfills.

small quantities this gas does not cause problems, but in highly populated areas the carbon monoxide levels can become dangerous. Other contaminants in the air include lead, CFCs (chemicals from air conditioners and plastics), and sulfur and nitrogen oxides. These chemicals cause various problems for people and wildlife. Smoke and other particles in the air can cause breathing problems for people and for animals, and can interfere with photosynthesis.

Most people realize that clean air is important for all life on Earth. Governments have passed regulations limiting the amounts of pollutants companies and automobiles can release into the air. And new technologies have made many industries much cleaner. Because of these changes, the air today is much cleaner than it was in the 1970s. Lead is down 93%, carbon monoxide is down by 60%, and sulfur dioxide is down by 70%. We need

🧪 Measuring your trash

One of the major sources of land pollution is trash. Every week people dump millions of tons of trash into landfills. Some of this trash will eventually decompose, meaning it is **biodegradable**; other items will last for hundreds of years. Reducing the amount of non-biodegradable trash will help keep landfills from filling up.

Purpose: To become aware of the amount of trash your family puts into the dump each year

Materials: rubber gloves, newspaper, bathroom scale, one week of family trash, "Our Family's Trash" worksheet

Procedure:

1. Collect all of your family's trash for one week.
2. Spread newspaper on the floor. You may want to do this activity in the garage or outside as the garbage is probably going to be pretty smelly.
3. Using rubber gloves, separate the trash into five piles: metal, glass, paper, plastic, other. If you recycle, separate your recycling materials into the same five piles.
4. Carefully weigh each pile and record the weight on the "Our Family's Trash" worksheet.
5. Multiply each number by 52 to calculate the yearly amount of each kind of trash or recycling your family is likely to generate. Record these numbers on your worksheet.
6. Clean up your mess and throw your trash away.

Conclusion:

Think about how much trash your family generates each year. There are millions of families around the world that are also generating trash. This can become a problem for the environment. Think of some ways your family can reduce the amount of trash it generates. One good way to reduce your trash is to recycle. Many areas have recycling programs that recycle aluminum, glass, plastics, and newspapers.

to keep working to reduce air pollution, but great progress has already occurred.

Land pollution is a third area that needs to be addressed. Chemicals that are used in farming not only get into the water, but also get into the soil. Factories which produce unwanted chemicals or by-products must get rid of their waste and in the past have been known to bury it underground. Radioactive waste from nuclear power plants is a real problem. The radioactivity does not go away, so radioactive waste must be disposed of in a way that will not harm people or animals. One way that people are dealing with radioactive waste is to store it in underground caves in the desert. This keeps it away from people and from most wildlife as well.

Although pollution is a real problem, in the past 40 years people have become much more aware of how their actions affect the world around them. Many programs have been implemented to help reduce or eliminate many sources of pollution. People are taking more responsibility for the pollution that they generate. There is still work to be done, but things are much better now than they were just 40 years ago.

What did we learn?

- What is pollution?
- What are some natural sources of pollution?
- What are some sources of man-made pollution?
- What are three major areas of the environment that can become polluted?

Taking it further

- What are some ways that people can reduce water pollution?
- What are some ways that people can reduce air pollution?
- What are some ways that people can reduce land pollution?
- Do you think that water, air, and land are cleaner or dirtier today than they were 40 years ago?

Blocking UV radiation

Purpose: To appreciate how ozone blocks ultraviolet radiation

Materials: clear plastic sheet protector, sunscreen lotion, newspaper, modeling clay

Procedure:

1. Completely coat one side of a clear plastic sheet protector with sunscreen lotion.
2. Make several marble-sized balls from modeling clay.
3. Place a sheet of newspaper in a sunny location.
4. Use the balls of clay to suspend the sheet protector above half of the newspaper by placing the balls under the edges of the sheet protector. Be sure that air can move between the sheet protector and the newspaper.
5. After several hours in the sun, remove the sheet protector. Compare the color of the newspaper that was under the sheet protector to the paper that was not under it.

Conclusion:

Newspaper is bleached to make it white. As ultraviolet rays strike the paper, the radiation causes oxygen in the air to chemically react with the paper, turning it yellow. The sunscreen blocks much of the UV radiation thus slowing the reaction with the paper. You should see a clear outline of the sheet protector where the newspaper remains whiter than the paper around it. Ozone blocks much of the UV radiation, just as the sunscreen did.

Ozone depletion

Near the surface of the Earth, most of the oxygen in the air is O_2. This means that two oxygen atoms are bonded together. This is the oxygen that people and animals need to breathe. But in the upper atmosphere, much of the oxygen has combined to form ozone which is three oxygen atoms together, O_3. We cannot breathe ozone; it is poisonous when it is down near the surface, but ozone is critical for the survival of life on Earth. Ozone blocks much of the **ultraviolet radiation** (UV) from the sun.

We need some UV radiation to reach the surface of the Earth to help warm us up, but too much UV radiation damages plant and animal tissues. Have you ever spent too much time outside on a sunny day and gotten a sunburn? The tissues of your skin were damaged by the ultraviolet radiation from the sun. Ozone in the upper atmosphere helps to limit the amount of UV radiation that reaches the surface.

In the 1970s it was discovered that substances called **chlorofluorocarbons** or **CFCs** were causing problems. CFCs are molecules that were used in air conditioning and refrigeration, aerosol sprays, and plastic foams. It is believed that the CFCs, which are very stable molecules, rise into the upper atmosphere where the UV light breaks them apart. The chlorine then bonds with the ozone, breaking off one of the oxygen atoms and turning it into O_2. The O_2 cannot protect the Earth from UV radiation like ozone can. It is believed that if enough CFCs reach the upper atmosphere, enough ozone can be removed to cause problems to plants, animals, and humans. Therefore in the late 1980s CFCs were banned in most countries.

CFCs have been largely replaced with different substances called **hydrochlorofluorocarbons**, or **HCFCs**. The presence of hydrogen in the molecule makes it less stable so HCFCs react with other molecules in the atmosphere before they reach the ozone layer, thus preventing the chlorine from reaching the ozone most of the time. Although HCFCs are much better than CFCs, some molecules still reach the upper atmosphere where they break apart and the chlorine bonds with oxygen, breaking down ozone. Replacing CFCs with HCFCs has greatly slowed down the ozone depletion but has not completely stopped it. Therefore, most governments now require that HCFCs no longer be used and be replaced with **hydrofluorocarbons**, or **HFCs**, which are similar molecules but do not have any chlorine in them. Since HFCs do not have chlorine atoms, they cannot cause ozone depletion. This will allow the ozone that God created to continue to protect us from UV radiation.

31

Acid Rain

Does it burn?

Why is acid rain harmful?

Words to know:

acid rain   buffering capacity

One of the most damaging forms of pollution in recent decades has been acid rain. Pure water does not have any acids or bases, it is neutral. But water in the atmosphere combines with some of the carbon dioxide in the atmosphere to form carbonic acid. Thus, all rainwater is slightly acidic. Most plants and animals can use slightly acidic water. The problem is that when fossil fuels such as coal, oil, and gasoline are burned, they release sulfur dioxide and nitrogen oxides. When these substances get into the air, they react with the water to form sulfuric acid and nitric acid. These acids are much stronger than carbonic acid and form what is known as **acid rain**.

When acid rain falls on an ecosystem it can damage the plant and animal life there. If lakes become too acidic the water will kill hatching fish, insects, and amphibians. This impacts the whole food chain. Acid rain in the ground reacts with other substances to release aluminum and mercury. These metals can wash into the water and do more damage to the fish and other animals that live in the water.

Acid rain also damages plants. Some plants are weakened by the acid causing them to be more easily attacked by insects and more easily blown over by the wind. Acid rain can also react with

This forest was killed by acid rain.

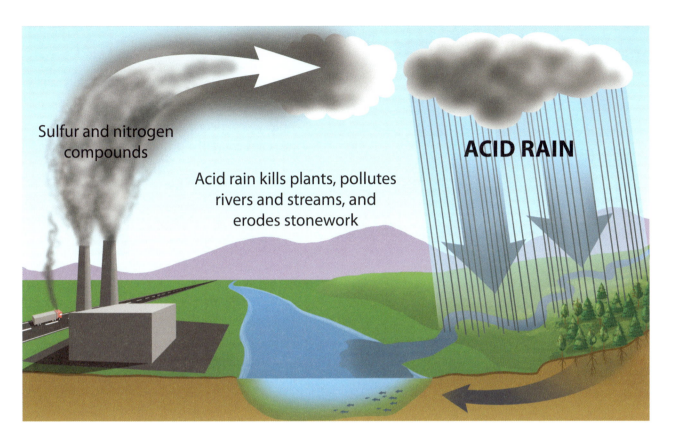

nutrients in the soil, washing them away so that plants do not have enough nutrients. This can cause plants to die. The red spruce tree is one plant that is especially sensitive to the effects of acid rain.

Although automobiles and the power plants that burn coal and oil can be found in many areas, acid rain is not a problem everywhere. Many soils have basic compounds in them which help to neutralize the acid in the rain water. This ability to neutralize acid is called the **buffering capacity** of the area. Many parts of the world have adequate buffering capacity to handle the acid rain. But areas such as the Adirondack and Catskill Mountains in New York have low buffering capacity so the acid rain is able to do considerable damage to ecosystems in that area.

Most countries have placed limits on the amount of sulfur dioxide and nitrogen oxides that power plants can release into the air. Thus many power plants have switched to low sulfur coal, wash their coal, and use scrubbers in their smokestacks to eliminate most of the sulfur and nitrogen compounds from the smoke they release. This has greatly reduced the amount of acid rain around the world.

Fun Fact

Acid rain does not necessarily have to be rain. Acid rain really refers to any acidic precipitation including snow, sleet, and hail, as well as rain.

Fun Fact

Not all acid rain is manmade. Erupting volcanoes release large amounts of sulfur and water vapor which combine to form acid rain.

LESSON 31 **Properties of Ecosystems**

What did we learn?

- Why is rain naturally slightly acidic?
- What is acid rain?
- What are the main causes of acid rain?
- What is buffering capacity?

Taking it further

- What are some ways to help reduce acid rain?
- If the buffering capacity were the same, would you expect acid rain to be more of a problem or less of a problem in areas with high population densities? Why?

Effects of acid rain

Purpose: To observe the effects of acid rain

Materials: two identical houseplants, two spray bottles, vinegar, water, "Acid Rain" worksheet

Procedure:

1. Place two houseplants in a sunny location. Label one plant with the word "water" and the other plant with "acid rain."
2. Fill one spray bottle with tap water and label it as "water." Fill a second spray bottle half full of tap water and then fill it the rest of the way with vinegar. Vinegar is an acid. Label this bottle as "acid rain."
3. Each day use the liquid in the spray bottles to water the plants. Squirt the same amount of liquid on the soil of each plant. The soil should be kept moist but not soggy.
4. Observe the plants each day for seven days. Record your observations on the "Acid Rain" worksheet.

Conclusion:

You should see a difference in the two plants after a week of watering one plant with tap water and the other with acid water. Some plants can handle acid better than others so the extent of the damage will vary with the type of plants you have.

Alternative energy sources

There are several forms of energy that do not create acid rain. These include solar, hydroelectric, nuclear, and wind energy. Also, electric and hydrogen cars are much less likely to produce nitrogen oxides than gasoline or diesel engine cars. Choose one of these alternative energy sources and find out all you can about it. Make a presentation of what you learned and share it with your family or class. Include what you learned in your notebook. Be creative in your presentation.

Wind and solar energy are clean energy alternatives.

32

Global Warming

Is it really heating up?

Is global warming man-made?

Words to know:

greenhouse effect

deforestation

global warming

Challenge words:

non-renewable resource

You can hardly watch the news or read a newspaper today without hearing about global warming or climate change. But you may not really understand what global warming is. Let's take a look at how the Earth's temperature might change.

Energy from the sun reaches the Earth. Some of the energy is reflected back into space by the atmosphere. This is because of the ozone and other molecules in the upper atmosphere. Some of the energy is absorbed by the atmosphere itself. And some of the energy reaches the surface of the Earth. The energy that reaches the surface of the Earth is either absorbed or reflected. Some of the reflected energy passes back through the atmosphere, but some of it is trapped by the gases in the atmosphere and some is reflected back to the surface. The gases in the atmosphere that trap energy are primarily water vapor and carbon dioxide. Water vapor is by far the most important gas and traps the majority of the heat. Carbon dioxide only accounts for about 5% of all the heat that is trapped in this way.

The trapping of solar energy is called the **greenhouse effect**. Have you ever been in a greenhouse? The glass or plastic panels trap solar energy, keeping the temperature warmer inside the greenhouse than it is outside. The greenhouse effect on Earth is very important. Without it, the average temperature on Earth would be about -100°F (-73°C). But because the gases in the atmosphere trap some of the sun's energy, the average temperature on Earth is a comfortable 50°F (10°C).

When the amount of energy that reaches the Earth is equal to the amount of energy that is reflected back into space, the Earth is in equilibrium, this means that the average temperature worldwide stays about the same. This does not mean

Fun Fact

Some people are trying to change the name global warming to global climate change to reflect not only changes in temperature, but changes in precipitation and other weather patterns.

that any particular area on Earth will be exactly 50 degrees all the time. It just means that the average is staying pretty much the same.

The concern is that in the last 130 years, the average surface temperature of the Earth has risen about 1.2°F, and thus the name **global warming**. Many scientists believe that this warming is primarily due to increased carbon dioxide in the atmosphere. They assert that the carbon being released by burning fossil fuels is the main cause of this global warming. They also claim that **deforestation**, the cutting of large areas of forest trees, is contributing to increasing carbon dioxide levels since plants use carbon dioxide for photosynthesis.

Some people who are concerned about global warming claim that if the current trend continues there could be severe consequences to global ecosystems. They claim that glaciers and ice sheets are melting at an alarming rate and that this could lead to coastal flooding and loss of habitat for polar bears. They claim that there are more severe droughts and more severe storms than in the past. They also claim that increased carbon dioxide in the oceans is causing the ocean to become more acidic. They also predict that more species will become extinct and diseases will spread. This is a very grim prediction of the future.

Politicians around the world have become very concerned about these predictions and have started passing laws to limit how much carbon dioxide people can put into the air. They are also passing laws to limit how many trees can be cut in the rainforests and requiring new trees to be planted. They believe that these changes are necessary to stop the dire consequences that are looming in the future. Many of these laws will be very harmful to people because it will require companies to change the way they do many things causing products to become much more expensive. This will be especially harmful if energy costs increase and poor people cannot afford to adequately heat their houses in the winter.

Are these laws really necessary? Should we be alarmed about global warming? Let's take a look at some of the facts.

First, not all scientists agree that global warming is a problem. Although the media makes it sound like every scientist on the planet thinks global warming is a major problem, this is simply not true; many scientists are urging caution, not panic. It is true that the average temperature of the Earth has increased by about 1.2°F in the past 130 years. It is also true that the amount of carbon dioxide in the atmosphere has increased by about 30% in the past 130 years. It is almost certain that much of the increase in carbon dioxide levels is due to the burning of fossil fuels and deforestation. However, not everyone agrees that this is a serious problem.

Some glaciers are melting. But others are growing. Many of the glaciers that are shrinking have been shrinking since the 1700s so their decrease is likely not related to the recent increase in global temperatures or to increased carbon dioxide levels. On the other hand, the Briksdal glacier in Norway has been growing at a rate of 200 feet (60 m) per year. In 2005 *National Geographic* reported that glaciers in the Himalayas were shrinking. But in 2006 the same magazine reported that glaciers in the Himalayas were growing. So it is unclear the true effect that global warming is having on the glaciers overall.

Many of the dire predictions that are circulating about global warming are based on computer models of what could happen. These computer models have not been able to accurately predict what we have seen in the past 25 years, so it is unrealistic to rely on them to predict what will happen in the next 25, or even 100 years. One of the major problems with the computer models is how they deal with the effect of water vapor on global temperatures. Most models say that the water vapor will magnify the effects of the carbon dioxide making the global temperature continue to rise. But experience has shown that when temperatures rise on Earth there is more evaporation which results in more clouds. The clouds reflect more of the sunlight thus cooling the surface of the Earth. This appears to be God's design for keeping the temperature of the Earth in check.

The Environmental Protection Agency, the organization in the United States most concerned with protecting the environment says, "Scientists are certain that human activities are changing the composition of the atmosphere, and that increasing the concentration of greenhouse gases will change the planet's climate. But they are not sure by how much it will change, at what rate it will change, or what the exact effects will be." The honest scientist will urge cautious objective research and not panic. There is simply not enough evidence to support most of the claims that you hear about global warming. Scientists continue to study other factors, including the amount of energy released by the sun, and to understand the very complex systems that control Earth's climate. More time is needed to study this important issue. In 1 Thessalonians 5:2, the Apostle Paul urges believers to "Test all things; hold fast what is good." This is what Christians must do in the global warming debate.

What did we learn?

- What is the greenhouse effect?
- Why is the greenhouse effect important on Earth?
- What is global warming?
- What do many scientists claim are the two main causes of global warming?

Taking it further

- What are some ways that people might reduce the amount of carbon dioxide they are putting into the atmosphere?
- Why is it inappropriate to panic about global warming?

The greenhouse effect

Purpose: To understand how the greenhouse effect raises temperatures

Materials: two thermometers, glass jar with a lid, "The Greenhouse Effect" worksheet

Procedure:

1. Place a thermometer inside a glass jar and seal the jar.
2. Place the jar in a sunny location. Place a second thermometer next to the jar.
3. Record the initial temperature of each thermometer on "The Greenhouse Effect" worksheet.
4. Record the temperature of each thermometer every 5 minutes for 30 minutes.

Conclusion:

The radiation from the sun passes through the glass of the jar. Some of the rays are absorbed by the glass and turned to thermal energy. This traps some of the energy inside the jar and increases the temperature of the air in the jar. This is the greenhouse effect and is essentially what happens to energy passing through the atmosphere and being absorbed by the surface of the Earth.

More alternative energy sources

You should be able to see by now that the use of fossil fuels, such as coal, oil, and natural gas, causes many problems. Not only are fossil fuels possibly linked to global warming, but they put many pollutants in the air, and are linked to acid rain. In addition, fossil fuels are considered **non-renewable resources**. This means that there is a limited amount of fossil fuels and they are not being replaced. This is because most fossil fuels were formed as a result of the Great Flood so they are not being formed in any large amount today.

It is important that we find alternatives to fossil fuels if we are to have reliable, clean energy in the future. Make a poster, encouraging people to use alternative energy sources. Include this poster in your notebook.

Carbon dioxide: a greenhouse gas

All scientists agree that carbon dioxide is a greenhouse gas and traps some of the sun's energy close to the surface of the Earth. You can demonstrate this by doing the following experiment. You are going to put air in one zipper bag and put carbon dioxide in a second bag. When you place both bags in a sunny location the gas in each bag will trap some of the heat from the sun. Would you expect the bag with just air to trap more or less heat than the bag with carbon dioxide in it?

Purpose: To demonstrate that carbon dioxide traps heat and is thus a greenhouse gas

Materials: Copy of "Carbon Dioxide: A Green House Gas" worksheet, two 1-gallon plastic zipper bags, two thermometers, 2-liter soda bottle, vinegar, baking soda

Procedure:

1. Label one zipper bag "Air" and label the second zipper bag "Carbon Dioxide." Complete the hypothesis on the worksheet.

2. Fill the "Air" bag with air by opening it wide and pulling it quickly through the air. Do not blow into the bag. It does not have to be very full. Place a thermometer in the bag and zip it shut.

3. Fill a 2-liter bottle ⅓ full of vinegar. Keep in mind that the space in the bottle above the vinegar is not empty it is actually full of air. You want to get rid of the air and replace it with carbon dioxide. To do this, pour 1 teaspoon of baking soda into the bottle. The reaction between the baking soda and vinegar produces carbon dioxide. Carbon dioxide is heavier than air so as it bubbles up, it pushes the air out of the bottle.

4. Now you want to trap some carbon dioxide in your second bag. To do this, close the zipper all except the last 2 inches. Pour another spoon of baking soda into the bottle and quickly place the bag over the top of the bottle. Wait for the bubbles to stop. This will push carbon dioxide into the bag. The bag does not need to be really full. Quickly remove the bag, slip a thermometer into the bag and seal it.

5. Record the temperature in each bag on your worksheet.

6. Place both bags side by side in a sunny location. Wait 15 minutes then record the final temperature in each bag on your worksheet and answer the questions.

Conclusion: You should find that the temperature in the bag with carbon dioxide was higher than the temperature in the bag with only air. This demonstrates that carbon dioxide traps more heat than just regular air. The atmosphere contains some carbon dioxide which contributes to the greenhouse effect. However, water vapor is the main greenhouse gas; carbon dioxide only accounts for about 5% of all warming from greenhouse gases.

33

What can you do?

How can I help?

What are the 3 Rs?

Challenge words:

polymer   virgin resin
recycled resin   photodegradable

Now that you have seen some of the negative impacts that human activity can have on the environment, you might want to do something to improve the situation. In fact, because God has given people the responsibility of caring for His creation, we should do what we can to be good stewards.

Many of the issues relating to man's impact on the environment are things that have to be dealt with by adults. Children cannot develop new technologies or enforce hunting laws. But you can do a small part to reduce the amount of energy and water that you use and reduce the amount of trash that you send to the landfill.

A simple way to help you plan ways to help the environment is to think of the 3 Rs: reduce, reuse, recycle. You can reduce the amount of water you use by not running the water when you brush your teeth and by taking a short shower rather than a long bath. You can reduce the amount of energy you use by replacing incandescent light bulbs with LED bulbs that use less energy while providing the same amount of light, and you can always turn off the lights when you leave a room. You can also reduce your energy usage by carpooling places with friends and neighbors or riding a bus or a bicycle.

Reusing is another way to help the environment. When you reuse something it does not go into the trash and a new one does not need to be made. This helps reduce the impact on the environment. What things can you reuse around your house? Can you fix something that is broken instead of buying a new one? Is there someone who can use things you no longer need? Give it away instead of throwing it away.

Finally, most communities have recycling programs. You can often sign up for curbside recycling for a small fee. You can place recyclable items such as steel and aluminum cans, glass containers, newspapers, and many plastics in a special bin. Then place the bin on the curb on recycling

Fun Fact

Americans drink an average of 167 liters of bottled water per person per year. Only about 23% of these bottles are recycled. Approximately 38 billion water bottles are sent to landfills each year.

LESSON 33 Properties of Ecosystems

🧪 Making a plan

Think about all the ways your family uses energy and water and think about all of the trash your family throws away. Use the ideas in this lesson and ask for ideas from your family to make a plan to help you Reduce, Reuse, and Recycle. Write out your plan on the "3 Rs of Conservation" worksheet and place it in a visible location so everyone in your family will remember to do them. Make a second copy of your plan and put it in your notebook.

day and the recycling company will pick them up and send them to companies that will use them to make new cans, bottles, and other items.

You can also recycle food scraps by starting a compost bin in your backyard. Fruit and vegetable peels and other food items, other than meat and dairy products, can be placed in a pile outside. Add strips of newspaper and grass clippings to the pile. Bacteria and worms will decompose these materials and turn them into compost, which can be used to fertilize your garden. Even as a kid, you can do a small part to take care of God's creation. ✾

Composting is a great way to recycle organic waste.

🧠 What did we learn?

- What are the three Rs of conservation?
- List two ways you plan to do each of these things.

🚀 Taking it further

- Why is it important to be concerned about how humans impact the environment?

Fun Fact

It actually takes more energy and raw materials to make a paper cup than it does to make a Styrofoam cup.

Plastic recycling

Plastic is something you use every day. Think about all the ways you use plastic. Your toothbrush, milk carton, refrigerator door handle, even your clothes are likely to be made from some sort of plastic. Plastic is useful because it can be made to be light and flexible, like the plastic wrap you use to cover your food or a grocery bag. Plastic can also be hard and strong like a coffee mug or door handle.

Plastic is made from petroleum, better known as oil. So in a sense it is a form of fossil fuel, although it is usually not used as fuel. The oil molecules are put together in such a way that they form long chains called **polymers**. Different types of plastics have different polymers. The type of plastic something is made from is usually marked on the bottom of the item with a number inside a triangle of arrows. This is a quick way to identify the type of plastic something is made from.

Many plastic items are semi-permanent parts of your life. Plastic furniture and rayon, nylon, and polyester clothes are things you are likely to keep around for a long time. Other plastics are meant to be used and thrown away, such as milk cartons or water bottles. These disposable plastic items are the ones that most people think of when they think of recycling. Products marked with a 1 or 2 on the bottom are usually picked up at curbside recycling. Plastic bags, which are a number 4, are often collected by grocery stores for recycling.

Most aluminum and glass containers that are recycled are used to make new containers. But most plastic that is recycled is made into something different. Soda bottles are often used to make carpet, T-shirts, car parts and plastic lumber. Some recycled plastic is used to make new bottles. It is too difficult and thus too expensive to recycle some types of plastic so some plastics are used as fuel for power plants. Remember, plastic is made from oil so it contains a fair amount of energy that is released as it burns.

There is a seven step process for recycling plastic. Once collected plastic reaches the recycling plant, it is first inspected and sorted. The plastic must all be the same type to be recycled together because different types of plastic melt at different temperatures and have different molecular structures. Next, the plastic is chopped into tiny flakes and washed. After this, the flakes are put into a flotation tank. Some plastic floats and other plastic sinks, so this is a second way to sort the different types of plastic.

The sorted flakes are then dried in a tumbler. After this they are melted at high temperature and pressure. The melted plastic is then pressed through a fine filter to remove any impurities. This turns the plastic into long strands. Finally, the strands are chopped into uniform sized pellets, called **recycled resin**. The pellets are then sold to manufacturers for various uses. In the past few years, several plastic recycling plants have closed because there has not been a good market for recycled plastic pellets. Pellets made directly from oil, called **virgin resin**, are often cheaper than recycled resin. However, the market for recycled plastic in China is expanding and much of the plastic that is sent to recycling plants in the United States is then shipped to China where it is made into new products.

One of the biggest concerns about putting plastics into landfills is that they do not break down or degrade quickly. So some manufacturers are trying to make biodegradable plastic, which will break down more quickly. Biodegradable plastic is made with 5% cornstarch or vegetable oil. It is believed that bacteria in the landfills will eat the cornstarch or vegetable oil, causing the plastic to lose its structure and break down. Other plastics are **photodegradable**, meaning they break down when exposed to sunlight. These plastics are strong for a short period of time, but lose their strength when exposed to sunlight for a few days or weeks. Many of the 6 ring plastic holders that hold soda cans together are made from photodegradable plastics.

Today only about 6.5% of all plastic is recycled. Another 7.7% is burned as fuel. This is a very small amount compared to the total amount of plastic that is being manufactured. The EPA has a goal that 25% of all American waste should be recycled. You can do your part in helping recycle the plastics that you use.

34

Reviewing Ecosystems: Final Project

Completing your book

In this book you have learned about the world that God created. You have learned about niches and habitats, food chains, and food webs. You have learned about various ecosystems and the plants and animals that live in each one. The beauty of how everything in nature works together is a testimony to God's care for His creation.

Now is the time to finish your notebook to display all that you have learned. Make a title page and a table of contents. Decorate the cover and make it look attractive. God's beauty is everywhere so your book should reflect that beauty.

Be sure to finish all of the pages that belong in your book. Review each lesson's activity to make sure that you included each item that should be in the book. Then, if you would like to add more pages for animals or ecosystems that you are particularly interested in, feel free to do so. The goal is that someone reading your book will appreciate the wonderful creation we call Earth.

Final projects

Choose one or more of the projects below to add to your notebook. Be creative and make your book a unique expression of what you have learned.

1. Make a diorama in a box of your favorite ecosystem. Take pictures and include them in the appropriate section of your notebook.
2. Cut pictures of plants and animals from magazines and make scenes of appropriate ecosystems. Only include plants and animals that are likely to share a particular ecosystem. Add your scenes to the appropriate sections of your notebook.
3. Paint a picture of how you imagine the Garden of Eden might have looked. Add your painting to the first section of your notebook.
4. Make a chapter page for each section of your notebook.
5. Take photographs of plants and animals from as many different ecosystems as you can visit. Add them to your notebook.

35

Conclusion

Appreciating our beautiful but cursed world

Thank God for this amazing planet.

When God created the world he pro-nounced it "very good" (Genesis 1:31). It was a perfect world without death. But man sinned and God cursed the Earth. This changed the way the world works. Death is now the rule and thorns and weeds grow rampant. Romans 8:22 tells us that the Curse did not just change life for men—all of nature was affected.

Even with the Curse, creation still declares the glory of God (see Psalm 19:1 and Romans 1:20–21). When we look at the diversity among all living things, the amazing interactions among plants and animals, and the interdependence of different species, we see the marks of design—they are not an accident. Your study of ecosystems should have helped to open your eyes to the wonder of God's creation.

God is concerned about everything in His creation. In Matthew 6:28–29 we read, "And why do you worry about clothes? See how the lilies of the field grow. They do not labor or spin. Yet I tell you that not even Solomon in all his splendor was dressed like one of these." Matthew 10:29 says, "Are not two sparrows sold for a copper coin? And not one of them falls to the ground apart from your Father's will." If God cares for the smallest animal and the plants of the field, we should care for them, too. We are to fulfill the commission given to Adam and Eve to take care of every living thing.

Most importantly, Matthew 10:31 goes on to say, "Do not fear therefore; you are of more value than many sparrows." Man is infinitely more valuable to God than any animal or plant He created. That is why Christ came to die for us. So we need to first dedicate ourselves to God, and then we can care for His creation as He intended.

 **A poem**

Write a poem expressing your love for the Creator and the beauty of His creation. Include your poem as the final entry in your Ecosystems Notebook.

Properties of Ecosystems — Glossary

Abiotic   Nonliving

Acid rain   Precipitation that has a higher than normal acid level

Adaptation   Physical trait or behavior due to inherited characteristics that gives an organism the ability to survive in a given environment

Algae bloom   Sudden rapid growth of algae

Alpine tundra   Tundra on high mountains

Antarctic tundra   Tundra below the Antarctic Circle

Arboreal   Living life primarily in trees

Arctic tundra   Tundra above the Arctic Circle

Atoll   Coral reef formed around a sunken volcano

Bactrian camel   Two-humped camel

Balance of nature   Condition in which the producers and consumers are in equilibrium

Barrier reef   Coral reef formed away from the shore

Beach   Shore where the ocean meets the land

Benthos   Plants and animals that live on the ocean floor

Biodegradable   Able to be decomposed by natural means

Biome   Many connected ecosystems that share a similar climate

Biosphere   Area of the Earth containing life

Biotic   Living

Buffering capacity   Ability of the soil to neutralize acid

Camouflage   How an animal blends in with its surroundings

Canopy   Roof of forest, formed by tops of mature trees

Captive breeding   Breeding of endangered species in captivity

Carnivore   Organism that eats only animals

Cave   Cavern in a mountain or underground

Chaparral   Area on hot dry mountain slopes with dense shrubbery

Chemosynthesis   Conversion of chemicals into food

Climate   The general or average weather conditions of a certain region

Cold desert   Desert with daytime temperatures below freezing in the winter

Commensalism   Relationship in which one species benefits and the other is unaffected

Community   All the populations living together in a given area

Competition   Relationship in which two species compete for scarce resources; can be harmful to both species

Coniferous forest/Boreal forest/Taiga   Ecosystem dominated by coniferous trees

Coniferous tree   Tree with needle-like leaves that bear seeds in cones

Consumer   Organism which obtains energy by eating other organisms

Coral reef   Formation built from exoskeletons of coral polyps

Deciduous forest   Ecosystem dominated by deciduous trees

Deciduous tree   Tree with broad, flat leaves that are shed in autumn

Decomposer   Organism that breaks down dead plants and animals into simpler elements

Decomposition   The act of breaking down dead tissues into simpler elements

Deforestation   Cutting of large numbers of trees without replacing them

Desert   Area that receives less than 10 inches of rain per year

Dromedary camel   One-humped camel

Dynamic equilibrium   The amount of debris deposited is equal to the amount removed

Ecology   Study of the interaction of the environment and living things

Ecosystem   Community of living things and the environment in which they live

Emergent layer   Top layer of a forest

Endangered   A species is in danger of becoming extinct

Ephemeral   Plant with an accelerated life cycle

Epiphyte   A plant that grows on another plant without harming it

Estivation   Summer sleep in which the animal's metabolism is very slow

Estuary   Ecosystem where fresh water flows into the ocean

Evergreen tree   Tree with needle-like leaves that are not shed in autumn

Extinct   There are no living specimens of a particular species

Fauna   Animals in an ecosystem
Floor   Ground layer of forest
Flora   Plants in an ecosystem
Food chain   Diagram of the flow of energy from one organism to another
Food web   Diagram showing the interconnection of the food chains within an ecosystem
Forest   Ecosystem dominated by trees
Fringing reef   Coral reef attached to land

Global warming   Increase in the average surface temperature of the Earth
Grassland   Ecosystem dominated by grass with few trees or shrubs
Greenhouse effect   The warming of the Earth due to trapped solar radiation
Guano   Bat or bird droppings

Habitat reduction   Destruction or elimination of natural habitat
Habitat   The environment in which an organism lives
Herb layer   Layer of forest containing grass, flowers, and other small plants
Herbivore   Organism that eats only plants
Hibernation   Winter sleep in which the animal's metabolism is very slow
Hot desert   Desert that does not experience freezing temperatures

Inter-tidal zone   The part of the shore that is covered with water at high tide and uncovered at low tide

Lake   Large body of freshwater
Law of conservation of mass   Mass/matter cannot be created or destroyed by any natural means; it can only change form

Mangrove forest   Estuary dominated by mangrove trees
Midnight/Aphotic zone   Layer of water that sunlight is unable to penetrate
Migration   Moving from one area to another for a particular season
Mirage   An optical illusion that reflects the sky onto the ground making it appear as water
Mutualism   Relationship in which both species benefit

Natural selection/Survival of the fittest   Ability of an organism to survive better than others of its kind because of a particular characteristic
Nekton   Animals that freely move throughout the ocean
Neutralism   Relationship in which two species do not significantly benefit nor harm one another
Niche   Role of an organism within its environment
Nocturnal   Active at night
Northern polar region   Area of Earth north of the Arctic Circle
Northern temperate zone   Area of Earth between Tropic of Cancer and Arctic Circle

Oasis   An area in the desert watered by a spring
Omnivore   Organism that eats both plants and animals
Overturn   Rapid exchange of cold and warm water regions within a lake
Oxygen cycle   Process through which oxygen is recycled

Pampas   Grasslands of South America
Parasitism   Relationship in which one species benefits and the other is harmed
Permafrost   Layer of ground that does not defrost even in summer
Photosynthesis   Chemical reaction in which water and carbon dioxide are changed into glucose and oxygen using sunlight
Phytoplankton   Microscopic aquatic organisms that perform photosynthesis
Plankton   Plants and animals that move with the ocean currents
Pollution   Any contaminant that harms an ecosystem
Pond   Lake that is too shallow to have an aphotic zone
Population   Number of a species in a given area
Prairie   Grassland of North America
Predator   Animal that hunts other animals
Prey   Animal that is hunted by other animals
Predator-prey feedback   Process in which a change in population among the prey produces a change in population among the predators and vise-versa
Producer   Organism which produces its own food

Respiration   Chemical reaction in which glucose and oxygen are changed into water and carbon dioxide
Riparian zone   Land along the banks of a river or stream
River/Stream   Freshwater that is flowing

Salt marsh   An estuary flooded and drained by seawater; dominated by rushes

Salt meadow   A meadow that may be flooded by seawater; dominated by sea grass

Savannah   Grassland of Africa

Scavenger   Organism that eats dead plants or animals

Semi-arid   Climate with more rain than a desert but less rain than temperate areas (usually 10–20 inches of rain per year)

Shrub layer   Layer of forest containing shrubs and other short plants

Snow line   Altitude above which the snow does not completely melt each year

Southern polar region   Area of the Earth south of the Antarctic Circle

Southern temperate zone   Area of Earth between Tropic of Capricorn and Antarctic Circle

Steppe   Grassland of Europe and Asia

Stomata   Tiny holes in leaves for exchange of gases and releasing water

Succulent   Plants with fleshy stems for storing water

Sunlit/Euphotic zone   Layer of water that sunlight is able to penetrate

Symbiosis   Close relationship between two different species

Territoriality   An animal's defense of its territory for breeding or other purposes

Tide pool   Area along shore that fills with water during high tide

Timberline   Altitude above which no trees can grow

Transpiration   The release of water from leaves

Tributary   Smaller stream or river that flows into a larger stream or river

Troglobites   Animals that spend their entire lives in caves

Troglophiles   Animals that like to live in caves but can be found living outside caves as well

Trogloxenes   Animals that visit caves but don't spend their whole lives there

Tropical rainforest   Forest in tropical region receiving more than 80 inches of rain per year

Tropical zone   Area of Earth between Tropic of Cancer and Tropic of Capricorn

Tundra   Treeless area with long cold winters and cool summers

Twilight/Disphotic zone   Layer of water that only a small amount of sunlight is able to penetrate

Understory   Layer consisting of immature trees and shorter species

Water cycle   Process through which water is recycled

Zooplankton   Microscopic aquatic consumers

Properties of Ecosystems — Challenge Glossary

Adaptive radiation   Development of several species from one common ancestral kind

Animal courtship   Animal behaviors performed to attract a mate

Biogeographic realms/Ecozones   Large areas of land separated by natural barriers

Bioluminescence   An organism's ability to produce its own light

Cambium   Area inside a tree where new bark and wood cells are produced

Carrying capacity   Maximum population an area can support

Chlorofluorocarbons/CFCs   Molecules used in refrigeration and plastics that degrade ozone

Climax ecosystem   Final, stable ecosystem

Coral bleaching   Condition in which coral expels its symbiotic algae

Dormant   An inactive, non-growing state

Dune system   Ecosystem that develops as you move inland from the shore

Echolocation   Use of high pitched sound waves to detect objects

Ecotone   Transitional area between two different ecosystems

Fire cue   Condition caused by fire which triggers seed germination

Genetically modified organism/GMO An organism whose DNA has been intentionally modified by man to produce a particular trait

Heartwood Center of tree; dead wood cells which provide support and strength
Hydrochlorofluorocarbons/HCFCs Molecules used to replace CFCs in many applications
Hydrofluorocarbons/HFCs Molecules similar to HCFCs without the chlorine atoms, used to replace HCFCs in many applications

Maritime forest Area inland from the shore containing shrubs and trees

Nitrogen cycle Process through which nitrogen is recycled
Non-renewable resources Natural resources that are not being replaced

Outer bark Protective outer layer of a tree trunk or branch

Papillae Tiny projections
Phloem/Inner bark Carries food down from leaves
Photodegradable Able to be decomposed by sunlight

Pioneer plants First plants to move into an area after a drastic change
Polymer Very long chain of molecules used to form plastics and other materials
Primary dune Area along the beach that is dominated by grass

Recycled resin Plastic pellets made from recycled plastic

Secondary dune Area inland from the primary dune that is covered with shrubs and grass
Succession Change over time from one ecosystem to another

Ultraviolet radiation/UV radiation Energy rays from the sun that can cause damage to plant and animal tissues

Virgin resin Plastic pellets made directly from oil

Watershed All the land drained by a particular river or body of water

Xylem/Sapwood Carries water and nutrients from roots to the rest of the tree

Properties of Ecosystems • 273

Properties of Atoms & Molecules

UNIT 1

Atoms & Molecules

1 Introduction to Chemistry
2 Atoms
3 Atomic Mass
4 Molecules

◊ **Identify** and **describe** the parts of an atom using diagrams.
◊ **Use** the periodic table to determine the characteristics of atoms.
◊ **Describe** the relationship between atoms and molecules.

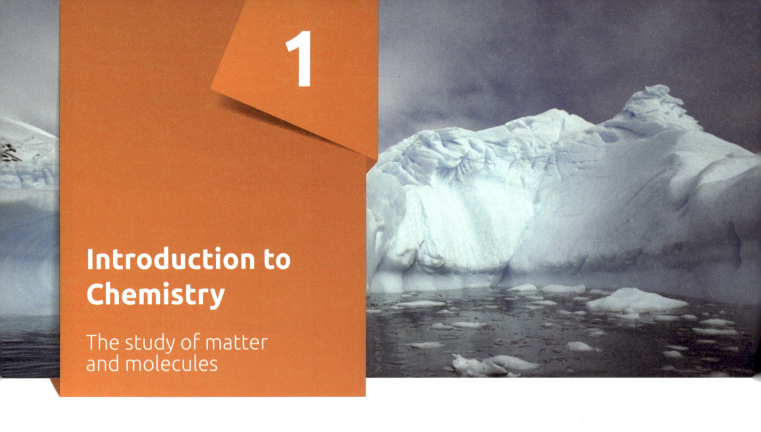

1

Introduction to Chemistry

The study of matter and molecules

What is chemistry?

Words to know:

chemistry

chemist

matter

Chemistry may sound like a big word and a difficult subject to study, but it's not. **Chemistry** is simply the study of matter, and **matter** is anything that has mass and takes up space. Some examples of matter are water, wood, air, food, paper, your pet skunk, or your little brother. So if you are interested in learning more about anything around you, then you are ready to learn about chemistry.

Chemists are scientists who study what things are made of, how they react to each other, and how they react to their environment. Chemistry is the study of the basic building blocks of life and the world.

In chemistry you will learn about atoms and molecules. You will learn about how substances combine to make other substances. You will find out how a substance changes form and you will discover that God created our world with such intricate designs that we may never fully understand how everything works.

God has established laws that govern how chemicals react and how matter changes. Many of these laws seem mysterious because they happen on an atomic level. Although these changes cannot be seen with the naked eye, the results of these laws can be seen all around us. As you study atoms and molecules you will begin to understand these laws and appreciate the beauty of God's design on the atomic level.

What did we learn?

- What is matter?
- Does air have mass?
- What do chemists study?

Taking it further

- Would you expect to see the same reaction each time you combine baking soda and vinegar?

Chemistry is fun

As you will learn in the upcoming lessons, some materials are very stable and do not change easily. Other materials are very reactive and easily combine with other substances to make a new substance.

Purpose: To see a chemical reaction

Materials: baking soda, drinking cup, vinegar

Procedure:

1. Place 1 teaspoon of baking soda in a drinking cup.
2. Pour 1 tablespoon of vinegar into the cup. Now watch the reaction!

Conclusion: Vinegar is an acid and baking soda is a base. Acids and bases easily combine together to form salts. In this reaction they also produce a gas. Can you guess what that gas might be? It is carbon dioxide.

Soda fountain

For an even more impressive reaction, you can make a Mentos® and diet soda fountain. This chemical reaction is very messy so this experiment must be done outside. This experiment happens quickly so you want to have everything ready before you start. Read through the directions below before you try the experiment so you know what to do.

Purpose: To make a diet soda fountain

Materials: 2-liter bottle of diet cola, heavy paper, tape, toothpick, Mentos® mints

Procedure:

1. Remove the cap from a 2-liter bottle of diet cola.
2. Make a tube to hold the mints: roll a piece of heavy paper into a tube that just fits around the mouth of the soda bottle. Tape the paper so it stays rolled up.
3. Use a toothpick to punch holes through the bottom of the tube just above the mouth of the bottle so that the toothpick goes through the tube and holds the mints in place.
4. Load up your tube with four or more mints.
5. Quickly remove the toothpick and step back so you don't get sprayed. You should see a fountain of soda. Be sure to clean up your mess when you are done.

Conclusion: This reaction is partially a chemical reaction and partially a physical reaction between the mints and the soda. Soda contains a gas called carbon dioxide. This gas is trapped between the liquid molecules. The mints have many tiny pits on their surfaces which allows the gas to collect very quickly and escape the liquid. There is also a chemical reaction between the mints and soda that further allows the gas to escape quickly producing a fountain of foam. Now, don't you think chemistry is fun?

2

Atoms

Basic building blocks

What are the basic building blocks of matter?

Words to know:

atom
proton
neutron
electron
nucleus
electron energy level
valence electron

Everything around you is made of matter. But what is matter made of? This is a question that has interested scientists for thousands of years. It is obvious that water is a different kind of substance than a rock and that a person is very different from a tree. But what makes each thing unique? As scientists considered this question, they began to try to separate and break down different substances to understand what they were made of. Eventually, scientists have discovered that everything in the universe is made of very small particles called atoms. **Atoms** are the smallest part of matter that cannot be broken down by ordinary chemical means. Atoms are so small that we cannot see them, even with the best microscope. But we can see how different types of atoms behave and see how they combine with other atoms.

Because atoms are so small, scientists have had to develop models to describe what an atom is like. Have you every played with a toy truck or airplane? That toy was a model of the real thing. It allowed you to see the basic parts of the vehicle, but it was not the same size or as complex as the real thing. In the same way, models of atoms help us to understand the basic parts of an atom, but they are not the same size or as complex as a real atom.

The earliest written ideas showing that matter was made of atoms come from the Greeks around 400 BC. The Greek scientists believed that matter was made of very small particles. But they did not try to describe those particles. Work on an actual atomic model did not really begin until the 1700s when experimental science became more popular. Early experiments showed that different atoms had different masses. In 1897 it was discovered that atoms consisted of electrically charged particles

Fun Fact

The models used to represent atoms do not accurately show the size relationship between the nucleus and the electrons. If the nucleus of the atom was the size of a tennis ball, the electrons would be orbiting about 1 mile away.

Neils Bohr

Ernest Rutherford

and that some particles in the atom were smaller than others. By 1911 a scientist named Ernest Rutherford discovered that atoms consisted of a positively-charged nucleus with negatively-charged particles whirling around it. And finally, Neils Bohr discovered that the electrons whirling around the nucleus had different energy levels.

All of these discoveries have helped in the development of the current atomic model. Today scientists describe an atom as having three parts: protons, neutrons, and electrons. **Protons** are positively charged particles. **Neutrons** are neutral; they do not have a positive or negative charge. And **electrons** are negatively charged particles. All of the parts of an atom are extremely small; however, electrons are much smaller than protons and neutrons. Protons and neutrons are approximately 1,800 times more massive than electrons.

The protons and neutrons in an atom are combined together in a tight mass called the **nucleus**. The electrons move very quickly around the nucleus. Some electrons orbit more closely to the nucleus than others. It is believed that the electrons in an atom occupy different levels, or distances, from the nucleus depending on how much energy they have. These levels are often referred to as the **electron energy levels**. The electrons that are in the level farthest away from the nucleus are called the **valence electrons**.

The model of a lithium atom below shows its nucleus containing four neutrons and three protons. It also shows three electrons orbiting the nucleus. Two electrons orbit closer to the nucleus, and the third electron orbits farther away, thus lithium has one valence electron.

The number of protons in the nucleus of an atom determines what kind of atom it is. If an atom loses or gains a neutron, or loses or gains an electron, it is still the same type of atom. But if the atom loses or gains a proton, it becomes a different type of material. Regardless of the number of neutrons or electrons that a lithium atom may have, a lithium atom always has three protons.

As research into the structure of atoms continues, scientists continue to gain more understanding of the complexity of the atom. It is believed that protons, neutrons, and electrons are made of smaller particles called quarks, but because of their extremely small size, they are difficult to study. This complexity continues to amaze scientists and shows God's mighty hand in the design of the universe.

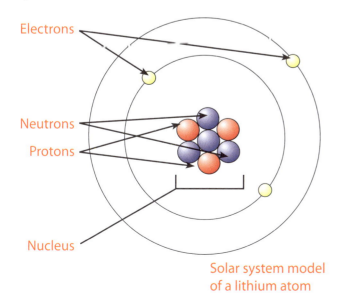
Solar system model of a lithium atom

What did we learn?

- What is an atom?
- What are the three parts of an atom?
- What electrical charge does each part of the atom have?
- What is the nucleus of an atom?
- What part of the atom determines what type of atom it is?
- What is a valence electron?

LESSON 2 **Properties of Atoms & Molecules** • 281

🚀 Taking it further

- Why is it necessary to use a model to show what an atom is like?
- On your worksheet, you colored neutrons blue and protons red. Are neutrons actually blue and protons actually red in a real atom?

🧪 Atomic models

Color the parts of the atoms on the "Atomic Models" worksheet.

🏅 Energy levels

In all atoms, the lowest electron energy level, that which is closest to the nucleus, is filled with electrons first. If a level is full, electrons will occupy the next level. The first level can hold up to two electrons. If an atom has more than two electrons, two electrons will orbit close to the nucleus, and the others will begin to fill the next layer. The following chart shows how many electrons scientists believe that each energy level can hold. Note, however, that although scientists believe that certain levels can hold more electrons, the highest number of electrons that has been determined to be in any energy level is 32. Other than in level one, an inner energy level does not have to be full before the next level begins to fill up. For example, even though the fourth level can hold 32 electrons, the fifth level begins filling up after the fourth level has only 8 electrons in it.

The electrons orbiting in the energy level farthest from the nucleus of an atom are called valence electrons. The lithium model on the previous page shows that lithium has one valence electron. Valence electrons play a vital role in how an element behaves. Neutral atoms are ones in which the number of electrons equals the number of protons. However, atoms can gain or lose valence electrons. The ability to gain, lose, or share valence electrons is what allows atoms to bond with each other to form new substances.

Look at the periodic table of the elements on page 294. The small numbers in the bottom of each box show the electron configuration for each element. For example, look at the box for lithium (Li), element number 3. The numbers at the bottom of the box are 2, 1. This means there are 2 electrons in the first energy level and 1 electron in the second energy level. This corresponds to the model earlier in this lesson.

Let's look at another example. Look at Potassium (K), which is element number 19. The numbers for the electron configuration are 2, 8, 8, 1. This means that there are 2 electrons orbiting close to the nucleus. There are 8 electrons orbiting in the next level out. There are 8 electrons orbiting in the third level out, and there is 1 electron in the outermost layer. We will learn more about why these electron configurations are so important as we learn about how atoms bond with each other.

Complete the "Energy Levels" worksheet to help you better understand how electrons are distributed in atoms.

Chart of Maximum Electrons in Each Energy Level

Energy Level	Maximum Number of Electrons
1	2
2	8
3	18
4	32
5	50
6	72
7	98

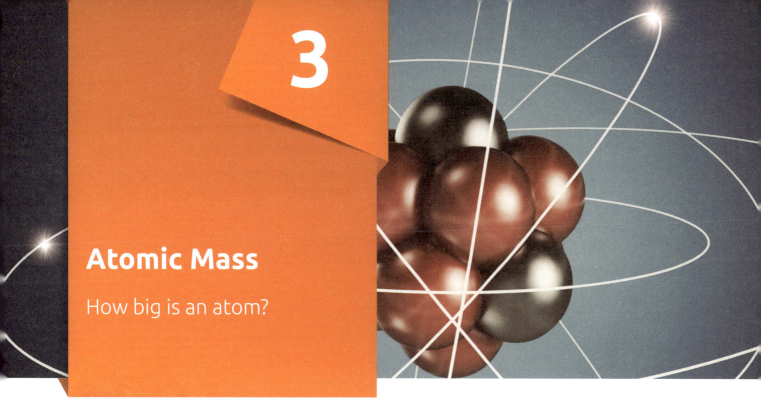

3

Atomic Mass

How big is an atom?

How do you measure an atom's mass?

Words to know:

atomic number   atomic mass
mass number   atomic mass unit (amu)

Challenge words:

isotope

As you learned in the previous lesson, an atom consists of three parts: protons, neutrons, and electrons. You also learned that the number of protons determines the type of element the atom will be. Therefore, the number of protons in an atom is called the **atomic number**. Hydrogen, which only has one proton, has an atomic number of 1. Oxygen, which has eight protons in its nucleus, has an atomic number of 8. The atomic number of an atom is very useful in identifying the type of atom.

The number of electrons in an atom typically equals the number of protons. However, an atom may lose or gain electrons. When this happens it is called an ion. You'll learn about ions in Lesson 11.

The mass of an atom is also an important characteristic to know about the atom. The mass of an atom is determined by the number of protons and neutrons in the atom. Electrons are so small and contribute such a tiny amount to the mass that their mass can be ignored. The **mass number** or **atomic mass** of an element is found by adding the number of protons and the number of neutrons in the atom. For example, hydrogen has only one proton and no neutrons so its atomic mass is one. Oxygen, which usually has eight protons and eight neutrons, has an atomic mass of 16.

On the other hand, if you are given the atomic number and atomic mass for an element you can figure out how many protons, electrons, and neutrons that element has. The atomic number given at the top of the square on a periodic table tells you how many protons the element has. The number of electrons is equal to the number of protons. Then to calculate the number of neutrons you subtract the number of protons from the atomic mass.

Because the mass of a proton or neutron is so small, it would not make sense to measure an atom's mass in grams. Therefore a special unit has been defined for measuring the mass of an atom. This unit is an **atomic mass unit**, or **amu** (it is also called *unified atomic mass unit* and abbreviated as u). An amu is defined as ¹⁄₁₂ the mass of a

LESSON 3   **Properties of Atoms & Molecules • 283**

carbon atom. A carbon atom has six protons and six neutrons, and thus has an atomic mass of 12 amu. Protons and neutrons have nearly identical masses, so for most applications an amu can be used to describe the mass of either type of particle. The mass of an electron is 1,800 times smaller than that of a proton or neutron, so we usually say its mass is negligible—it can be ignored.

What did we learn?

- What are the three particles that make up an atom?
- What is the atomic number of an atom?
- What is the atomic mass of an atom?
- How can you determine the number of electrons, protons, and neutrons in an atom if you are given the atomic number and atomic mass?

Taking it further

- What does a hydrogen atom become if it loses its electron?
- Why are electrons ignored when calculating an element's mass?

Learning about atoms

Complete the "Learning About Atoms" worksheet.

Isotopes

If you look up the atomic mass for an element on the periodic table, you find that most are not listed as whole numbers. This does not mean that an atom has only part of a proton or part of a neutron. It means that even though all atoms of a particular element have the same number of protons, some have different numbers of neutrons. Each variety of atom is called an *isotope* of that element.

To help you understand this better, let's look at carbon. On most periodic tables, the atomic mass of carbon is listed as 12.01 amu. This is an average mass for carbon atoms that naturally occur. All carbon atoms have 6 protons or they would not be carbon atoms. Ninety-nine percent of all carbon atoms also have 6 neutrons giving them a mass of 12 amu. However, a little less than 1% of all carbon atoms have 7 neutrons so they have a mass of 13 amu; and a very small percentage of carbon atoms have 8 neutrons and a mass of 14 amu. When you average the mass for all isotopes of carbon the average mass is 12.01 amu. Some elements have only a few known isotopes while others have many. Chlorine has 24 known isotopes, but only two are common. The most common has an atomic mass of 35 amu, and the second most common has a mass of 37 amu.

Now that you have a better understanding of what atomic number, atomic mass, and isotopes mean, use a periodic table to fill in the "Understanding Atoms" worksheet. Round the atomic mass from the periodic table to find the most common number of neutrons, in other words the most common isotope, for each element. You may use the periodic table on page 24.

Madame Curie
1867–1934

SPECIAL FEATURE

Atoms, isotopes, and radioactive decay are the things that Marie Curie is best known for. But who was she? When she was born in 1867, her last name was Sklodowska. She was born in Warsaw, Poland, an area that was controlled by the czar of Russia at that time. Because of their pro-Polish leanings, Marie's parents lost their jobs and her father was forced into a series of lower academic posts. The family was poor and took in students as boarders to help pay the rent. When Marie was eight, her oldest sister died, and less than three years later her mother also died. This made the family turn to each other for strength.

As they were growing up, their father read them classics and exposed them to science. Marie graduated from high school at the age of 15, at the top of her class. But, women were not allowed to attend the University of Warsaw, so Marie went to a floating university, named so because it changed locations frequently to hide it from the Russian authorities. This schooling was not a high quality education, so Marie made a pact with her older sister. Marie would work and help send her sister to Paris for medical school, and then her sister would work to send her to school. For two years, Marie worked as a teacher and then, to make more money, she became a governess and sent as much money as she could to her sister.

Eventually, Marie went back home and because of her father's new job she was able to leave for Paris in 1891, when she was 24 years old. There, life was hard for her. In the winters, she would wear every piece of clothing she had to keep herself warm. And sometimes she would get so absorbed in her studies she would forget to eat and she would pass out. In later years, Marie said it was very common for the Polish students to be poor.

In Paris, Marie found that she was ill prepared for college. She was lacking in both math

and science, plus her technical French was behind where it needed to be. She overcame this by working hard, and it paid off. She finished first in her class for her master's degree in physics and second in her class in math the following year.

In 1894 Marie began sharing lab space with a man named Pierre Curie. Their work drew them together, and in July 1895 the two were married in a simple ceremony. In September of 1897 their first child was born—a baby girl. Pierre's father delivered the baby. A few weeks later Pierre's mother died and Pierre's father, along with Marie, Pierre, and baby Irene moved into a house together. Marie kept working in the lab and found her father-in-law to be the perfect babysitter.

About six months after Marie and Pierre were married, a German scientist name Wilhelm Conrad Roentgen discovered X-rays. He discovered that X-rays could travel through wood and flesh and produce an image on photographic paper. A few months later a French physicist named Henri Becquerel discovered that uranium produced similar rays.

These discoveries prompted Marie and Pierre to start working with uranium. They soon discovered other materials that also emitted strong rays and they called this characteristic *radioactivity*. One element they discovered was polonium, named for Marie's home country of Poland, but the most important radioactive material discovered by the Curies was radium. It wasn't long before radium was in demand. In cheap novels, it was touted as "a magical substance whose rays could cure all ills, power wondrous machines, or destroy a city at one blow." This obviously was quite an exaggeration; however, the damaging effects radioactivity has on tissues was soon used on cancer cells. These damaging effects also took their toll on both Marie and her husband. Pierre developed sores on his body and was constantly fatigued. Marie lost 20 pounds and her fingertips were scarred from the radiation, but they had no knowledge of the long-term effects. While they noted Pierre's loss of good health and the severe pains he experienced, they did not link this to their work.

In 1903 both Pierre and Marie were invited to England to be honored for their work at the Royal Institution. Because it was not customary for women to speak there, Lord Kelvin showed his support for Marie by sitting next to her as her husband gave his speech. Later, when Pierre was nominated for the Nobel Prize in Physics for his and Marie's discovery of radium, he said it would be a travesty if his wife was not also included; so she was. In 1911 Marie received a second Nobel Prize, this one in chemistry, for the discovery of the atomic mass of radium.

After Pierre was killed by a horse-drawn wagon in 1906, Marie continued to carry on their work. A little while later, she was offered and accepted her husband's academic post at the Sorbonne, becoming the first woman to teach at this prestigious French college. Over the next few years, with the help of some wealthy friends and the French government, she was able to found the Radium Institute where research into the uses of radium in treating cancer and other illnesses was to be conducted.

When war came to France in 1914, the Radium Institute was complete, but Marie had not moved in yet. The other researchers who worked there were drafted to fight the Germans, and Marie also wanted to help. She knew X-rays could help save soldiers' lives by showing the doctors where the bullets or shrapnel were located, and they could see how the bones were broken. So she helped design 20 radiology vans to be taken into the field to treat the wounded. Since no one else was trained to use the X-ray equipment, Marie learned how to drive and she and her very mature 17 year old daughter, along with a doctor, made the first trip to the front lines in the fall of 1914. By 1916 Marie was training other women to work in the 20 mobile units and at the 200 stationary units.

After the war, Marie went back to work at the Radium Institute, and between 1919 and her death from leukemia in 1934, the Institute published 483 works, including 31 papers and books by Marie. Both of her daughters also achieved distinction. Irene and her husband won a Nobel Prize, and her daughter Eve was recognized for her writings. But the Curies will continue to be best known for their discovery of radioactivity.

4

Molecules

Putting atoms together

How do atoms combine?

Words to know:

molecule
compound
diatomic molecule

Atoms seldom exist by themselves. Instead, most atoms bond with other atoms. A group of chemically connected atoms is called a **molecule**. How atoms will connect with each other is determined by the number of valence electrons each atom has. Atoms do not usually react if they have eight valence electrons. This is considered a full outer layer and the atom will be stable and unlikely to bond with other atoms. There are only six elements that naturally have eight valence electrons and are thus stable by themselves. All other atoms will try to combine with other atoms to create a full outer layer of electrons.

Some atoms combine with other atoms of the same element. For example, oxygen atoms are seldom found by themselves. Instead, two oxygen atoms usually combine together. When two atoms of the same material combine to form a molecule the result is called a **diatomic** (meaning *two atoms*) **molecule.** Oxygen is the most common diatomic molecule, but several other gases exist as diatomic molecules as well, including hydrogen, nitrogen, and fluorine.

When atoms of different elements combine together to form a molecule, it is called a **compound**. The most common compound on earth is water, which is a combination of two hydrogen atoms and one oxygen atom. God has created the elements in such a way that multiple atoms can combine together in nearly innumerable ways. So far, over three million compounds have been identified.

When atoms combine to form compounds, the new substance that is formed has completely different characteristics from the elements that form it.

A model of the diatomic molecule nitrogen, N_2

LESSON 4   **Properties of Atoms & Molecules • 287**

🧪 Understanding molecules

To help you better understand the differences between atoms, diatomic molecules, and compounds, complete the "What Am I?" worksheet.

For example, sugar is formed from carbon, hydrogen, and oxygen atoms. Yet the sweet compound we use in so many baked goods has no resemblance to carbon, which is what diamonds are made of, or hydrogen or oxygen, which are both colorless gases.

It is also important to understand the same groups of elements can combine in different amounts to form different substances. For example, if two hydrogen atoms and one oxygen atom combine they form water. But if two hydrogen atoms and two oxygen atoms combine together, they form hydrogen peroxide, which is a clear liquid, but is very different from water. Similarly six carbon atoms, twelve hydrogen atoms, and six oxygen atoms can combine together to form glucose which is a simple sugar. But if only one carbon atom, two hydrogen atoms, and one oxygen atom combine together it forms formaldehyde, which is a substance used to preserve dead animals. So you can see that even though the same types of atoms combine together, they form very different substances if the numbers of atoms are different.

Sugar is formed from carbon, hydrogen, and oxygen, yet it bares no resemblance to carbon.

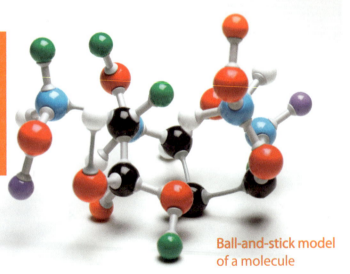

Ball-and-stick model of a molecule

🧠 What did we learn?

- What is a molecule?
- What is a diatomic molecule?
- What is a compound?

🚀 Taking it further

- What is the most important factor in determining if two atoms will bond with each other?
- Table salt is a compound formed from sodium and chlorine. Would you expect sodium atoms and chlorine atoms to taste salty? Why or why not?

🏅 Molecule puzzle pieces

Atoms bond together primarily based on the configuration of their valence electrons. Atoms like to have eight electrons in their outermost layer. So an atom with six valence electrons will bond easily with an atom with two valence electrons. It will also bond with two atoms that each have one valence electron. This is why water is made up of one oxygen atom, which has six valence electrons, and two hydrogen atoms which each have one electron. Together they make eight electrons in the outermost layer.

Check your understanding of this idea by using the "Molecule Puzzle Pieces" to try to form molecules.

Cut out each piece of the "Molecule Puzzle Pieces" worksheet. The dots on each piece represent the number of valence electrons that each element has. For example, the oxygen piece (O) has six dots because it has six valence electrons and the sodium piece (Na) has only one dot because it has only one valence electron. Atoms are stable when they have eight valence electrons with the exception of hydrogen and helium. Hydrogen and helium only need two valence electrons to be stable. Try fitting the various pieces together to form stable molecules.

Try the following combinations:

- Two hydrogen pieces together—this forms the diatomic molecule H_2
- Two hydrogen pieces and one oxygen piece—this forms a water molecule
- One sodium and one chlorine—this forms table salt

Neon cannot combine with any of the other pieces because it already has a full set of eight electrons in its outer level. Neon is on the far right side of the periodic table. Elements on the far right side are unlikely to bond with other elements.

LESSON 4 **Properties of Atoms & Molecules**

UNIT 2

Elements

5 Periodic Table of the Elements
6 Metals
7 Nonmetals
8 Hydrogen
9 Carbon
10 Oxygen

◊ **Describe** how the periodic table can be used to classify elements.
◊ **Distinguish** between the properties of metals and nonmetals.
◊ **Explain** the importance of hydrogen, carbon, and oxygen.

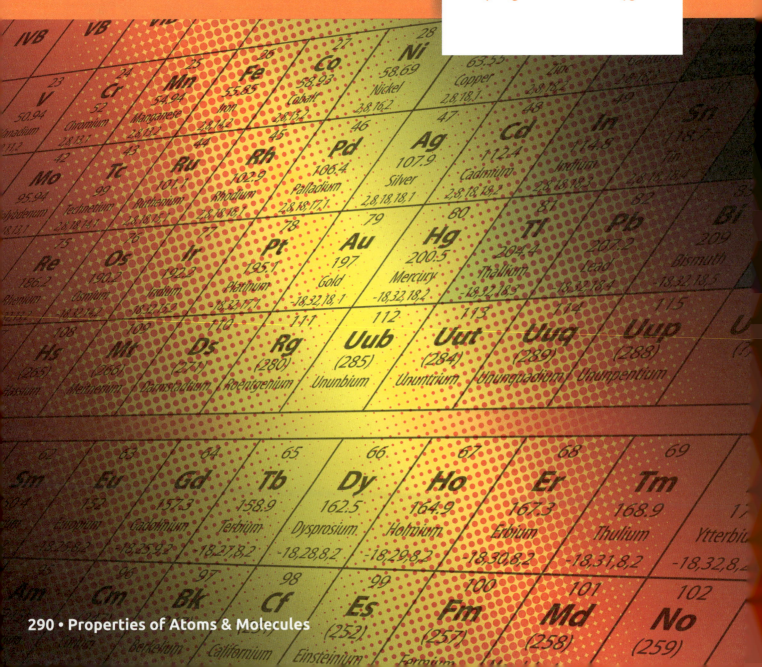

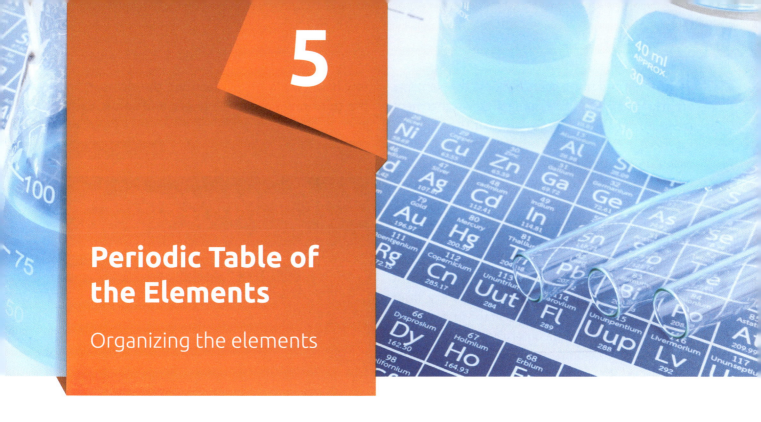

5

Periodic Table of the Elements

Organizing the elements

How are elements classified?

Words to know:

transition elements period
family

There are 98 naturally occurring elements. And scientists have been able to produce another 20 elements in the laboratory. Each of these elements has unique properties. Keeping track of information for all of the elements that have been discovered can be a big task. Fortunately for you, much of that information can be found in a periodic table of the elements. Periodic tables sometimes have additional information, but all periodic tables have at least the following information. Each square will show the name of the element, the symbol for the element, the atomic number, and the atomic mass. Remember that the atomic number is equal to the number of protons in the nucleus and the atomic mass is equal to the average of the number of protons plus the number of neutrons for all isotopes of the element. The symbol is one to three letters derived from the name of the element.

Some of the symbols for the names of the elements are obvious while others may not make much sense to you. It is obvious why hydrogen has the symbol H and oxygen is O. However, what would you expect the symbol to be for lead? Did you think it should be L or Le? That would make sense, but the symbol for lead is Pb. This may seem strange, but the Latin name for lead is *plumbum* so lead is designated at Pb. Several other symbols on the periodic table are also derived from their Latin names, while we are familiar with the English names for those elements. Look at the periodic table and see which other symbols do not match the English version of the element's name. Did you find Na for sodium, K for potassium, Fe for iron, and Au for gold? If you look up the Latin names for these elements you will find that the symbols make more sense.

You know that the periodic table gives information about all of the known elements on earth. Look

Fun Fact

The Latin term for lead is *plumbum*. Because pipes that carried water were originally made of lead or plumbum, the people who installed the pipes were called plumbers.

LESSON 5 **Properties of Atoms & Molecules • 291**

at the table and try to figure out how the elements are arranged on the table. Some of the arrangement is easy to figure out. It is obvious that the elements are listed in order of their atomic numbers. It starts with number 1 and goes up to number 118. But it might not be as obvious why certain elements are in each column or each row. The arrangement of the electrons in an atom determines which row and which column it will be in.

The elements are put into columns according to the number of electrons in the outermost layer, those that are orbiting farthest away from the nucleus. All of the elements in the first column have one electron in the outer layer. The elements in the second column have two electrons in the outer layer. The columns in the middle are called **transition elements** and have special rules for how they are arranged. These are the elements in the yellow squares below. If you skip over them, to the column labeled IIIA, you find elements that have three electrons in their outermost layer and so on.

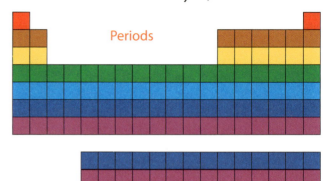

Periods

row have two layers of electrons, the elements in the third row have three layers, and so on. Each

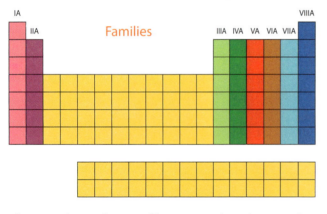

Families

Elements in a column will react to other elements in a similar way as other elements in that column. The elements in a column are called a **family** or group.

The elements are put into rows according to how many layers of electrons they have. Hydrogen and helium are the only elements in the first row because they are the only elements that have only one layer of electrons. The elements in the second

Fun Fact

When lighter-than-air ships were first designed, many were filled with hydrogen. Hydrogen is the lightest known element, thus making the ships float easily. However, after a terrible fire in the *Hindenburg* airship in 1937, it was decided that hydrogen was too dangerous. It was believed at the time that since hydrogen has only one electron and is highly reactive, that the fire was caused by lightning, or some other spark, causing the hydrogen to react with oxygen leading to the tragedy. After this tragic accident, it was decided to change from hydrogen to helium. Helium is a noble gas. It is very stable and does not react easily with other elements, yet it is still much lighter than air and allows blimps to be used in many areas still today.

Many researchers today believe that the fire in the *Hindenburg* was not a result of the hydrogen gas. It is now believed that the aluminum powder varnish used to coat the outside of the ship built up a huge electrostatic charge and caught fire when the electricity was discharged. They believe this fire would have started regardless of the gas used inside the ship.

🧪 Using the periodic table

You will become more comfortable with the periodic table of the elements as you use it more, so now you have a chance to use it. Use the periodic table on page 294 to answer the questions on the "Learning About the Elements" worksheet.

292 • Properties of Atoms & Molecules   LESSON 5

row in the periodic table is called a **period**, hence the name periodic table.

Once you understand more about how the elements are arranged on the periodic table, you will be able to tell a lot about each element just by where it is on the table.

What did we learn?

- How many valence electrons do the elements in each column have?
- What four pieces of information are included for each element in any periodic table of the elements?
- What do all elements in a column on the periodic table have in common?
- What do all elements in a row on the periodic table have in common?

Taking it further

- Atoms are stable when they have eight electrons in their outermost energy level. Therefore elements from column IA will react easily with elements from which column?
- Elements from column IIA will react easily with elements from which column?

Synthetic elements

Twenty of the elements on the periodic table do not occur naturally; they have only been produced by scientists in a laboratory or as a result of a nuclear explosion. These elements are called synthetic elements. Elements 99–118 are all synthetic. Nine other elements have been discovered in trace amounts in nature but were first observed by scientists who were making elements in a laboratory. All synthetic elements are very unstable. The nucleus of a synthetic element quickly breaks apart in a process called radioactive decay. The time for half of a sample to decay is called the half-life of that element. Synthetic elements have a half-life of between a microsecond and 1 year. Most synthetic elements only exits for a very short period of time.

Synthetic elements are created in one of two ways, either through a nuclear reaction such as a nuclear bomb explosion or in a nuclear reactor, or by experiment using a particle accelerator, which causes atoms to collide with each other at very high speeds. When two atoms collide, sometimes their nuclei fuse together. This forms a new element because the new nucleus now has more protons. Because these elements exist for such a short period of time, it is difficult to verify their existence, and there has been some controversy over whether some elements have actually been discovered or not.

The first element to be created in a laboratory was technetium in 1936. This is element 43 on the periodic table. It was created to fill a hole or space left in the periodic table. Scientists knew that there had to be an element that fit in that spot on the table, but no one had ever found any naturally occurring. Since that time trace amounts of technetium have been discovered on earth, and it is believed that it exists in larger amounts in red giant stars.

The first truly synthetic elements were discovered in 1952 when scientists were analyzing the results of the detonation of the first hydrogen bomb. Two new elements were discovered and were given the names einsteinium and fermium. They are numbers 99 and 100 on the periodic table. Synthetic elements are usually named for famous scientists or the locations where they were discovered. Although it is believed that elements 113, 115, 117, and 118 have been created, they have not completed the confirmation process and remain officially unnamed.

Naturally occurring elements usually have several different isotopes so the atomic mass listed on the periodic table is an average of these various masses for each element. However, since there are no naturally occurring isotopes for synthetic elements, there is no way to find an average mass. So the mass listed on the periodic table is the mass for the isotope with the longest half-life.

Because synthetic elements do not last very long they do not have any practical uses other than for scientific experimentation. Many other naturally occurring elements, such as technetium, which occur only in trace amounts, also have no commercial or other practical uses.

Periodic Table of the Elements

	IA	IIA												IIIA	IVA	VA	VIA	VIIA	VIIIA
1	1 **H** 1.008 Hydrogen 1																		2 **He** 4.0026 Helium 2
2	3 **Li** 6.941 Lithium 2,1	4 **Be** 9.012 Beryllium 2,2												5 **B** 10.81 Boron 2,3	6 **C** 12.01 Carbon 2,4	7 **N** 14.01 Nitrogen 2,5	8 **O** 16 Oxygen 2,6	9 **F** 19 Fluorine 2,7	10 **Ne** 20.18 Neon 2,8
			IIIB	IVB	VB	VIB	VIIB	VIII	VIII	VIII	IB	IIB						
3	11 **Na** 22.99 Sodium 2,8,1	12 **Mg** 24.31 Magnesium 2,8,2											13 **Al** 26.98 Aluminum 2,8,3	14 **Si** 28.09 Silicon 2,8,4	15 **P** 30.97 Phosphorus 2,8,5	16 **S** 32.07 Sulfur 2,8,6	17 **Cl** 35.45 Chlorine 2,8,7	18 **Ar** 39.95 Argon 2,8,8
4	19 **K** 39.1 Potassium 2,8,8,1	20 **Ca** 40.08 Calcium 2,8,8,2	21 **Sc** 44.96 Scandium 2,8,9,2	22 **Ti** 47.9 Titanium 2,8,10,2	23 **V** 50.94 Vanadium 2,8,11,2	24 **Cr** 52 Chromium 2,8,13,1	25 **Mn** 54.94 Manganese 2,8,13,2	26 **Fe** 55.85 Iron 2,8,14,2	27 **Co** 58.93 Cobalt 2,8,15,2	28 **Ni** 58.69 Nickel 2,8,16,2	29 **Cu** 63.55 Copper 2,8,18,1	30 **Zn** 65.39 Zinc 2,8,18,2	31 **Ga** 69.72 Gallium 2,8,18,3	32 **Ge** 72.59 Germanium 2,8,18,4	33 **As** 74.92 Arsenic 2,8,18,5	34 **Se** 78.96 Selenium 2,8,18,6	35 **Br** 79.9 Bromine 2,8,18,7	36 **Kr** 83.8 Krypton 2,8,18,8
5	37 **Rb** 85.47 Rubidium 2,8,18,8,1	38 **Sr** 87.62 Strontium 2,8,18,8,2	39 **Y** 88.91 Yttrium 2,8,18,9,2	40 **Zr** 91.22 Zirconium 2,8,18,10,2	41 **Nb** 92.91 Niobium 2,8,18,12,1	42 **Mo** 95.94 Molybdenum 2,8,18,13,1	43 **Tc** 99 Technetium 2,8,18,14,1	44 **Ru** 101.1 Ruthenium 2,8,18,15,1	45 **Rh** 102.9 Rhodium 2,8,18,16,1	46 **Pd** 106.4 Palladium 2,8,18,17,1	47 **Ag** 107.9 Silver 2,8,18,18,1	48 **Cd** 112.4 Cadmium 2,8,18,18,2	49 **In** 114.8 Indium 2,8,18,18,3	50 **Sn** 118.7 Tin 2,8,18,18,4	51 **Sb** 121.8 Antimony 2,8,18,18,5	52 **Te** 127.6 Tellurium 2,8,18,18,6	53 **I** 126.9 Iodine 2,8,18,18,7	54 **Xe** 131.3 Xenon 2,8,18,18,8
6	55 **Cs** 132.9 Cesium -18,18,8,1	56 **Ba** 137.3 Barium -18,18,8,2	57 **La** 138.9 Lanthanum -18,18,9,2	72 **Hf** 178.5 Hafnium -18,32,10,2	73 **Ta** 180.9 Tantalum -18,32,11,2	74 **W** 183.9 Tungsten -18,32,12,2	75 **Re** 186.2 Rhenium -18,32,13,2	76 **Os** 190.2 Osmium -18,32,14,2	77 **Ir** 192.2 Iridium -18,32,15,2	78 **Pt** 195.1 Platinum -18,32,17,1	79 **Au** 197 Gold -18,32,18,1	80 **Hg** 200.5 Mercury -18,32,18,2	81 **Tl** 204.4 Thallium -18,32,18,3	82 **Pb** 207.2 Lead -18,32,18,4	83 **Bi** 209 Bismuth -18,32,18,5	84 **Po** (209) Polonium -18,32,18,6	85 **At** (210) Astatine -18,32,18,7	86 **Rn** (222) Radon -18,32,18,8
7	87 **Fr** (223) Francium -18,32,18,8,1	88 **Ra** (226) Radium -18,32,18,8,2	89 **Ac** (227) Actinium -18,32,18,9,2	104 **Rf** (261) Rutherfordium	105 **Db** (262) Dubnium	106 **Sg** 262.94 Seaborgium	107 **Bh** (264) Bohrium	108 **Hs** (265) Hassium	109 **Mt** (266) Meitnerium	110 **Ds** (271) Darmstadtium	111 **Rg** (280) Roentgenium	112 **Cn** (285) Copernicium	113 **Nh** (284) Nihonium	114 **Fl** (289) Flerovium	115 **Mc** (289) Moscovium	116 **Lv** (293) Livermorium	117 **Ts** (294) Tennessine	118 **Og** (294) Oganesson

58 **Ce** 140.1 Cerium -18,20,8,2	59 **Pr** 140.9 Praseodymium -18,21,8,2	60 **Nd** 144.2 Neodymium -18,22,8,2	61 **Pm** (145) Promethium -18,23,8,2	62 **Sm** 150.4 Samarium -18,24,8,2	63 **Eu** 152 Europium -18,25,8,2	64 **Gd** 157.3 Gadolinium -18,25,9,2	65 **Tb** 158.9 Terbium -18,27,8,2	66 **Dy** 162.5 Dysprosium -18,28,8,2	67 **Ho** 164.9 Holmium -18,29,8,2	68 **Er** 167.3 Erbium -18,30,8,2	69 **Tm** 168.9 Thulium -18,31,8,2	70 **Yb** 173 Ytterbium -18,32,8,2	71 **Lu** 175 Lutetium -18,32,9,2
90 **Th** 232 Thorium -18,32,18,10,2	91 **Pa** 233 Protactinium -18,32,20,9,2	92 **U** 238 Uranium -18,32,21,9,2	93 **Np** (237) Neptunium -18,32,22,9,2	94 **Pu** (244) Plutonium -18,32,24,8,2	95 **Am** (243) Americium -18,32,25,8,2	96 **Cm** (247) Curium -18,32,25,9,2	97 **Bk** (247) Berkelium -18,32,26,9,2	98 **Cf** (251) Californium -18,32,28,8,2	99 **Es** (252) Einsteinium -18,32,29,8,2	100 **Fm** (257) Fermium -18,32,30,8,2	101 **Md** (258) Mendelevium -18,32,31,8,2	102 **No** (259) Nobelium -18,32,32,8,2	103 **Lr** (262) Lawrencium -18,32,32,9,2

Legend:
- Alkali metals
- Alkali-earth metals
- Transition metals
- Poor metals
- Metalloids
- Nonmetals
- Noble gases
- Hydrogen nonmetal

Key:
- Atomic number
- Symbol
- Atomic Mass
- Name
- Electron structure by energy level

Example: 12 **Mg** 24.31 Magnesium 2,8,2

Note: the lowest electron levels are not shown for rows 6 and 7, instead they are indicated by a -, which means 2, 8.

Development of the Periodic Table

SPECIAL FEATURE

The periodic table of the elements, which is probably the most important tool a chemist has, is a table that lists some of the most important properties known about each element. But like all tools, someone had to invent it. So how did this useful tool come into existence?

In the early 1800s many scientists tried to find relationships between the various elements, but this was a daunting task, somewhat like sorting the pieces of a jigsaw puzzle without knowing what the picture should be. In 1866 an English scientist named John Newlands began sorting the then-known elements according to their atomic masses. He believed that the properties of the elements repeated every eighth element and he called this the law of octaves. However, there were many elements that did not fit into this pattern and eventually it was discarded. Only three years later, in 1869, a Russian chemist named Dmitri Mendeleev came up with what is considered by most people to be the first periodic table of elements.

Dmitri Mendeleev was born in Siberia, Russia, in 1834, and was the youngest of 14 children. About the time Dmitri finished high school, his father died and his mother moved to St. Petersburg, Russia, where she worked hard to earn the money needed to send Dmitri to college. This sacrifice paid off, not only for Dmitri, but also for all scientists to follow.

After college, Mendeleev began to catalog all the data he could find on the 63 elements known at the time. He was sure that the elements had repeating or "periodic" properties. He wrote the properties of each element on a card, and then he arranged the cards according to their similar properties and in order of their increasing atomic masses.

A draft of Mendeleev's periodic table

Dmitri Mendeleev

Mendeleev found that this method worked very well for most elements, and for the few that did not fit, he did a very unusual thing; he moved the element over one space and left a hole. He was convinced that where holes appeared, elements would be found in the future to fill these holes. He was so sure of this that he even predicted what the properties of the elements would be. And over the next several years, these predicted elements were indeed discovered.

The shape of the table has changed several times as scientists have discovered new elements, but the concept developed by Mendeleev has remained the same. The biggest advance to the periodic table after Mendeleev was in 1914, when Henry Moseley decided to arrange the elements according to their atomic numbers instead of their atomic masses. This arrangement solved some of the problems in Mendeleev's table and led to the periodic table we use today.

Properties of Atoms & Molecules

6

Metals

Silver and gold have I none ...

What are the properties of metals?

Words to know:

malleable
ductile
reactive
metalloid
semiconductor

Challenge words:

noble metals
poor metals
reactivity series

What do you think of when you hear that something is made of metal? Do you think of something hard, heavy, strong, and shiny? That would be a pretty good description of most metals. Do you think of cars, washing machines, silverware, or coins? Those are just a few of the many uses for metal.

Three quarters of all elements on earth are metals. Most are hard, strong, and heavy. However, a few pure metals are weak or soft. The majority of metals have the following six characteristics:

1. Silvery luster on the surface
2. Solid at room temperature (except for mercury, which is a liquid)
3. **Malleable**—can be hammered into shape
4. **Ductile**—can be drawn into a wire
5. Good conductor of electricity
6. Tend to be very **reactive**—easily combines with other elements

Most metals have one to three valence electrons. They easily give up their electrons, which is why they are reactive with many other elements. This is also why metals are good conductors of electricity. Because of their unique characteristics, metals are used in a variety of ways. Their strength and malleability allow them to be used for making cars and trucks, bridges, appliances, aluminum siding, and soda pop cans. Metal's conductivity of electricity is why metal is used for electrical wiring in nearly every house and building in America.

As you go across the periodic table from left to right, the elements become less metallic; this means that they are less malleable, less ductile, and less conductive. The dividing line between the metals and the nonmetals is the diagonal line of elements shaded dark green on the periodic table.

Conducting electricity

Purpose: To demonstrate the electrical conducting ability of metals

Materials: copper wire, flashlight, battery, electrical tape (or duct tape)

Procedure:

1. Cut two 12-inch pieces of copper wire. Strip about 1 inch of plastic off of each end of the wires.
2. Remove the light bulb and battery from a flashlight.
3. Using electrical tape or duct tape, attach one end of a wire to the positive terminal of the battery and the other end of the wire to the side of the metal contact on the light bulb.
4. Tape one end of the second wire to the negative terminal of the battery.
5. Touch the other end of the second wire to the end of the light bulb and watch the bulb light up.

Conclusion: Electrons are flowing from the battery, through the first wire, through the light bulb, through the second wire, and finally back to the battery. Because copper conducts electricity so well, it is used for most electrical wiring in buildings.

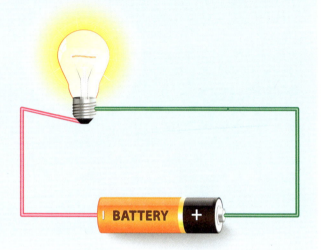

These seven elements are called **metalloids**. They act partially like metals and partially like nonmetals. The metalloids are called **semiconductors** because they conduct electricity only under certain circumstances. This is an important characteristic for the technological world. Semiconductors are used to provide low power yet high speed in electronic devices used in many products such as computers, digital watches, and cell phones.

Fun Fact

Manganese is the only metal that is not silvery, ductile, or malleable, which may make it seem like a nonmetal. However, manganese acts like a metal when it is alloyed, or added to other metals.

What did we learn?

- What are the six characteristics of most metals?
- How many valence electrons do most metals have?
- What is a metalloid?

Taking it further

- What are the most likely elements to be used in making computer chips?
- Is arsenic likely to be used as electrical wire in a house?

LESSON 6 **Properties of Atoms & Molecules** • 297

Reactive metals

Metals can be grouped according to how reactive they are. Based on what you have learned so far, which metals would you expect to be the most reactive? If you said the metals in the first column, you would be correct. The alkali metals are the most reactive metals. Sodium and potassium are both very reactive with air and water as well as other metals. Therefore, they must be stored in oil to keep them from reacting with the air.

The alkali-earth metals, those elements in column IIA, are also highly reactive, but not as reactive as the alkali metals. Because these metals react easily with other metals they readily form compounds. Calcium, one of the alkali-earth metals, is found in large amounts in the crust of the earth in the form of calcite, which is the main ingredient in limestone. Calcium is also found in seashells in the form of calcium carbonate. Alkali metals and alkali-earth metals are seldom found in their pure form because they are so reactive.

All of the metals in the center of the periodic table are called transition metals. Transition metals are what you might consider more typical metals. These are the metals that are used in many of the applications we mentioned earlier in this lesson. A few of the transition metals, those to the far right of the group, including silver, gold, copper, palladium, and platinum are often grouped together as the noble metals. Do you remember which column was called the noble gases? The noble gases are the elements in the far right column. They are the gases that do not easily react with other elements. Similarly, the **noble metals** do not react easily with other metals and are often found in their pure form in the crust of the earth.

Finally, the metals to the right of the transition metals, including aluminum, gallium, indium, tin, antimony, lead, and bismuth are called the **poor metals**. These metals are colored light green on the periodic table in lesson 5. These metals are generally more reactive than the transition metals but less reactive than the alkali and alkali-earth metals. Poor metals are very soft and usually are not very useful by themselves. However, when mixed with other elements, they become very useful.

In order to give people a good idea of how reactive various metals are, scientists have developed the "**Reactivity Series**" which lists a selected group of metals from most reactive to least reactive. This list helps people know what to expect from several common metals when they come in contact with various substances. The most reactive metals will react with water, acids, oxygen and many other substances. These metals are at the top of the list and are found in columns IA and IIA. Metals that react with acids but do not react with water are next on the list. These metals are mostly found in columns IIA, IIIA and IVA. Many of the transition metals (IB – VIIB) react only with oxygen and are near the bottom of the list. Finally, metals that do not easily react with anything (the noble metals) are at the bottom of the list.

Reactivity Series

- Potassium
- Sodium
- Calcium
- Magnesium
- Aluminum
- Zinc
- Iron
- Tin
- Lead
- Copper
- Silver
- Gold
- Platinum

7

Nonmetals

The rest of the elements

What are the properties of nonmetals?

Words to know:
halogens
noble gases
inert

The vast majority of the elements on earth are metals or metalloids. There are only 18 elements that are nonmetals. These include hydrogen, plus the 17 elements to the right of the metalloids. Although hydrogen is listed in the left column of the periodic table because it has only one valence electron, it often acts more like the elements in column VIIA because it only needs one electron to have a full outer shell and is classified as a nonmetal.

Nonmetals have very different characteristics from metals. They generally do not have a silvery luster or shiny appearance. Because the nonmetals need only one to three electrons to fill their outer shells, they do not easily give up electrons but share or gain electrons when they combine with other elements. Because they do not give up electrons, nonmetals are poor conductors of electricity. At room temperature, some nonmetals are solid, some are liquid, but most are gases. Those that are solid are usually brittle and shatter easily.

The top four elements in column VIIA are called **halogens**. These are very reactive elements. These elements can be very dangerous in large quantities, but in small quantities they are very useful. Chlorine is added to drinking water and swimming pools to kill bacteria. Fluorine (in the ionic form of fluoride) is added to drinking water and toothpaste to prevent tooth decay. Iodine can also be used to kill germs and is an essential nutrient in our diets. Because hydrogen is so reactive, it is sometimes grouped with the halogens.

The elements in column VIIIA are called the **noble gases** because they do not easily combine with any other element. Because these elements have eight

Chlorine is added to swimming pools to kill bacteria.

LESSON 7 **Properties of Atoms & Molecules • 299**

electrons in their outer shells, they are very stable. They are referred to as **inert** gases because they do not react. Their inability to react makes noble gases very useful for certain applications. Noble gases are sometimes used to fill a space instead of air to prevent a reaction from occurring. For example, in the process of making semiconductor chips, the space around the circuitry may be filled with argon gas instead of air to prevent a reaction from occurring.

Noble gases do not easily react with other elements. However, they have another special characteristic that makes them very useful. When a noble gas is in an enclosed container and an electrical current is passed through it, the gas turns into a plasma

The noble gas in a neon sign glows brightly when high voltage is applied.

and a colored light is given off. You have probably seen this phenomenon many times without realizing it. Most of the lighted signs that look like

🧪 Protecting your teeth

As we mentioned earlier, fluoride is often added to drinking water and toothpaste to help prevent tooth decay. Fluoride reacts with the calcium in your tooth enamel to make it stronger. The shell of an egg is made of calcium carbonate so it will react with fluoride in a similar way to your teeth. Do the activity below to see how these two elements react with each other. This experiment will take several days.

Purpose: To demonstrate how calcium carbonate and fluoride react together

Materials: fluoride toothpaste, two cups, uncooked egg, nail polish or permanent marker, vinegar

Procedure:

1. Squirt a bunch of fluoride toothpaste into a cup.
2. Make a small mark on one side of an uncooked egg with nail polish or a permanent marker.

3. Push the marked side of the egg into the toothpaste until half of the egg is covered. Let the egg sit in the toothpaste for 3 days.
4. At the end of 3 days, remove the egg and gently wash off the toothpaste and dry the egg.
5. Place the egg in an empty cup.
6. Pour enough vinegar in the cup to completely cover the egg. Observe what happens.
7. Allow the egg to sit in the vinegar for 8 to 12 hours. At the end of this time, carefully remove the egg. It will be very fragile on one side so be very careful.
8. Use a pen to gently tap each side of the egg.

Questions/Conclusion:

- What happened to the egg when you added the vinegar? You should see small bubbles coming off of half the egg shell.
- Did the bubbles form on the half with the mark or on the unmarked half? The bubbles will be forming on the half of the egg that did not have any toothpaste (the unmarked half).
- What do you think is causing the bubbles? Vinegar is an acid. This acid reacts with the calcium in the egg shell and creates carbon dioxide. This is what forms the bubbles. Acid in your mouth reacts with your teeth. If you have acid in your mouth too long, the reaction will make small holes in your teeth. These holes are called cavities.
- When you tapped the egg shell, what did you notice about the difference between the two sides of the egg? The side that soaked in the toothpaste should be much harder than the side that did not have the toothpaste. This is how toothpaste helps to make your teeth stronger so they do not react as easily with the acids in your mouth.

glass tubes that have been formed into the shape of words or symbols (like the sign on the previous page) are referred to as neon signs. When these tubes are filled with neon gas, the electrical current produces a red/orange colored glow. Not all lighted signs are neon signs. Other noble gases will produce different colors of light. This same trait of inert gases is used in plasma ball toys and plasma TV screens.

What did we learn?

- What are some common characteristics of nonmetals?
- What is the most common state, solid, liquid, or gas, for nonmetal elements?
- Why are halogens very reactive?
- Why are noble gases very non-reactive?

Taking it further

- Hydrogen often acts like a halogen. How might it act differently from a halogen?
- Why are balloons filled with helium instead of hydrogen?

Incandescent light bulbs

Halogens and noble gases both play important roles in the light bulb industry. A regular incandescent light bulb is a glass bulb with a tungsten filament in the middle. Electricity flows through the filament and heats it up. As it gets hot, as much as 4,500°F (2,500°C), the filament begins to glow and gives off light. Some of the tungsten gets hot enough to evaporate.

The bulb is filled with a noble gas, usually argon. The noble gas prevents the tungsten molecules from reacting with anything, which helps the filament last longer than it would if the bulb were filled with air.

One disadvantage of the incandescent light bulb is that it gives off a large amount of heat as well as light. This is considered an inefficient use of electricity. So scientists have developed other types of light bulbs that give off the same amount of light but use less electricity to do it.

One type of more efficient light bulb is the halogen bulb. A halogen bulb is a special kind of incandescent light bulb that contains a filament surrounded by a small quartz envelope. This envelope is much smaller than the glass bulb of a regular light bulb. Quartz is used because the heat given off by the filament would melt glass.

The envelope is filled with a halogen gas. As you just learned, halogens are very reactive. When the tungsten on the filament vaporizes, it reacts with the halogen gas inside the tube. This new molecule is heavy and falls back onto the filament. The tungsten is then redeposited on the filament. This helps the filament last much longer than it does in a regular

Incandescent, halogen, florescent, and LED bulbs

incandescent light bulb. This design also allows the filament to burn at a much higher temperature, which gives off more light for the amount of electricity. Because it burns at a much higher temperature, a halogen bulb is still very hot, but it lasts much longer than a regular incandescent bulb and produces more light with less electricity.

In this lesson you already learned that noble gases are used in neon lights. And now you know that noble gases and halogen gases are used in incandescent lights as well. The special properties of halogen and noble gases allow us to see in the dark. However, incandescent lights turn most of the energy they use into heat rather than light so other sources of light have been developed. Mercury vapor is used in fluorescent lights, and semi-conducting materials are used in LED lights to provide more efficient and longer lasting light bulbs. By understanding the various properties of the different elements, scientists are able to find just the right material for each purpose.

LESSON 7 **Properties of Atoms & Molecules** • 301

8

Hydrogen
Very reactive

What is special about hydrogen?

Words to know:
reduction

hydrogenation

dehydrogenation

Challenge words:
hydrogen fuel cell

It's the first element listed in the periodic table and it's the smallest and simplest atom. What is it? It's hydrogen. Hydrogen has an atomic number of 1 because it has one proton and no neutrons in its nucleus. It has one electron in orbit around the nucleus. This is the simplest possible atom. Hydrogen needs only two electrons to make it stable, and since it already has one, it needs only one more. Therefore, hydrogen is often classified as a halogen, because it reacts like a halogen. However, hydrogen can also give up one electron, so it sometimes acts like an alkali metal.

Hydrogen is the lightest element in the universe. If you had a swimming pool full of hydrogen, all the molecules together would only weigh about two pounds. Hydrogen has no smell, taste, or color. At normal room temperature and pressure, hydrogen is a gas. Hydrogen's boiling point is -423.17°F (-252.87°C) and its freezing point is -434.45°F (-259.14°C).

Hydrogen is the most abundant element in the universe. Hydrogen is believed to be the main element comprising the sun, as well as Jupiter and Saturn. Nearly 90% of all atoms in the universe are believed to be hydrogen atoms. Yet on earth, hydrogen is only the tenth most abundant element. God made the earth different from other planets, with additional elements necessary for life being more abundant than hydrogen.

Most of the hydrogen on earth does not exist as hydrogen gas. Most of the hydrogen is combined with other elements to form compounds. The most common compound containing hydrogen is water. Hydrogen is also found in sugars, amino acids, proteins, cellulose, and fossil fuels such as oil and gasoline. And hydrogen can combine with nitrogen to form ammonia.

Fun Fact
About 1 out of every 6,000 hydrogen atoms has a neutron in its nucleus.

Because hydrogen is so reactive it has many uses. It combines explosively to form water, H2O. This makes liquid hydrogen and liquid oxygen ideal as rocket fuel. Hydrogen is also being explored as an alternative form of energy for cars. Some hybrid cars now have engines that can use either compressed hydrogen or gasoline to power them.

Hydrogen is used in many chemical processes as well. Hydrogen can be used to remove oxygen from metal oxide ores in a process called **reduction**. A process called **hydrogenation** forces hydrogen molecules through a substance to change its molecular structure. For example, vegetable oil is hydrogenated to become margarine and crude oil is hydrogenated to produce gasoline. **Dehydrogenation** is the process that removes hydrogen atoms from a substance.

What did we learn?

- What is the most common element in the universe?
- What is the atomic structure of hydrogen?
- What is the atomic number for hydrogen?
- Why is hydrogen sometimes grouped with the alkali metals?
- Why is hydrogen sometimes grouped with the halogens?

Taking it further

- Why is hydrogen one of the most reactive elements?
- Margarine contains only partially hydrogenated oil. What do you suppose fully hydrogenated oils are like?

Hydrogenation

Hydrogenation is a process where hydrogen is added to vegetable oil at high temperature, forcing the hydrogen to bond with the oil molecules. This process causes the oil to become thicker. This allows vegetable oil to become margarine. When peanut butter is hydrogenated the peanut oil stays mixed into the peanut butter. Hydrogenated or partially hydrogenated foods are very common.

Purpose: To see which foods contain hydrogenated products

Materials: vegetable oil, margarine, peanut butter, crackers, cookies, dry soup, other pre-packaged meals

Procedure:

1. Read the list of ingredients for vegetable oil, margarine, and peanut butter.
2. Compare the thickness of each of these substances. How does the thickness of the margarine and the peanut butter compare to that of the oil? Which products are hydrogenated?
3. Look at other food labels for hydrogenated oils. You may be surprised at how many products have these substances. Possible places to look include crackers, cookies and other snack foods, dry noodle soups, and many pre-packaged meals.

Hydrogen fuel cells

Hydrogen plays a vital role in society today, but it is likely to become even more important in the future. Today nearly all vehicles are powered by gasoline or diesel fuel, both of which come from petroleum. However, **hydrogen fuel cells** are being developed, which could power these vehicles using hydrogen instead of gasoline.

Hydrogen fuel cells combine hydrogen gas and oxygen gas to form water. This process releases electricity, which is used to power an electrical motor in the vehicle. This process has many advantages over the current gasoline engine. It produces less air pollution and reduces dependence on foreign oil. It is likely that cars in the future may use this type of technology.

Today there are several obstacles to switching from gasoline to hydrogen. First, there needs to be a cost effective way to obtain the hydrogen. We can't just go find a pocket of hydrogen in the ground and pump it out like we do petroleum for gasoline.

Hydrogen is very reactive and is almost always chemically bonded with other elements. Therefore, the hydrogen must be separated from the other atoms before it can be used to power vehicles.

Second, there needs to be a system set up for drivers to get more hydrogen when they are traveling. Now, a driver can fill up his gas tank nearly anywhere, but there are few hydrogen stations available.

Third, hydrogen is difficult to store and transport. Hydrogen is very light and very small so it must be stored in a container that will keep it from escaping. Also, it must be compressed or cooled so that a large amount will fit in a small space inside the car.

As scientists and engineers work on these problems, it is likely that hydrogen fuel cell cars will become more readily available in the future. To learn more about how hydrogen fuel cells work, visit How Stuff Works or other Internet sites on hydrogen fuel cells.

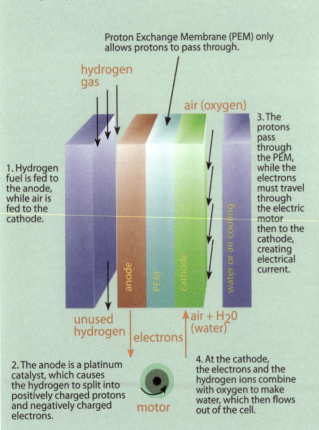

A London bus that runs on hydrogen fuel cells

9

Carbon
Graphite and diamonds

What is special about carbon?

Words to know:
carbon cycle

Challenge words:
allotrope
buckminsterfullerene
buckyballs
nanotechnology
carbon nanotubes

Carbon is one of the most important and interesting elements on earth. It can exist as a soft slippery powder called graphite. It can also be found in the form of a diamond, which is the hardest substance on earth. How can the same atoms form such very different substances? It depends on how the atoms are arranged. The carbon atoms in graphite line up in long chains that easily slip over each other. But the carbon atoms in diamond are arranged in a lattice network or crystalline structure that holds each atom tightly in place (see the diagrams later in this lesson). How the carbon atoms line up is greatly affected by temperature and pressure.

The atomic number of carbon is 6. Carbon has six protons and usually has six neutrons in its nucleus. It also has six electrons. Two electrons are in the inner shell and four electrons are in the outer shell. Since elements are most stable when they have eight electrons in their outer shell, carbon needs to either lose four or gain four electrons. This is not easy, so instead, carbon shares its electrons with other elements to form what are called covalent bonds. In this way, carbon will combine with many other elements to form many different compounds.

Carbon is one of the most important elements in all living things. Therefore, carbon compounds are called organic compounds. All plant and animal cells are made from organic compounds. Because carbon is essential to all living things, God has designed a way for carbon to be recycled in what is called the **carbon cycle**. First, plants absorb carbon in the form of carbon dioxide gas from the air. This carbon dioxide is used in the photosynthesis process to form sugar. Next, animals eat the plants containing sugar and absorb the carbon through digestion. Much of the carbon is released back into the atmosphere through respiration when the animal breathes, exhaling carbon dioxide. Some carbon remains in the animal's body. When an animal

LESSON 9 **Properties of Atoms & Molecules • 305**

Drawing the carbon cycle

Draw a picture demonstrating the carbon cycle. Be sure to include plants performing photosynthesis, animals eating plants and exhaling carbon dioxide, and animals and plants decaying. You may also want to include coal being formed in the earth and/or being mined and burned to return carbon to the air. Finally, draw arrows showing which direction the carbon is moving.

dies, its body decays and the carbon enters the soil. Finally, bacteria and fungi in the soil absorb the carbon from the soil, convert it into carbon dioxide, and release it into the air to begin the cycle again.

Many plants are not eaten by animals, but this does not mean that the carbon in those plants is lost. When a plant dies, it decays and the carbon enters the soil. Again, bacteria and fungi absorb the carbon and release it into the air as carbon dioxide. Also, some plants that have been buried under a large amount of mud or rock and have experienced great pressure have turned into coal. When coal is mined and then burned, it releases carbon dioxide back into the air to be used by plants again. So you can see that God designed a wonderful way to allow carbon atoms to be used over and over again to sustain life on earth.

What did we learn?

- What is the atomic number and atomic structure of carbon?
- What makes a compound an organic compound?
- Name two common forms of carbon.
- What is one by-product of burning coal?

Taking it further

- How does the carbon cycle demonstrate God's care for His creation?
- What is the most likely event that caused coal formation?
- What would happen if bacteria and fungi did not convert carbon into carbon dioxide gas?

Examining carbon

Purpose: To examine a carbon sample

Materials: candle, ceramic plate, knife

Procedure:

1. Hold a ceramic plate about 5 inches above a burning candle.
2. Slowly lower the plate until a black film forms on the bottom of the plate. This film is composed of carbon atoms.
3. Scrape the carbon from the bottom of the plate and feel it.
4. Use the carbon to write/smear a message on a piece of paper.

Questions:

- How does the carbon look? How does the carbon feel?
- Do you think these carbon atoms are more like graphite or diamond?

Carbon allotropes

Carbon atoms can link with other carbon atoms in several different ways. When atoms of the same element link together in different ways to form substances with different properties, the different formations are called **allotropes**. For centuries scientists have known about several common allotropes of carbon. The most common form of carbon is coal. The carbon atoms in coal do not have a specific pattern. Carbon, in the form of diamonds, has a crystalline structure where each carbon atom is linked to four other carbon atoms forming a tetrahedron. These tetrahedrons bond with other tetrahedrons to form large crystals. This allotrope of carbon is the strongest naturally occurring substance on earth.

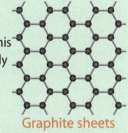
Diamond crystal

Graphite, on the other hand, is formed when carbon atoms link together to form sheets of hexagons. This is another allotrope of carbon. These sheets stack on top of each other and have weak forces holding the sheets together. This is why graphite molecules easily slip over one another, making graphite a good lubricant. Graphite is also mixed with clay to form pencil lead. The more graphite in the lead, the softer it is.

Graphite sheets

In 1985 a new allotrope, or molecular structure, for carbon molecules was discovered by three scientists, Harold W. Kroto, Robert F. Curl, and Richard E. Smalley. These scientists were studying the composition of carbon-rich stars. In their experiments they discovered a new form of carbon that always consisted of 60 atoms. After further experimentation, they found that these molecules were shaped like spheres with the atoms connecting together with hexagons and pentagons just like a soccer ball. They named these balls after the famous architect Buckminster Fuller who developed the geodesic dome. They are called **buckminsterfullerene** or **buckyballs** for short. The discovery of this form of carbon has opened up a whole new field of chemistry. In 1996 Harold Kroto, Richard Smalley, and Robert Curl shared the Nobel Prize in Chemistry for their discovery of buckminsterfullerene.

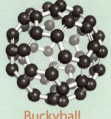
Buckyball

Several other related allotropes have been discovered in recent years. Graphene is similar to graphite as it forms a sheet of hexagons. However, graphene is only one atom thick; it only has one sheet of atoms instead of many sheets. Graphene has unusual electrical properties, and studies are being conducted to see if it can be used to replace semiconductors in many applications.

One of the newest fields of chemistry is called **nanotechnology**, which is the manipulation of matter on the atomic or molecular level for technological uses. Carbon atoms are being used to form structures called nanotubes. **Carbon nanotubes** are formed when a sheet of graphite or a sheet of graphene is formed into a cylinder, usually with a half of a buckyball at one end. These cylinders have special properties that make them very useful. First they are extremely strong. Carbon nanotubes can be 100 times stronger than reinforced steel. There are obvious uses for very strong fibers. For example, the bicycle ridden by Floyd Landis in the 2006 Tour de France used carbon nanotubes to reinforce the carbon frame, making it very strong but extremely light.

A second important property of nanotubes is their ability to conduct electricity. Depending on how they are made, some nanotubes conduct electricity better than silver or copper. Other nanotubes are semiconductors. Although they are not yet being used in production, it is believed that nanotechnology has the potential to replace much of the semiconductor technology being used today.

A third property of nanotubes is their ability to slide inside each other with nearly no friction. Scientists have been able to make tiny motors and rotors that are almost frictionless. This has great potential for tiny machines in the future. Although nanotechnology is very new, it has great potential in many areas. Keep your eyes and ears open to the news, and you will likely hear more about nanotechnology in the future.

10

Oxygen
A very essential element

What is special about oxygen?

Words to know:

oxidation

The most abundant element on earth is oxygen. Oxygen is also believed to be the fourth most abundant element in the universe. Oxygen is element number 8 on the periodic table of the elements. It has 8 protons and usually has 8 neutrons in its nucleus. Oxygen is in column VIA because it has six valence electrons. This means that oxygen needs two electrons added to its outer shell to be stable.

Most often oxygen atoms are found in the atmosphere as O_2, where two oxygen atoms have bonded together to form what is called a diatomic molecule. These two atoms share electrons so they are said to have covalent bonds. A small percentage of oxygen atoms combine in groups of three atoms, O_3, also known as ozone. Most of the ozone is high in the atmosphere and protects the earth from harmful radiation coming from the sun. The fact that oxygen near the surface of the earth is O_2 and not O_3 shows God's provision for life; the type of oxygen necessary for breathing is near the surface where people and animals are, and the type of oxygen that would be poisonous is high in the atmosphere where it can help shield the earth without harming us.

Oxygen is also abundant on earth in the form of water. Every water molecule has an oxygen atom in it. So between water and air, oxygen is perhaps the most critical element for life. Oxygen is also found combined with many other elements to form oxides, which are generally rocks. For example, oxygen combines with silicon to form silicon oxide, which is better known as quartz. Adding oxygen to a molecule is called **oxidation**.

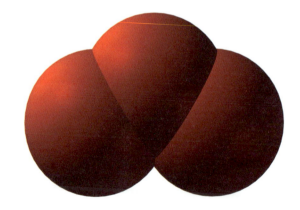

Ozone is a triatomic molecule, consisting of three oxygen atoms.

308 • Properties of Atoms & Molecules   LESSON 10

Another vital function of oxygen is in the releasing of energy. Oxygen is necessary for most burning processes. This may not seem very vital for life; however, just as oxygen is necessary to keep a wood fire burning, oxygen is also necessary to "burn" the food you eat. Oxygen is a key element in the process of converting food into energy. This is why animals need to continually breathe oxygen.

Earth is the only planet in our solar system with an abundant supply of oxygen both in the atmosphere and in the form of water. God designed our planet to be the perfect place for life to exist.

What did we learn?

- What is the atomic structure of oxygen?
- How is ozone different from the oxygen we breathe?

Taking it further

- Why does the existence of ozone in the upper atmosphere show God's provision for life on earth?
- How do animals in the ocean get the needed oxygen to "burn" the food they eat?
- Why are oxygen atoms nearly always combined with other atoms?

Oxygen—needed for burning

Oxygen is a necessary element in the combustion process. Whether you are burning wood for a campfire, a candle for a birthday cake, or the food you eat for energy, oxygen is necessary.

Purpose: To demonstrate the necessity for oxygen in combustion

Materials: small candle, glass cup, gloves, dry ice

Activity 1—Procedure:

1. Light a small candle.
2. Cover the candle completely with a glass cup. After a few seconds the candle will go out. Why?

Conclusion: The flame has used up the oxygen in the air and if no new air can reach the flame the burning will stop.

Activity 2—Procedure:

1. Using gloves, place a small piece of dry ice in an open container.
2. Remove the glass and relight the candle from the first activity.
3. Scoop a cup of gas from the container with the dry ice in it and pour the gas above the lighted candle. What happened to the candle? It went out.

Conclusion: Dry ice is frozen carbon dioxide so the gas in the container is carbon dioxide gas. Carbon dioxide gas is heavier than air so when you pour it over the candle it pushes the air molecules out of the way; it moves the oxygen away from the flame and the flame dies. So you can see that oxygen in the air is a very important element.

Oxidation

You have just demonstrated that oxygen is needed for burning, which is rapid oxidation. However, oxygen is also needed for slow oxidation processes. Cellular respiration is the "burning" of food for energy. This is a slow oxidation process. In cellular respiration a glucose molecule is combined with six oxygen molecules (O_2) to form six carbon dioxide molecules and six water molecules. The chemical equation for this process is:

$$C_6H_{12}O_6 + 6\,O_2 \rightarrow 6\,CO_2 + 6\,H_2O + \text{Energy}$$

All animal and plant cells must perform cellular respiration to obtain the necessary energy for growth and activity. Oxygen must be present for this to take place.

You also learned that oxygen combines with many other elements in the earth's crust. This often produces useful compounds. However, sometimes this chemical reaction can cause problems. One oxidation reaction that is not helpful is rusting. Oxygen slowly combines with iron to produce iron oxide, which we commonly call rust. Iron is strong, but rust is brittle and easily breaks down. We often coat things made from iron with paint or other coatings to keep the oxygen away from the iron. This helps to slow down the rusting process.

Water is an important part of the rusting process. Although rust is the combination of oxygen and iron, this is a very slow reaction. Water is a catalyst, which is a substance that speeds up a reaction but is not used up in the process. So rust occurs in areas with high humidity more than in areas of low humidity. Also, salts and other substances in the water can also speed up the rusting process.

Purpose: To observe the use of oxygen in the rusting process

Materials: steel wool, water, dish soap, two test tubes, pencil, dish

Procedure:

1. Cut two strips of steel wool, approximately the same size.
2. Wash one strip with dish soap to remove any oil on the steel wool. Rinse away all the soap, but leave the steel wool moist. Keep the other strip dry.
3. Place each strip in a test tube. Use a pencil to gently push the steel wool to the bottom of the test tubes.
4. Place 1–2 inches of water in a small dish and place the dish in a location where it will not be disturbed.
5. Turn the test tubes upside down and place them in the water.
6. After 12–24 hours, check the level of the water in each tube.

Questions:

- What do you observe?
- Which sample of wool has the most rust?
- Is the water higher in one tube than in the other tube? Why?

Conclusion: The moistened steel wool had the water needed to speed up the reaction so the iron in this tube was able to quickly react with the oxygen in the air to produce more rust than in the test tube without water. As the oxygen in the tube is used up it creates a vacuum in the tube. This draws water into the tube. The air is approximately 21% oxygen. If all of the oxygen in the tube has been used up, the water should fill approximately ⅕ of the tube. You should observe that the water level in the tube with the moistened steel wool is much higher than the water level in the tube with the dry steel wool.

UNIT 3

Bonding

11 Ionic Bonding
12 Covalent Bonding
13 Metallic Bonding
14 Mining & Metal Alloys
15 Crystals
16 Ceramics

◊ **Describe** the differences between ionic and covalent bonds.
◊ **Demonstrate** different bonds using models.
◊ **Describe** the properties of crystals.

Properties of Atoms & Molecules • 311

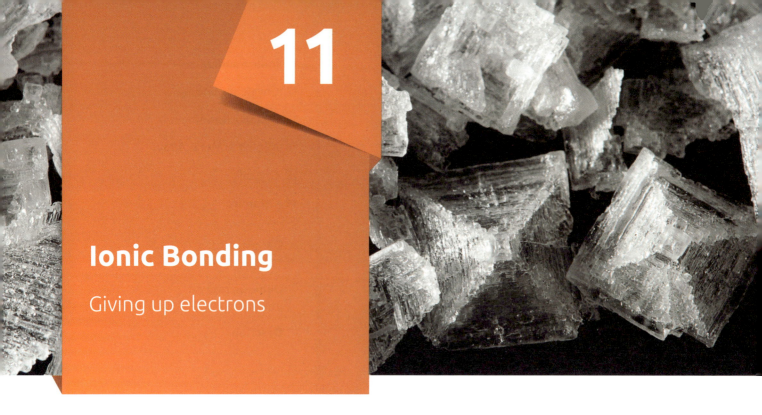

11

Ionic Bonding

Giving up electrons

How do atoms bond together?

Words to know:

chemical formula
electronegativity
low electronegativity
high electronegativity
ion
ionic bond
ionic compound

Challenge words:

cation
anion
valence

In general atoms do not exist long as single atoms. Most of them bond with other atoms to form molecules. Some molecules are small and consist of only two atoms. Others are large and can contain hundreds of atoms. To easily describe which atoms make up a particular molecule scientists write out its chemical formula. A **chemical formula** contains the symbol for each type of atom in the molecule followed by a subscripted number showing how many of that type of atom are in the molecule. If there is only one of a particular atom in the molecule there is no number after the symbol. You are probably familiar with the chemical formula for water, H_2O. This formula shows us that a water molecule consists of two hydrogen atoms and one oxygen atom. The chemical formula for methane is CH_4, one carbon

Linus Pauling established the concept of electronegativity in 1932.

312 • Properties of Atoms & Molecules   LESSON 11

and four hydrogen atoms. This is a quick and easy way to describe a molecule.

Atoms chemically connect with other atoms based on the number of valence electrons each atom has. Remember that valence electrons are the electrons in the outermost energy level of the atom. Scientists have determined that each atom is most stable when its outermost level is filled with eight electrons. The only exceptions to this are hydrogen and helium, which only have two electrons in their outermost level.

The ability of an atom to attract electrons to itself is called **electronegativity**. Electronegativity increases as you go from left to right on the periodic table. Atoms with one or two valence electrons easily give up those electrons when they bond with other atoms, allowing the next level in to become the outermost level, so that their outermost level will be full. For example, sodium has one valence electron in the third energy level. If a sodium atom gives up that electron, the second level is now the outermost level with electrons. The second level already has eight electrons so the sodium atom is now stable. Atoms that do not strongly attract electrons from other atoms are said to have a **low electronegativity** and easily give up their valence electrons. Sodium has low electronegativity and easily gives up its one valence electron.

Atoms with six or seven valence electrons easily pull electrons away from other atoms when they bond. These atoms are said to have **high electronegativity**; they hold tightly to their electrons. Chlorine has seven valence electrons. It has high electronegativity and strongly attracts electrons to itself while holding on tightly to its own valence electrons.

When an atom gains or loses electrons, it becomes electrically charged and is called an **ion**. When electrons are transferred as atoms bond, the bond that is formed is called an **ionic bond**. One of the most common substances that is formed by ionic bonding is table salt—sodium chloride (NaCl). Sodium is in the alkali metal family and has one valence electron. Chlorine is in the halogen

🧪 Atomic models

Purpose: To make models of a lithium atom and a fluorine atom, and to use those models to demonstrate ionic bonding

Materials: colored mini-marshmallows, toothpicks, glue

Procedure:

1. To make each model, use different colored mini-marshmallows for each part of the atom. If you have these colors, use green marshmallows to represent protons, yellow to represent neutrons, and orange to represent electrons.
2. Lithium has three protons and four neutrons in its nucleus. So glue together three green and four yellow marshmallows.
3. After these have dried, break a toothpick in half and put an electron (orange marshmallow) on the end of each half. Insert these toothpicks into the nucleus.
4. Place an electron on the end of an unbroken toothpick and insert it in the nucleus as well. This is a model of a lithium atom. Notice how two of the electrons orbit closely to the nucleus and one electron is farther away, in the outer shell.
5. Repeat the process to make a fluorine atom. Fluorine has 9 protons and 10 neutrons in its nucleus, so glue together 9 green and 10 yellow marshmallows. The marshmallows for the nucleus can be stacked together to form a ball.
6. After the glue has dried, again break a toothpick in half and use these short pieces to add two electrons to the atom.
7. Finally, add 7 full-length toothpicks with electrons to the nucleus. You now have an atom that has 9 electrons, with 7 of those electrons in the outer shell.
8. Now demonstrate how a lithium atom and a fluorine atom would combine by removing the valence electron from the lithium atom and adding it to the fluorine atom. Now the lithium atom has a positive charge, since it lost an electron, and the fluorine atom has a negative charge, since it gained an electron. These atoms will be attracted to each other and form an ionic bond.

LESSON 11 **Properties of Atoms & Molecules**

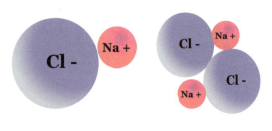
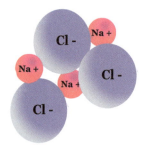
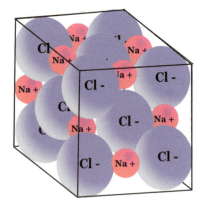

Salt molecules combine with ionic bonds

family and has seven valence electrons. In order to be stable, sodium must lose its one valence electron and chlorine must gain one electron. When sodium and chlorine atoms combine, the chlorine pulls one electron away from the sodium.

The chlorine atom now has one more electron than protons so it has a negative charge. It is now a negative ion. The sodium atom now has one less electron than protons so it is a positively charged ion. These two atoms stay bonded together by their opposite charges and now form the compound sodium chloride. Salt crystals are formed when the positively charged sodium side of a salt molecule lines up with the negatively charged chlorine side of another salt molecule. These opposite charges hold the molecules together. Salt molecules line up to form crystal lattices as shown in the diagram above.

All **ionic compounds**, those formed by exchanging electrons, have similar characteristics. First, ionic compounds are formed from elements that have very different electronegativities. One element always has a high electronegativity, the other has a low electronegativity. Ionic compounds are also brittle and have high melting points. Because they are only held together by their opposite charges, ionic compounds are easily dissolved in water. The oxygen side of a water molecule is slightly negative and the hydrogen side is slightly positive; therefore, water easily pulls ionic molecules away from each other, allowing them to dissolve. Ionic compounds also conduct electricity easily when they are melted or dissolved in water.

What did we learn?

- What is the main feature in an atom that determines how it will bond with other atoms?
- What kind of bond is formed when one atom gives up electrons and the other atom takes the electrons from it?
- What is electronegativity?
- Why are compounds that are formed when one element takes electrons from another called ionic compounds?
- What are some common characteristics of ionic compounds?
- Which element has a higher electronegativity, chlorine or potassium?

Taking it further

- Which column of elements are the atoms in column IA most likely to form ionic bonds with?
- Use the periodic table of the elements to determine the number of electrons that barium would give up in an ionic bond.

Ions

As you just learned, ionic bonding occurs when two ions are attracted to each other. Ions are formed when an atom either loses or gains electrons. If an atom loses one or more electrons it becomes positively charged. A positively charged ion is called a **cation** (KAT-ī-on). In table salt the sodium loses an electron so it is the cation.

An ion that is formed by gaining electrons is called an **anion** (AN-ī-on). Because an anion has more electrons than protons it has a negative charge. The chlorine atom in salt gains an electron so it is the anion in the compound.

The charge of an ion is shown by a superscript + or -. For example, since sodium in table salt has a positive charge, it is shown as Na^+. Similarly, the chlorine in salt is shown as Cl^-. If an ion has gained or lost more than one electron, the magnitude of the charge is shown as well. If calcium were to bond with sulfur, the calcium would lose two electrons and the sulfur would gain two electrons. They would then be designated as Ca^{2+} and S^{2-}.

The number of electrons that an atom is willing to give up or gain is determined by the number of valence electrons in that atom. Thus, the number of electrons an element is willing to give up or accept when forming a compound is called its **valence**. Sodium and chlorine both have a valence of 1 and calcium and sulfur both have a valence of 2. Atoms will form ionic bonds with other atoms that have the same valence, but very different electronegativities.

Look at the periodic table of the elements. Which types of elements have low electronegativities? If you said metals, or alkali and alkali-earth metals, you are correct. Which kinds of elements have high electronegativities?

Table salt is an ionic compound of sodium cations and chloride anions.

Nonmetals have high electronegativities. Thus, ionic compounds are only formed between metals and nonmetals.

In order to help people remember when a substance is an ionic compound, scientists change the ending of the name of the nonmetal. You have probably heard salt called sodium chloride. It is an ionic compound formed from sodium and chlorine. To show that it is an ionic compound the ending of chlorine is changed to "ide" and is thus called chloride.

You can practice naming ions by completing the "Name that Ion" worksheet.

12

Covalent Bonding

Sharing electrons

What is another way that atoms bond together?

Words to know:

covalent bonding

Elements that give up electrons when they bond with other elements form ionic bonds. Ionic bonds occur between elements with very different numbers of valence electrons, however not all compounds are formed by ionic bonding. Sometimes atoms have a similar number of valence electrons and do not easily give them up. In this case, the elements share electrons when they bond. This type of bonding is called **covalent bonding**.

Compounds made by covalent bonding have very different characteristics from those formed by ionic bonding. Covalent compounds have low melting points. They are usually strong and flexible. They are also lightweight, and many do not easily dissolve in water. Because covalent compounds do not form ions, they do not conduct electricity very well. Also, because covalent compounds do not form ions, these molecules have only a slight attraction for each other compared to ionic compounds.

One common type of compound formed by covalent bonding is a diatomic molecule. For example, oxygen gas almost always occurs as O_2—two oxygen atoms bonded together. Each atom of oxygen has 6 valence electrons. None of these atoms will easily give up its electrons. However, when two oxygen atoms bond, they each share two of the other atom's electrons, thus making each atom seem to have a full 8 electrons in its outer shell. This allows the O_2 molecule to be stable because each atom is stable.

Covalent bonding does not just occur between two identical atoms. Bonds between nonmetals are usually covalent. The most common covalent compound on earth is water. Hydrogen has one valence electron and oxygen has six. You might think that the oxygen would pull the electron away from each of the hydrogen atoms to form ionic bonds. However, each hydrogen needs only one additional electron to have a full outer shell so it does not give up its electron as easily as other elements with only one valence electron. Therefore, two hydrogen atoms share their electrons with one oxygen atom and the oxygen shares one of its electrons with each of the hydrogen atoms to form a water molecule. In this way, each atom feels like it has a full outer shell, so the compound is stable.

Scientists have developed a visual way to show the sharing of valence electrons. They write the atomic symbol with dots representing

More atomic models

Purpose: To demonstrate covalent bonding
Materials: colored mini-marshmallows, toothpicks, glue
Procedure:

1. Make marshmallow models for one oxygen and two hydrogen atoms. Use the same color of marshmallows as you did in lesson 11. Hydrogen atoms are extremely easy to make because they have only one proton and one electron. An oxygen atom has 8 protons and 8 neutrons in its nucleus and 8 electrons orbiting the nucleus. Break a toothpick in half and use the two shorter pieces for the first two electrons, thus showing that the 6 remaining electrons are in the outer shell.

2. Once the models are complete, set the hydrogen atoms close to the oxygen atom in such a way that the electrons of all three atoms form a group of 8 electrons in the outer layer around the nucleus of the oxygen atom. This demonstrates a covalent bond. Do not remove any electrons from any of the atoms.

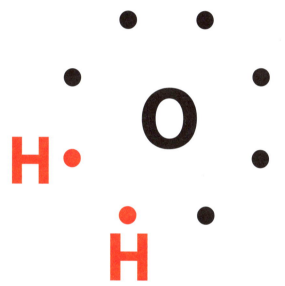

A dot diagram showing how hydrogen and oxygen atoms share electrons

the valence electrons. Above is a dot diagram showing how the hydrogen and oxygen atoms share electrons. The electrons from the hydrogen atoms are shown in a different color from the electrons from the oxygen atom. You can see that the two electrons from the hydrogen atoms give the oxygen atom a full eight electrons and two electrons from the oxygen atom make the hydrogen atoms feel like they have a full outer shell as well since they need only two valence electrons.

Covalent compounds are vital to life. Not only is water a covalent compound, but most of the compounds that make up our bodies are covalent compounds. These include proteins, fats, and carbohydrates.

What did we learn?

- What is a covalent bond?
- What are some common characteristics of covalent compounds?
- What is the most common covalent compound on earth?

Taking it further

- Why do diatomic molecules form covalent bonds instead of ionic bonds?
- Would you expect more compounds to form ionic bonds or covalent bonds?

Ionic vs. covalent

Ionic compounds and covalent compounds have very different characteristics. Because ionic compounds are formed from ions, which are electrically charged particles, they readily conduct electricity when they are dissolved in water. However, because covalent compounds are sharing electrons, they do not easily give up electrons so they do not conduct electricity.

You can conduct an experiment to see which substances are ionic compounds and which substances are covalent compounds. Because water is a covalent compound we know that pure water does not conduct electricity. However, the water from your tap is not pure water. There are small amounts of minerals dissolved in tap water. Therefore, you need to use distilled water for this experiment.

Purpose: To determine which compounds are ionic and which are covalent

Materials: "Bonding Experiment" worksheet, four paper cups, distilled water, copper wire, 9-volt battery, baking soda, sugar, salt, olive oil

Procedure:

1. Look at the chemical formula for each of the substances we will be testing. These are listed on the "Bonding Experiment" worksheet. Next to each substance write whether it is composed of all metals, all nonmetals, or both metals and nonmetals. Use a periodic table if you need to see which elements are metals and nonmetals.

2. Based on what the compounds are composed of, make a reasonable guess, called a hypothesis, about whether you think each substance will conduct electricity. Write your guesses on the worksheet. Now you are ready to test your hypotheses.

3. Fill four paper cups with distilled water.

4. Strip off at least 1–2 inches of insulation from each end of 2 copper wires.

5. Attach a length of copper wire to each terminal of a 9 volt battery.

6. Place the ends of the two wires into one of the cups of water. Watch to see if anything happens at the ends of the wires. If electricity is being conducted, you should see small bubbles forming on the ends of the wires. This is because the energy from the moving electrons breaks apart some of the water molecules and the hydrogen gas will collect on one wire and the oxygen gas will collect on the other wire. If you see bubbles, then electricity is being conducted. If you do not see any bubbles, then no electricity is being conducted. Write your observations on your worksheet. After a few seconds, remove the wires from the water.

7. Now, dissolve one teaspoon of baking soda in the first cup, one teaspoon of sugar in the second cup, one teaspoon of salt in the third cup, and one teaspoon of olive oil in the fourth cup. The oil will not really dissolve, but stir it into the water.

8. Use your battery and wire set-up to test for the conduction of electricity in each solution. Be sure to wipe off the wires after each test. Write your observations on your worksheet.

9. Finally, complete the worksheet by writing your conclusions about each compound. Decide if each compound is ionic or covalent.

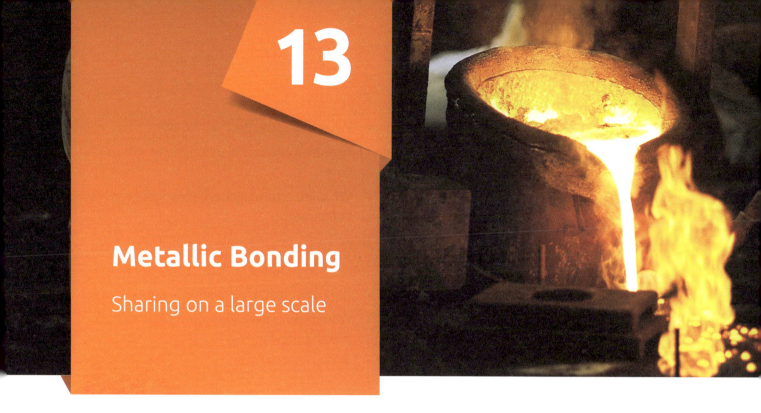

13

Metallic Bonding

Sharing on a large scale

How do metal elements bond together?

Words to know:

free electron model   metallic bonding

It is easy to see how metals and nonmetals such as sodium and chlorine exchange electrons to form ionic bonds. Sodium gives up one electron so its outermost shell has 8 electrons and chlorine accepts one electron to make 8 electrons in its outermost shell. It is also easy to understand how nonmetals can form covalent bonds by sharing electrons. Two oxygen atoms can share two electrons so that each atom feels that it has 8 electrons in its outer shell. However, it is more difficult to understand how metals can bond with each other. For example, aluminum, with three valence electrons, cannot form ionic bonds with other aluminum atoms. If one aluminum atom gave up its three valence electrons, the other atom would then have six valence electrons and would not be stable. If one atom gave up its three valence electrons to two other atoms there would still not be enough electrons to make the atoms stable. So you can see that metals do not form ionic bonds with other metals.

Similarly, metals do not form covalent bonds. Since metals usually have only one, two, or three valence electrons, two or three atoms together would not have enough electrons to share to make all of the atoms stable. The best explanation for how metals form bonds is called the **free electron model**. This model states that metals share electrons on a grand scale. Thousands of atoms join together and electrons freely move from one atom to another to form stable atoms. This type of bonding is called **metallic bonding**.

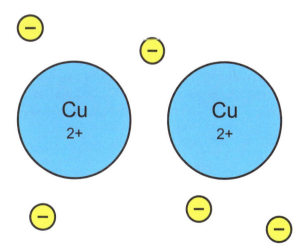

Metallic bonds are found in metals like copper, with free electrons among a lattice of positively-charged metal ions.

LESSON 13 **Properties of Atoms & Molecules • 319**

This free movement of electrons explains why most metals are good conductors of electricity. Compounds formed by metallic bonds also have other similar characteristics. The free movement of electrons allows metals to conduct heat and gives metals their shiny appearance. They also have high melting points and are insoluble in water.

You can see that because elements can form ionic bonds, covalent bonds, and metallic bonds, God has created elements that can produce a nearly endless variety of compounds. This is one reason why our world is so wonderful and so complex.

What did we learn?

- What is the free electron model?
- How many valence electrons do metals usually have?
- What are common characteristics of metallic compounds?

Taking it further

- Why don't metals form ionic or covalent bonds?
- Would you expect semiconductors to form metallic bonds?

Metal models

Purpose: To demonstrate metallic bonding

Materials: colored mini-marshmallows, toothpicks, glue

Procedure:

1. Using marshmallows and toothpicks, make three or more beryllium models. Beryllium has 4 protons and 5 neutrons. It also has 4 electrons, two in its inner shell and two in its outer shell.

2. After making the models, place the models near each other. Note that none of the atoms has enough electrons to be stable.

3. Add some free electrons around the models. These represent the electrons that are shared freely among thousands of metal atoms in metallic bonds. Use the models made in the previous lessons as well as the ones made today to review the differences between ionic, covalent, and metallic bonds.

Bonding characteristics

Because bonding occurs on an atomic level, you cannot see if a compound is ionic, covalent, or metallic. However, you have learned that each type of compound has certain characteristics. If a substance has the characteristics of an ionic compound it probably has ionic bonds; if it has the characteristics of a metal, it probably has metallic bonds. Review the characteristics listed in the previous lessons, then fill in the "Bonding Characteristics" worksheet.

14

Mining & Metal Alloys

Making it stronger

How do we process and use metals?

Words to know:

smelting   alloy

electrolysis

Challenge words:

super alloy

Most metals will form metallic bonds with other metal elements, and they form ionic bonds with nonmetals as well. Most metal found in nature is not in a pure metal form. Most commonly, metal atoms combine with oxygen atoms to form metal oxides. To obtain pure metal from the metal ore, a chemical reaction must take place that will remove the oxygen from the metal ore. This type of reaction is called a reduction reaction. The process used to remove oxygen varies depending on the type of metal.

Many metal ores are purified through a process called **smelting**. For example, copper ore is smelted to reduce the amount of oxygen in it. During smelting the ore is crushed and heated. Then hydrogen is blown through the molten metal. The hydrogen combines with the oxygen in the liquid to form water; leaving nearly pure copper behind.

Further refining of copper is done by **electrolysis**. Carbon electrodes are used to pass an electrical current through the liquid copper. This allows any remaining oxygen atoms to combine with carbon atoms from the electrode to form carbon dioxide and allows the pure copper to collect on the other electrode. This results in nearly pure sheets of copper.

A similar process is used to produce pure aluminum. Bauxite is an ore that contains aluminum. The bauxite is dissolved in a cryolite solution (sodium aluminum fluoride), and then placed in an electrolysis set-up like the one shown on the next page. The liquid aluminum collects on the bottom when electricity is passed through the solution, and oxygen combines with the carbon in the electrodes becoming carbon dioxide and escapes from the solution.

Metals like copper and aluminum are very useful because they can be molded into pipes or cans, or drawn into wires. However, pure metals are not always the best choice for a particular job. Scientists have found that by adding a small amount

LESSON 14   **Properties of Atoms & Molecules • 321**

🧪 Polishing silver

When a metal combines with oxygen, an oxide is produced. Iron oxide is commonly called rust. Copper combines with oxygen to form a layer that is green instead of the shiny reddish-gold we commonly think of as copper. The Statue of Liberty is made of copper, but is green because the copper has oxidized.

Silver also oxidizes. We usually say that silver has tarnished when the silver combines with oxygen. This oxidation leaves a streaky black surface on our silverware and other silver items. Because people prefer silver to be shiny and silvery, scientists have developed tarnish remover. Tarnish remover is usually a liquid or cream that combines chemically with the silver oxide, leaving behind a shiny silvery surface.

Remove silver oxide from a piece of silver that is tarnished. Follow the directions on the tarnish remover. This will allow you to perform a chemical reaction and help restore the beauty of your silver at the same time.

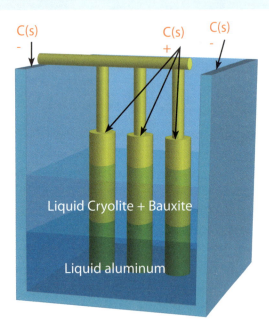

of another element to the molten ore, the resulting metal has superior qualities. When a small amount of one metal is added to another metal the result is called an **alloy**. Alloys are often stronger, more resilient, and easier to work with than the pure metal would be. For example, steel is iron with a small amount of carbon added. Steel is stronger and more flexible than iron.

To produce steel, iron ore is processed in a blast furnace. First, the ore is crushed. Then it is mixed with limestone and coke (a form of carbon—not the soft drink). This mixture is then heated to very high temperatures in a blast furnace. The resulting molten metal is called pig iron. This is iron with a significant amount of carbon in it. Pig iron can be cast into pots and other shapes, but it is brittle and is not useful for most other applications. To improve the quality of the iron, when the molten iron is removed from the furnace, oxygen is blown through the liquid where it combines with the carbon to form carbon dioxide, which bubbles out of the liquid. The remaining liquid is iron with just a small amount of carbon. This is called steel. Steel is pliable and strong and can be formed into rods, sheets, and other shapes that are useful for many applications.

Other elements besides carbon are sometimes added to steel to further improve its performance. For example, chromium is added to produce stainless steel. This metal does not easily corrode so it is preferable for many applications, such as making forks and spoons to eat with. Another element that is sometimes added to steel is tungsten. Tungsten makes steel very tough. Tungsten steel is often used to make saw blades that last longer than regular steel blades. ✹

Steel can be formed into many useful applications.

What did we learn?

- What element is combined with most metals to form metal ore?
- What must be done to metal oxides to obtain pure metal?
- What is an alloy?
- Why are alloys produced?

Taking it further

- Do you think chromium would be added to steel that is going to be used in saw blades? Why or why not?
- Is oxidation of metal always a bad thing?

Alloys

Pure metals are used in some applications. For instance, pure gold is used to cover the dome of the Colorado state capital building. However, in many applications metal alloys are used because they are stronger. But why are alloys stronger than pure metals? When pure metals bond together, they generally form straight lines of atoms. If something disrupts the atoms, they can break apart in sheets or crack in straight lines. However, if a small amount of another metal is added to the original metal while it is molten, the new atoms mix into the middle of the original metal atoms. This keeps the atoms from forming into long lines and actually makes the metal stronger.

Some alloys are very strong, even at high temperatures. These metals are called **super alloys**. Super alloys have nickel, iron, and cobalt added to make them strong.

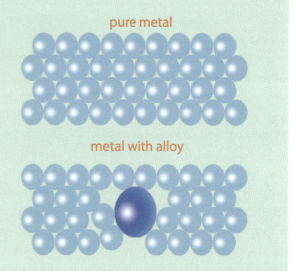

pure metal

metal with alloy

See what you can find out about some common alloys by completing the "Common Alloys" worksheet.

Charles Martin Hall
1863–1914

SPECIAL FEATURE

Aluminum, believed to be the most common metal on earth, was discovered by Friedrich Wohler in 1827. This discovery, however, did not mean that aluminum was immediately available for use. In its natural state, aluminum is always tightly bonded with other compounds; most often it is in a compound called bauxite. Without an economical method for extracting the aluminum from the bauxite, pure aluminum was very expensive. During most of the mid 1800s, aluminum was so valuable that it was mostly used in jewelry and for special projects, like capping the Washington Monument. Because the unique properties of aluminum made it ideal for many applications, a race was on to find a less expensive way to extract it from the ore. Two men, working independently from each other, won this race. These two men, Charles Martin Hall of the U.S. and Paul Heroult of France, were born the same year (1863), made their discoveries the same year (1886) and died the same year (1914).

Charles Martin Hall was born in Thompson, Ohio, to Rev. Heman Hall and Sophronia Brooks Hall. When he was 10 years old, Charles and his family moved to Oberlin, Ohio. There he did his preparatory work in high school and, in 1880, began his studies at Oberlin College.

Charles did not take a formal chemistry class until his junior year in college; however, his interest in chemistry began much earlier. Hall met Dr. Frank Jewett, a well-educated chemist, while buying some equipment and chemicals during his first year of college. Hall and Jewett spent many hours discussing chemistry, and it is believed that Jewett was instrumental in encouraging and helping Hall in his discovery of aluminum extraction. In class, Jewett talked about the challenge of finding

an economical method for extracting aluminum. Jewett said, "Any person who discovers a process by which aluminum can be made on a commercial scale will bless humanity and make a fortune for himself." Charles Hall took the challenge and told some of his fellow students, "I'm going for that metal."

Jewett, along with Charles's sister Julia, made many contributions to the discovery. In addition to working in a lab in the woodshed behind his house, Hall was allowed to use Jewett's personal laboratory. Jewett also supplied Hall with materials and up-to-date knowledge of chemistry. Jewett had gone to one of the best schools in Europe for his education, and before coming to Oberlin College, he taught at the Imperial University of Tokyo, so he was a valuable asset in Hall's quest for aluminum.

Charles's sister Julia had also gone to Oberlin

College and had taken most of the same science courses he had taken. She was very involved in his research and probably helped him prepare many of the chemicals that he used. When he finally made his successful experiment on February 9, 1886, he repeated the experiment for Julia the next day, after she returned from a trip to Cleveland.

The famous experiment in 1886 used electrolysis to remove the aluminum from aluminum oxide. Hall accomplished this by dissolving aluminum oxide in a cryolite-aluminum fluoride mixture, and then passing an electrical current through the liquid. The electricity caused aluminum to form and settle on the bottom of the vessel where it could not oxidize with the oxygen in the air.

Charles Hall applied for a patent for his aluminum reduction process in July, 1886, only to find that a Frenchman named Paul Heroult had already applied for a patent for the same process. How could two people in two different parts of the world come up with the same process at virtually the same time? These men were both very interested in solving this problem and had access to much of the same information and the same materials, so it is not surprising that they developed the same process. The patent dispute was resolved when it was confirmed that Hall had performed his successful experiment shortly before Heroult did.

Hall was very successful at overcoming obstacles and within a few years, he and his partners were making commercial quality aluminum. In 1888 he and his partners started the Pittsburgh Reduction Company, and in 1907 the name of the company was changed to the Aluminum Company of America, which is today known as ALCOA.

By 1914 Hall's new process had caused the price of aluminum to drop from $12.00/lb to $0.18/lb. As Jewett had predicted, Hall's discovery truly was a blessing to humanity and made a fortune for him. Today, a host of items are made from aluminum. However, Hall did not keep all of his fortune for himself. He donated over $10 million dollars to Oberlin College. He also donated substantial amounts of money to Berea College, to the American Missionary Association, and to educational programs in Asia and the Balkans.

15

Crystals

Sparkling like diamonds

How are crystals formed?

Words to know:

crystal

face

edge

Challenge words:

hydrated

water of crystallization

hydrate

anhydrous

dehydration

What do salt, sugar, sand, diamonds, and snowflakes all have in common? They are all solids that have a crystalline structure. Certain materials form crystals when the liquid form freezes or becomes a solid. **Crystals** are solids whose atoms are in an orderly pattern. Crystals have flat surfaces called **faces**, and **edges** where their faces meet. There are seven major types or shapes of crystals. These are shown on the next page.

Large, perfectly formed crystals can only form when liquids are allowed to cool slowly and are not disturbed. This allows the atoms to line up in the crystal formation. If a liquid cools rapidly, crystals will not form at all, or only very small crystals will form.

The most common place to find crystals is among rocks and minerals. These compounds are the most likely to form crystalline bonds. One very common type of crystal is quartz. Quartz always forms six-sided crystals. Another very interesting place to find crystals is in a geode. A geode is a rock in which crystals have formed in the center. A geode must be broken open to reveal the beauty of the crystals inside.

Some crystals are made of only one kind of element. Diamonds, for example, are pure carbon. But most crystals are made from two or more kinds of atoms.

326 • Properties of Atoms & Molecules   LESSON 15

The seven crystal systems

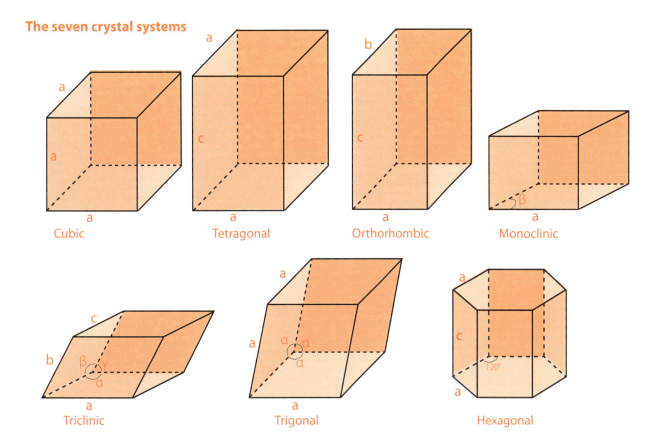

Cubic Tetragonal Orthorhombic Monoclinic

Triclinic Trigonal Hexagonal

Other crystals form when minerals which have been dissolved in water crystallize as the water evaporates. Salt crystals often form this way. Crystals that form this way are also common in limestone caves. As water seeps through the limestone, it dissolves small amounts of calcite and gypsum. When the water drips from the ceiling of the cave, it evaporates leaving behind the minerals as stalactites and stalagmites. These crystal formations can be very beautiful.

Not all crystals are formed naturally. The first artificial, or man-made, rubies were made in 1837 by a French scientist named Gaudin. Since that time, the process has been improved and artificial gems are now routinely made. Although some artificial gems, such as cubic zirconia, are crystals that are made to look like real gems but are made from different materials, many artificial gems are crystals made from the same chemicals as the naturally occurring gem. The elements are melted and then allowed to cool very slowly, sometimes under high pressure. Today over 44,000 pounds (20,000 kg) of artificial diamonds are manufactured each year, as well as artificial rubies, sapphires, spinets, and emeralds. Although artificial gems are chemically identical to naturally occurring gems, they are usually not as brilliant as the natural ones.

In addition to gems, crystals have many uses. Some crystals, such as salt and sugar, are part of

Fun Fact

Salt has been a valuable crystal throughout history. Not only is salt used to season food, it has been used for centuries as a natural preservative. In ancient times, salt was so valuable that it was used as a form of currency. At times, people preferred to be paid in salt rather than gold.

LESSON 15 **Properties of Atoms & Molecules** • 327

🧪 Growing crystals

You can make your own crystals by dissolving minerals in water and then allowing the water to evaporate slowly. This works best when you have a saturated solution. A saturated solution is one that cannot dissolve any more of the material being dissolved. The amount of material that can be dissolved increases with temperature, so you will want to heat the water before dissolving the minerals.

Purpose: To grow two different shapes of crystals

Materials: plate, water, black construction paper, scissors, small pan, stove, table salt, Epsom salt

Procedure:

1. Place a plate upside down on a piece of black or other dark construction paper. Trace around the edge of the plate. Remove the plate and cut out the circle. Place the paper on top of the plate.
2. Place ½ cup of water in a small pan and bring to a boil.
3. Dissolve as much table salt in the water as you can. Add the salt a teaspoon at a time until no more salt will dissolve.
4. Slowly pour the saltwater onto the paper on the plate until the paper is completely wet, but not soaked.
5. Place the plate in a place where it will not be disturbed.
6. Repeat steps 1-5 using Epsom salt instead of table salt
7. Allow the water to evaporate undisturbed for several days. After the water is gone, you should see crystals growing on the paper.

Conclusion: The table salt crystals will be cube-shaped and the Epsom salt crystals will be long needles. When the paper is completely dry, look under the paper. There may be some crystals that formed under the paper as well.

our food. Other crystals are used in the medical field as medications, and in hearing aids. Some crystals, like silicon, are used in the semiconductor industry. Diamonds that are not good enough to be gems have many uses because of their hardness. Diamonds are added to drill bits, saw blades, scalpels, and other cutting instruments to make them sharp and hard. There are many other uses for crystals as well. God has blessed us with an abundance of crystals for a variety of purposes.

🚀 Taking it further

- Why are naturally occurring gems more valuable than artificial gems when many are made from the same materials?
- Why is a saturated solution better for forming crystals?
- What are some ways you use crystals in your home?

🧠 What did we learn?

- What is a crystal?
- How do crystals form?
- What is an artificial gem?
- Where would you look to find crystals?

🧪 Opening a geode—optional

It is fun to crack open a geode and reveal the crystals inside. If you have access to a geode, open it up and enjoy the beauty hidden inside.

Hydrates

Many salts do not just dissolve in water to form a water solution. Some salts chemically bond with the water. When this occurs the crystals are said to be **hydrated**. The water that is stored in the molecules is called the **water of crystallization** and the substance itself is called a **hydrate**. Although hydrates contain water, they usually do not feel wet because the water molecules are bound to the crystals.

Hydrates have many important uses. One you are probably familiar with is keeping products dry. Have you ever opened a box with a new pair of shoes and found a little packet with crystals inside? The crystals are hydrates that can absorb more water. They are included to prevent your new shoes from getting moist when they are shipped. Other products including purses, wallets, and some food products also contain hydrate packets to absorb excess moisture.

Another important use of hydrates is in the making of fire resistant materials such as housing insulation. The materials used often have hydrates in them. This is important because if there is a fire in your house, the heat will cause the water in the hydrates to evaporate. This will use up energy from the fire, which keeps the fire from spreading. The more hydrates in the material the longer it will resist the spreading of fire.

Heating will usually remove the water from hydrates, leaving behind a dry solid. The dry solid is said to be **anhydrous**. The process of removing water is called **dehydration**. Anhydrous crystals are also used in many applications. You probably have concrete in the foundation of your house. Concrete is formed when cement, which is an anhydrous crystal, is mixed with water. The water causes a chemical reaction as it bonds with the crystals. This releases energy, which is why you might see steam coming off of newly poured cement. However, not all of the water is evaporated; some of it is bonded with the cement and actually gives strength to the concrete.

You can experience your own chemical reaction with an anhydrous crystal. Plaster of Paris is anhydrous calcium sulfate. When water is added to the powder, it hydrates the crystals and starts a chemical reaction. Some of the water evaporates, leaving behind a network of hydrated crystals that forms a solid.

A fun use of plaster of Paris is in casting animal tracks in the wild. Often a scientists or naturalist will pour a small amount of plaster into an animal track and allow it to dry. Then he can take the mold with him to identify at a later time.

Purpose: To make your own plaster track at home

Materials: modeling clay, water, a pet, plaster of Paris

Procedure:

1. Make a cookie-sized disk of modeling clay.
2. Press your cat's or dog's paw into the clay to make a print. Or make up your own track by making indentations in the clay with your fingers or other objects.
3. Mix ¼ cup of plaster of Paris with enough water to make a creamy liquid.
4. Carefully pour the liquid into the indentations.
5. Allow the plaster to dry for several hours.
6. When the plaster is dry, remove the clay and you will have a plaster cast of your animal track made from a hydrated salt.

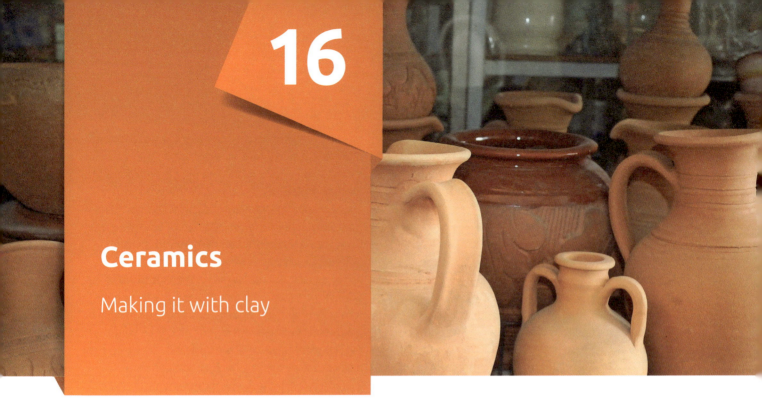

16

Ceramics

Making it with clay

What are ceramics, and how are they made?

Words to know:

ceramics

Challenge words:

bioceramics
inert ceramics
resorbable ceramics
active ceramics

Crystals are an essential part of our lives. From computers to jewelry, we use crystals every day. One very special type of crystal material is ceramic. **Ceramics** are inorganic nonmetallic materials which are formed by the action of heat. The word *ceramic* comes from the Greek word for *earthenware* and describes where ceramics came from. Traditional ceramics include pottery, brick, porcelain, and glass. And the common ingredient in each of these materials is clay, which comes from the earth, making them earthenware and thus the name, ceramic.

The clay molecules in ceramic materials are fused with other chemicals by heat. The toughness, look, and other characteristics of the ceramic material are determined by the crystal structures that are formed in the heating process. From the ancient Egyptians to the American Indians, people in many cultures have used the heating or firing process to strengthen their earthenware. People have been firing their pottery to make it stronger for thousands of years, long before anyone understood the chemistry behind it. Even at the Tower of Babel, they were baking bricks, more than 4,000 years ago (Genesis 11:3).

Today, with a better understanding of chemical bonding, scientists have developed advanced ceramics. These new ceramics are designed with specific, very pure substances that are fired in very

After a clay vessel is made, it is fired to make it strong.

specific ways to create very strong crystal structures. These new ceramics are replacing metal in many applications. The new ceramics are often stronger, harder, and more heat resistant than the metals they replace. Also, ceramics are more chemically stable. They do not react with oxygen to form rust as readily as metals often do.

New ceramics are engineered for specific purposes. For example, special ceramic material is used to make artificial joints used in medical procedures. This new ceramic material contains calcium that will fuse with the surrounding bone; allowing the new joint to become part of the body. Another special ceramic has been developed for use as heat-absorbing tiles on the underside of the space shuttle. Ceramics are also being used as tools such as scissors, knives, and blades for machines. As scientists learn more about chemistry, they will be able to continue developing more uses for special ceramics.

Close-up of space shuttle *Discovery's* ceramic thermal tiles

What did we learn?

- What is ceramic?
- What are some examples of traditional ceramics?
- What makes ceramics hard?
- What are some advantages of modern ceramics?

Taking it further

- Why are the tiles on the space shuttle made of ceramic?
- Why are crystalline structures stronger than noncrystalline structures?

Fun with clay

One of the most interesting new ceramics to be developed in recent years is polymer clay. Polymer clay is a material that is soft and pliable. It can be molded into any shape and then remolded as often as desired. However, when the clay is baked at a low temperature, a chemical reaction occurs and the clay becomes hard. This clay is fun for children, but has recently become an art medium for adults as well.

Make a sculpture, beads, pots, or other items using polymer clay such as Sculpey or Femo. Follow the manufacturer's directions for baking the finished masterpiece.

Bioceramics

Ceramics that are used in the human body are called **bioceramics**. We already mentioned a little about how ceramics can be used inside the body, but this is such an interesting topic we knew you would want to learn more about it. There are basically three different kinds of bioceramics: those that do not react at all with the body, those that break down inside the body, and those that combine with the tissues to become a part of the body.

Ceramics that do not react chemically inside the body are called **inert ceramics**. Alumina and zirconia are the most common inert ceramics used in the body. These materials are very hard and can be highly polished for low friction. This makes them ideal for joint replacements for hips and knees. Inert ceramics are sometimes used as spacers in bones where a small section of bone must be removed. The spacer is put in its place as a frame for new bone cells to grow on.

Another inert ceramic is pyrolytic carbon. This ceramic material is often used to replace valves inside a heart. The valves are very strong and can withstand the wear and tear of constant use. Even more importantly, they do not cause blood clots.

Ceramics that break down or dissolve inside the body are called **resorbable ceramics**. These ceramics can be used to deliver drugs, particularly radioactive particles, to a certain location in the body or to provide temporary strength to a bone while the body repairs itself. Most resorbable ceramics are made from calcium phosphates or from silica-based glass ceramics.

Finally, ceramics that react with the body but do not break down are called **active ceramics**. Active materials are often used to coat metal implants. Active ceramics are usually too brittle to be used as a load-bearing structure so titanium or stainless steel is used to replace leg bones or other structures, then the implant is coated with hydroxyapatite, which is an active ceramic.

Hydroxyapatite is made from a material that is chemically very similar to the bones in your spine. This material reacts chemically with the bone material in the patient's body to promote bone growth and to prevent infection in the area. It also helps to prevent rejection of the implant.

The use of bioceramics has helped millions of people to live better lives. This is a growing field of science and new ceramic techniques and devices are being designed every year.

UNIT 4

Chemical Reactions

17 Chemical Reactions
18 Chemical Equations
19 Catalysts
20 Endothermic & Exothermic Reactions

◊ **Use** equations to describe chemical reactions.
◊ **Identify** factors that effect chemical reactions.
◊ **Demonstrate** the first law of thermodynamics using chemical reactions.
◊ **Describe** what happens to heat during chemical reactions.

Properties of Atoms & Molecules • 333

17

Chemical Reactions

Changing from one thing to another

How do chemicals react with each other?

Words to know:

reactants
products
chemical reaction
first law of thermodynamics
composition reaction
decomposition reaction
reaction rate

As you learned in the past several lessons, elements bond in many different ways depending on their electron structures. When two or more different elements bond together, a chemical reaction takes place and a new substance is formed. In a chemical reaction, the beginning materials are called the **reactants** and the ending materials are called the **products**. Some common chemical reactions you are probably familiar with include photosynthesis, bread dough rising, a flame burning, or a firecracker exploding.

The formation of water is a very simple chemical reaction. Two hydrogen atoms bond with one oxygen atom to form a water molecule. Hydrogen and oxygen are the reactants and water is the product. Breaking water apart is also a chemical reaction. One water molecule can be broken apart to form two hydrogen atoms and one oxygen atom. In this reaction water is the reactant and hydrogen and oxygen are the products. A **chemical reaction** takes place whenever atomic bonds are formed or broken.

Simple chemical reactions occur when atoms combine to form molecules. But more complex chemical reactions take place when two or more molecules break apart and their atoms recombine to form new molecules. Let's look at a very important chemical reaction taking place all the time—photosynthesis. During photosynthesis molecules of carbon dioxide and water are broken apart and then combined inside a plant's leaf to form molecules of glucose (sugar) and oxygen. This diagram shows how this happens.

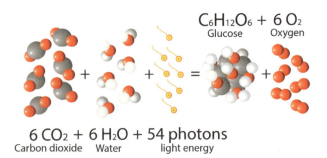

$C_6H_{12}O_6 + 6 O_2$
Glucose   Oxygen

$6 CO_2 + 6 H_2O + 54$ photons
Carbon dioxide   Water   light energy

The chemical reaction of photosynthesis

334 • Properties of Atoms & Molecules   LESSON 17

Inside a leaf, six carbon dioxide molecules and six water molecules (the reactants) are broken apart into carbon, hydrogen, and oxygen atoms. Then these atoms combine together to form one molecule of glucose, which uses all of the carbon atoms, all of the hydrogen atoms, and some of the oxygen atoms. The rest of the oxygen atoms combine to form oxygen gas. Glucose and oxygen gas are the products.

It is important to realize that when a chemical reaction takes place the atoms that are in the products are the same atoms that were in the reactants. Just because one substance went away and a new substance was formed does not mean that the original atoms disappeared and new atoms appeared. Once scientists began to understand chemical reactions they realized that there is no way to make new matter, only many ways to rearrange existing matter. This concept is the **first law of thermodynamics**, which says that matter and energy cannot be created or destroyed; they can only change forms. God is the only one who can create new matter or new energy.

Sometimes chemical reactions are reversible. If water is broken apart into oxygen and hydrogen gas, the gases can later be recombined to form water again. Other chemical reactions cannot be reversed. For example, if you cook an egg, the egg cannot be "uncooked." Some chemical reactions happen very easily. You notice an immediate reaction when you combine baking soda and vinegar. Other reactions are slow or may even require heat, light, or other stimuli to make them happen. For example, photosynthesis does not occur without sunlight and chlorophyll.

There are many different kinds of chemical reactions. If an element combines with oxygen, the reaction is called an oxidation reaction. If oxygen is removed from a substance, such as in the purification of metals, the reaction is called a reduction reaction. If elements other than oxygen combine to form a new substance, the reaction is a **composition reaction**, and if a substance is broken down into individual elements, the reaction is a **decomposition reaction**.

Some reactions happen very quickly. In fact, some are instantaneous, like the explosion of fireworks. Other reactions happen very slowly. The rate at which a reaction takes place is called its **reaction rate**. A piece of iron will eventually

🧪 Fire extinguisher in a jar

A flame is a chemical reaction that requires oxygen. Therefore, it is an oxidation reaction. If you are trying to build a campfire, you need to make sure that air, which contains oxygen, can reach the wood, paper, and other materials you may be using to build your fire. However, if a fire starts someplace you don't want a fire, one of the quickest ways to put out the fire is to deprive it of oxygen; the fire will then go out quickly. This is the way that many fire extinguishers work. The fire extinguisher sprays a chemical on the fire that keeps the oxygen away from the flames and allows the flames to go out.

Purpose: To build a fire extinguisher in a jar

Materials: modeling clay, birthday candle, jar, vinegar, baking soda

Procedure:
1. Using a piece of modeling clay, attach a birthday candle to the inside bottom of a jar.
2. Pour ¼ cup of vinegar into the jar. Be sure not to get the wick of the candle wet.
3. Light the candle.
4. Sprinkle a teaspoon of baking soda into the jar. Be sure not to sprinkle it on the candle.

Conclusion: The candle will quickly go out, even though none of the ingredients you added touched it. The carbon dioxide produced by the reaction of vinegar and baking soda pushes the air out of the jar and deprives the flame of oxygen.

LESSON 17 **Properties of Atoms & Molecules**

rust away, but depending on how much iron you start with, it may take years or even decades for the metal to all turn to rust. In order for a chemical reaction to take place, the reactants must be in contact with each other. So the speed of the reaction is not only affected by what kind of reaction is taking place, but also by the size and shape of the reactants. A cube of iron will rust much more slowly than a thin sheet of iron containing the same amount of material because the thin sheet has more surface area and the oxygen in the air can react with more of the iron molecules at one time.

Increasing the concentration of reactants will usually speed up the reaction. The more molecules of each type of reactant there are, the more likely they are to come in contact with each other and react together. So adding more reactants or pushing them closer together will speed up the rate of the reaction.

In other reactions, heat can speed up the rate at which the reaction takes place. Heat causes the molecules to move more quickly so the reactants come in contact with each other more often and the reaction speeds up. Another way to increase the reaction rate of some chemical reactions is to add a catalyst. A catalyst is a substance that is added that encourages the reaction to occur, but is not used up in the reaction. We will explore catalysts more in a later lesson.

It is important to remember that chemical reactions are taking place all around us and even inside us all the time. These reactions are necessary for life and are designed by God to happen in a very predictable way. So enjoy learning about chemical reactions.

What did we learn?

- What is a chemical reaction?
- What are the initial ingredients in a chemical reaction called?
- What are the resulting substances of a chemical reaction called?

Taking it further

- How might you speed up a chemical reaction?
- A fire hose usually sprays water on a fire to put it out. Water does not deprive the fire of oxygen, so why does water put out a fire?
- What chemical reaction do you think is taking place in the making of a loaf of bread?

Temperature & surface area

You just learned that reaction rate can be increased in several ways. What are some of those ways? Two of the most important things that affect reaction rate are temperature and the surface area of the reactants. Here is a fun way to see the effects of temperature and surface area on reaction rate. On a copy of the "Reaction Rate Experiment" worksheet, complete the hypotheses section before conducting the experiments below.

Activity 1

Purpose: To test the effects of temperature on reaction rate

Materials: water, stove, pan, three clear cups, ice, Alka-Seltzer® tablets, stop watch, "Reaction Rate Experiment" worksheet

Procedure:

1. Heat a small amount of water until it is boiling.
2. Carefully pour the hot water into a clear cup.
3. Fill a second clear cup with the same amount of room temperature water.
4. Fill a third cup to the same level with a combination of ice and water.
5. Drop an Alka-Seltzer® tablet into each cup at the same time and start your stop watch.
6. Time how long it takes for each tablet to completely dissolve.

Questions:

- What differences do you observe in each cup?
- Are there more bubbles in one cup than in another?
- Are there fewer bubbles in one cup?
- Which tablet dissolved the fastest?
- Which dissolved the slowest?
- Write your observations on the worksheet.

Activity 2

Purpose: To test the effects of surface area on reaction rate

Materials: water, three clear cups, Alka-Seltzer® tablets, paper, spoon, stop watch

Procedure:

1. Fill three clear cups with the same amount of room temperature water.
2. Break an Alka-Seltzer® tablet into several smaller pieces. Place all of the pieces together on a small sheet of paper.
3. Crush another tablet with the back of a spoon. Be sure to place all of the powder on another small sheet of paper.
4. Place a whole tablet in one cup, all of the pieces of the broken tablet in a second cup, and all of the powder of the crushed tablet in a third cup.
5. Begin your stop watch and again time how long it takes for each tablet to completely dissolve.

Questions:

- Which tablet dissolved first? Which tablet dissolved the slowest? Again, record your observations on the worksheet.
- Did your observations match your hypotheses? If not, try to figure out why you got an unexpected result.

LESSON 17 **Properties of Atoms & Molecules**

18

Chemical Equations

Describing how it works

How do we write a chemical equation?

Words to know:

chemical equation

Challenge words:

single-displacement reaction

double-displacement reaction

Chemical reactions are taking place all around us, but it may be difficult to understand or visualize what is happening in a reaction. Therefore, scientists have developed a method for describing what is happening in a chemical reaction. This method is called a **chemical equation**.

We could draw pictures of the atoms and molecules to help explain what is going on in a chemical reaction. This would certainly help us visualize what is going on. For example, it is easy to see what is happening in the reaction below. We see that two copper oxide molecules are combined with one carbon atom. The result is two copper atoms bonded together and one molecule of carbon dioxide. While pictures do help us to see what is going on, it quickly becomes difficult to draw pictures of all of the molecules that we know of. Do you remember what a simple sugar molecule (glucose) looks like? It has 6 carbon atoms, 12 hydrogen atoms, and 6 oxygen atoms. This is time consuming to draw and it is still a very small molecule. Some molecules have hundreds of atoms in them.

Because drawing reactions is sometimes difficult, scientists have created an easier way to describe what is going on in a reaction. They use a chemical equation.

Chemical equations work just like mathematical equations. When two quantities are added together in math, you can show that using an equation, such as $4 + 3 = 7$. This equation tells you that

This reaction would be written as:

$$2CuO + C \longrightarrow 2Cu + CO_2$$

if you add four apples to three apples you will end up with 7 apples. Similarly, a chemical equation tells you how different elements or compounds combine together to form new compounds. Chemical equations use a plus sign (+) to indicate which compounds are combined or added together, but instead of an equals sign, chemical equations use an arrow to show what the result is. The chemical symbol from the periodic table is used to represent each of the elements being combined. For example, below is the chemical equation for producing water:

$$2H_2 + O_2 \longrightarrow 2H_2O$$

Let's see how we got this equation. We know that water is made from hydrogen and oxygen. So we begin by writing the atomic symbols for hydrogen (H) and oxygen (O) on the left side of the equation with a plus sign between them.

$$H + O$$

This shows what types of atoms are being combined. Next, we draw an arrow to show that a reaction is taking place. Then we write the result on the right side of the arrow. You know that water is H_2O. This means that there are two atoms of hydrogen and one atom of oxygen in a molecule of water.

$$H + O \longrightarrow H_2O$$

There are a couple of problems with this equation. First, as you learned earlier, hydrogen and oxygen seldom occur as single atoms. They commonly occur as diatomic molecules. So let's rewrite the equation to reflect this.

$$H_2 + O_2 \longrightarrow H_2O$$

Finally, we have to make sure there are the same number of atoms on each side of the arrow. How many atoms are shown on the left side of the

Fun Fact

Two things that are necessary for photosynthesis to occur are not shown in the chemical equation. Sunlight and chlorophyll are needed but are not part of the equation. Chlorophyll is not shown in the equation because it does not permanently combine with any of the reactants. It only speeds up the reaction. Chlorophyll is called a catalyst and you will learn more about catalysts in the next lesson.

When photosynthesis occurs, energy from the sunlight is stored in the sugar molecule, but energy transfer is not shown in a chemical equation.

arrow? Four. How many atoms are shown on the right side of the arrow? Three. We have a problem with our equation; it is not complete yet. We need to show that we need two hydrogen diatomic molecules to react with every oxygen diatomic molecule and that when these molecules react they produce two molecules of water. So the complete equation would look like:

$$2H_2 + O_2 \longrightarrow 2H_2O$$

We now have four hydrogen atoms and two oxygen atoms on each side of the equation.

As we learned in the last lesson, the elements or compounds that are added together are called reactants, and the resulting compound is called the product of the reaction. In this case, the hydrogen and oxygen are the reactants and water is the product. This type of reaction, where two or more reactants are combined to form a single product is called a *composition reaction*. It has the general form of:

$$A + B \longrightarrow AB$$

Not all chemical reactions are composition reactions, however. Many reactions are just the opposite. If an electrical current is sent through a sample of water, some of the water molecules will break apart into separate hydrogen and oxygen gas molecules. This type of reaction is called a *decomposition reaction*. The general form of a decomposition equation is:

Chemical equations

Complete the "Understanding Chemical Equations" worksheet.

$$AB \longrightarrow A + B$$

Notice that this format is the opposite of the composition reaction. The compound on the left is still called the reactant, but in this type of reaction there are two or more products. In the water example, water is the reactant and hydrogen and oxygen gas are the products. The equation for the decomposition of water is:

$$2H_2O \longrightarrow 2H_2 + O_2$$

We will learn more about chemical reactions and chemical equations in the upcoming lessons.

What did we learn?

- What is a chemical equation?
- What are the elements or compounds on the left side of a chemical equation called?
- What are the elements or compounds on the right side of a chemical equation called?

Taking it further

- Why is it helpful to use chemical equations?

Reactants and products

So far we have looked at chemical equations for composition and decomposition reactions. Chemical equations can help us understand other types of chemical reactions as well. Sometimes a compound will combine with an element to form a new compound and a different element. This is shown by the chemical equation:

$$AB + C \longrightarrow AC + B$$

In this reaction, compound AB was broken apart. Then A combined with C leaving B by itself. This type of reaction is called a **single displacement reaction** or a single replacement reaction. Element B was displaced by element C in the chemical bonding. The chemical equation helps us to visualize this reaction. All metal with acid reactions are single displacement reactions. For example, when magnesium combines with hydrochloric acid the magnesium displaces the hydrogen as shown below:

$$Mg + 2HCl \longrightarrow MgCl_2 + H_2$$

Another type of chemical reaction is called a **double displacement reaction** or double replacement reaction. This is demonstrated by the equation:

$$AB + CD \longrightarrow AC + BD$$

In this type of reaction, elements B and C trade places, forming two new compounds. In the equation below you can see that iron (Fe) and hydrogen trade places.

$$Fe_2O_3 + 6HCl \longrightarrow 2FeCl_3 + 3H_2O$$

It is important to note that whatever elements you start with must also end up on the other side of the equation. For a mathematical equation to be true, both sides must be equal. For example, if you place 4 apples in a bowl then add 3 more apples to the bowl, you will have the same number of apples as in a bowl with 7 apples; you will not have 6 or 8 apples. Similarly, for a chemical equation to be true, the number of atoms of each type of element must be the same on each side of the arrow. In the water equation, there were four hydrogen atoms and two oxygen atoms on each side of the equation. Below is the chemical equation for photosynthesis:

$$6CO_2 + 6H_2O \longrightarrow C_6H_{12}O_6 + 6O_2$$

The 6 in front of the CO_2 indicates that 6 carbon dioxide molecules are needed for this reaction. Similarly, 6 water molecules are needed for this reaction. So, on the left side of the equation there are a total of 6 carbon atoms, 12 hydrogen atoms, and 18 (12 + 6) oxygen atoms. The carbon dioxide and water molecules are broken apart and the atoms combine to form one sugar molecule and 6 O_2 molecules. On the right side of the equation, there are 6 carbon atoms, 12 hydrogen atoms, and a total of 18 oxygen atoms, just like there were to begin with. The first law of thermodynamics says that matter cannot be created or destroyed; it can only change form. And chemical equations help us to see that even though the product does not look at all like what you started with, the atoms (or amount of matter) were not lost, their form was just changed.

Complete the "Reactants and Products" worksheet to get practice working with different kinds of equations.

19

Catalysts

Speeding things up

How can we speed up a reaction?

Words to know:

activation energy   enzyme

catalyst   inhibitor

Challenge words:

homogeneous catalyst   heterogeneous catalyst

As we have discussed in previous lessons, some chemical reactions are very quick and others are very slow. The rate at which a chemical reaction takes place depends on several things. What things have you already learned about that affected reaction rate? The temperature of the reactants, the concentration of reactants, and the surface area of the reactants all affect how quickly a reaction will occur. However, there are other factors that affect reaction rate as well. The reactivity of the substances affects how fast the reaction occurs. Reactions involving hydrogen are likely to go more quickly than reactions involving metals because hydrogen is very reactive.

Another factor affecting reaction rates is available energy. In order for many chemical reactions to take place, there must be a certain amount of energy available. Breaking atomic bonds requires energy and creating atomic bonds releases energy. Before a reaction can even begin there has to be enough energy available to break the necessary bonds. This energy is called the **activation energy**. When the activation energy of a reaction is low, the reaction occurs very quickly. When the activation energy is high, it occurs slowly.

The reactants have to reach a certain level of energy, or "height," before they can react with one another. Adding another substance to the mix can sometimes speed up this process. This type of substance is called a **catalyst**. Adding a catalyst is like finding a pass or shortcut over the mountain. It reduces the amount of energy necessary for the reaction to take place.

We already discussed one very important catalyst—chlorophyll. Chlorophyll is a necessary ingredient in plant cells that helps speed up the reaction rate between carbon dioxide and water in the photosynthesis reaction. However, as we learned in the fun fact in the previous lesson, chlorophyll does not show up in the chemical equation because it is not used up in the reaction. A catalyst is something that alters the rate of the reaction without being consumed in the reaction. It is important to remember that a catalyst does not make an impossible reaction possible, it just makes the reaction

LESSON 19 **Properties of Atoms & Molecules • 341**

easier. It can do this in a number of ways.

Some very important catalysts are called enzymes. **Enzymes** are found in living cells and are used in reactions involved in digestion, muscle contraction, cell construction, and reproduction. Without the many enzymes in our bodies, the chemical reactions necessary for life would occur so slowly that we would not be able to live. For example, in digestion starch is broken down into glucose. At normal body temperature, this reaction would take weeks to be completed. However, we cannot wait for weeks for our food to be digested. So God created the α-amylase (alpha amylase) enzyme to be part of our digestive systems. This enzyme makes it so the starch to glucose reaction takes only a few seconds.

Another common enzyme found in many living cells is catalase. Catalase allows the decomposition of hydrogen peroxide into water and oxygen (2 H_2O_2 $2H_2O + O_2$) to occur nearly ten billion times faster than it normally would without it. This is very important because hydrogen peroxide, H_2O_2, is a by-product of many cellular metabolic processes. This means that it is produced when your cells produce other needed chemicals. However, hydrogen peroxide is not a useful chemical in your body. So there needs to be a way to break it down into water and oxygen, two compounds that your body needs. Without catalase in your cells to break it down quickly, the levels of hydrogen peroxide would build up and poison your body. But God designed a way for the hydrogen peroxide to be quickly changed to useful compounds.

A catalyst can be very helpful if you wish to increase the rate of a reaction. But what if you want to slow down a reaction that is happening faster than you want it to? Food spoiling is a chemical reaction

Fun Fact

The metal nickel is a catalyst used in the making of margarine. It is needed in the hydrogenation process that turns liquid vegetable oil into solid margarine.

🧪 Catalysts & inhibitors

Activity 1

Purpose: To observe the catalytic effect of catalase, an enzyme that is present in many living cells

Materials: hydrogen peroxide, potato, drinking glass

Procedure:

1. Pour some hydrogen peroxide into a glass. Observe it for a few minutes. What do you observe happening? (Probably not much)
2. Make a hypothesis about what will happen if you place a slice of potato in the hydrogen peroxide.
3. Place a slice of potato in the peroxide and observe for a few minutes. What do you observe happening? (You should see little bubbles coming up off the potato.)

Conclusion: The potato contains catalase. The catalase is working as a catalyst to break the hydrogen peroxide into water and oxygen. The bubbles you see are the oxygen gas that is being produced. Is this what you predicted would happen?

Activity 2

Purpose: To observe the effects of inhibitors

Materials: apple, lemon juice, knife, brush

Procedure:

1. Slice an apple into quarters.
2. Brush two slices of apple with lemon juice.
3. What do you think will happen to the slices with the lemon juice? What do you think will happen to the slices without the lemon juice?
4. Wait 15 minutes. What differences do you see between the slices with the lemon juice and those without?

Conclusion: You should observe that the uncoated slices are turning brown and the coated slices are not. The acid in lemon juice acts as an inhibitor. It prevents the oxygen molecules from reacting with the apple molecules to produce the brown colored chemical. Other inhibitors are used in foods to prevent them from spoiling. They are often called preservatives on food labels.

that we all want to slow down as much as possible. In this case you need a "negative catalyst." A "negative catalyst" is called an **inhibitor**. An inhibitor prevents a reaction from occurring by either keeping the reactants apart, or by bonding with one of the reactants so that the chemical reaction cannot take place.

What did we learn?

- What is a catalyst?
- How does a catalyst work?
- What is an inhibitor?
- What is an enzyme?

Taking it further

- Why is it important that living cells have enzymes?
- Are catalysts always good?

Types of catalysts

There are two different kinds of catalysts and they work very differently. First, if the catalyst and the reactants are in the same phase, for example they are both gases or both liquids, then the catalyst is called a **homogeneous catalyst**. Homogeneous catalysts generally work by chemically combining with one of the reactants to form a new substance, which then quickly combines with the other reactant to form the product. An example will make this more clear.

The oxidation of sulfur dioxide is a very slow reaction. It is represented by the chemical equation:

$$2SO_2 + O_2 \rightarrow 2SO_3$$

All of these molecules are gases. To speed up this reaction we can add some nitrogen monoxide gas, NO, in with the reactants. Although the sulfur dioxide will react with the oxygen gas, the nitrogen monoxide reacts with the oxygen much more quickly to form nitrogen dioxide. Below is the chemical equation showing this reaction.

$$2NO + O_2 \rightarrow 2NO_2$$

Now the sulfur dioxide and nitrogen dioxide react very quickly to form the sulfur trioxide gas from the original equation. This reaction looks like this:

$$2NO_2 + 2SO_2 \rightarrow 2SO_3 + 2NO$$

Without the catalyst, you start with sulfur dioxide and oxygen and end up with sulfur trioxide, but it takes a long time for the reaction to occur. With the catalyst, you start with sulfur dioxide, oxygen, and nitrogen monoxide and end up with sulfur trioxide and nitrogen monoxide. The catalyst was used for an intermediate reaction, but was not part of the final product, and you got to the final product much more quickly.

A second kind of catalyst is a **heterogeneous catalyst**. This type of catalyst is one that is in a different phase from the reactants. Commonly the catalyst is a solid and the reactants are gases or liquids. This type of catalyst works in a very different way from the homogeneous catalyst. A heterogeneous catalyst attracts the reactants to itself, forcing them to come closer together. This encourages the reaction to occur more quickly because it increases the probability of the reactants interacting.

One very important use of heterogeneous catalysts is in the design of a catalytic converter. A catalytic converter reduces toxic emissions from your car.

The main reaction taking place inside your car's engine is a combustion reaction. Ideally, the gasoline mixes with the oxygen in the air to produce energy and carbon dioxide. However, the reaction takes place so quickly that there is not always enough oxygen to fully react with the carbon that is released in the combustion, so often carbon monoxide is produced instead of carbon dioxide. Carbon dioxide is harmless for people to breathe, but carbon monoxide is poisonous to breathe.

The catalytic converter is a device on a car through which the car exhaust must flow. The inside of the device is coated with platinum or rhodium. These solid metals attract carbon monoxide molecules and oxygen molecules in the exhaust. Since these molecules are attracted to the metal, they move to the metal surface. This causes them to come closer together than they were in the air. As they get close together the carbon monoxide combines with the oxygen to form carbon dioxide. The metal has reduced the energy required for the reaction by pulling the reactants together and increasing their concentration.

Both homogeneous and heterogeneous catalysts are important in many areas of our lives.

LESSON 19 Properties of Atoms & Molecules

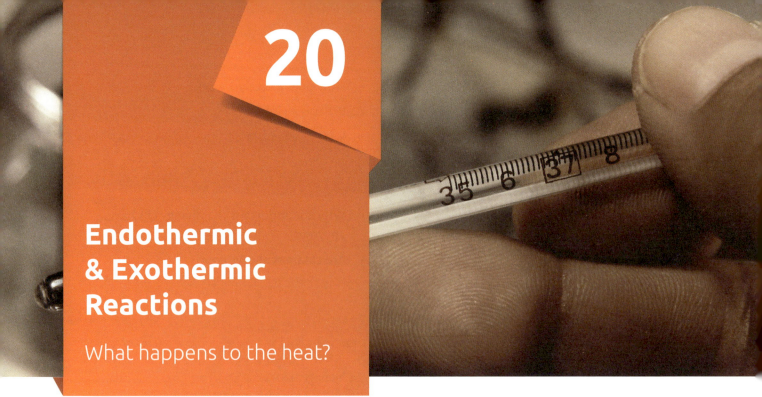

20

Endothermic & Exothermic Reactions

What happens to the heat?

What kinds of reactions create heat?

Words to know:

endothermic reaction   exothermic reaction

Challenge words:

enthalpy

Energy plays a very important role in chemical reactions. Without the energy from sunlight, photosynthesis cannot occur. The solar energy is stored in the sugar molecules and later released when an animal eats the plant. This storing and releasing of energy is a very important part of God's provision for life on earth.

A chemical reaction that stores energy is called an **endothermic reaction**. Another way to think of an endothermic reaction is to think of it as absorbing energy. The energy goes in; "in" sounds like "en," so endothermic has the energy going into the reaction. In photosynthesis, the energy that is absorbed is in the form of light. However, most endothermic reactions absorb energy in the form of heat. This energy must be added to the reactants in order for the reaction to occur.

One common endothermic reaction occurs when you are baking. Baking soda decomposes with the heat of the oven, releasing carbon dioxide gas. Similarly, yeast reacts much more quickly when heat is added. These are both endothermic reactions. One endothermic reaction you may not associate with heat is a chemical "ice pack." When an athlete is hurt, the doctor or trainer may use a special pack that contains two chemicals that are separated inside the pack. When these chemicals are combined, they absorb heat so the pack feels cool and helps reduce swelling around the injured area.

Not all chemical reactions absorb energy. Many chemical reactions release energy. These reactions are called **exothermic reactions**. You can remember this by thinking that "exo" sounds like "exit." The exit is how you leave a building, and energy is leaving an exothermic reaction. Energy can be released in the form of light or heat.

One of the most common examples of an exothermic reaction is combustion or burning. In one form of combustion, methane gas combines with oxygen to produce carbon dioxide, water, light, and heat. The chemical equation for this reaction is:

$$CH_4 + 2O_2 \longrightarrow CO_2 + 2H_2O$$

This is a similar reaction to the one between gasoline and oxygen in your car's engine.

344 • Properties of Atoms & Molecules   LESSON 20

Endothermic & exothermic reactions

Endothermic reaction

Purpose: To observe the effect of heat on an egg cooking

Materials: five pieces of paper, water, five uncooked eggs, small sauce pan, stove, timer or clock

Procedure:

1. Label five pieces of paper with the numbers 1–5.
2. Place five uncooked eggs in a small pan.
3. Cover them with water and bring to a boil over medium heat. Begin timing as soon as the water begins to boil.
4. After one minute remove one egg and place it on the paper marked with a 1.
5. One minute later, after a total two minutes of boiling, remove a second egg and place it on the number 2.
6. Continue removing an egg after each minute has passed, placing the egg on the paper showing its total cooking time.
7. After removing all of the eggs, remove the pan from the stove.
8. Break open each egg and you will be able to observe how the molecules in the egg have reacted as the heat was added to them.

Exothermic reaction

One reaction that produces or releases heat is the formation of rust, called oxidation. This occurs in small amounts and is not usually noticeable. However, you can measure the heat released in a sealed container.

Purpose: To measure heat released by oxidation

Materials: steel wool, jar with lid, thermometer, vinegar

Procedure:

1. Place a piece of steel wool, the kind without soap in it, in a jar.
2. Place a thermometer in the jar so that you can read the temperature in the jar.
3. Seal the jar and wait 5 minutes. Read the temperature inside the jar.
4. Next, open the jar and pour ¼ cup of room temperature vinegar over the steel. The acid in the vinegar and the oxygen in the air will react with the steel to form rust.
5. Reseal the jar and measure the temperature inside the jar every five minutes for twenty minutes. What did you observe? You should see the temperature in the jar rise as the chemical reaction releases heat.

The heat produced by combustion can be helpful if you are trying to heat your house, such as when you burn natural gas in your furnace. But in a car engine, too much heat can be harmful to the engine. Therefore, the engine must be cooled. This is most often done by running water from the radiator around the engine to absorb the heat.

Another very important exothermic reaction is the reaction that occurs as your food is digested. The energy released as food molecules are broken down is necessary for you to be able to function. The heat released helps your body regulate its temperature. Warm-blooded animals generally have to eat more food than cold-blooded animals in order to regulate their body temperatures.

What did we learn?

- What is an exothermic reaction?
- What is an endothermic reaction?

Taking it further

- If a chemical reaction produces a spark, is it likely to be an endothermic or exothermic reaction?
- How do photosynthesis and digestion reveal God's plan for life?
- If the temperature of the product is lower than the temperature of the reactants, was the reaction endothermic or exothermic?

Energy in a reaction

Endothermic and exothermic reactions are all about energy: energy entering the substance that is formed or energy exiting the substance that is formed. But where does this energy come from? With an endothermic reaction it may be somewhat obvious where the energy is coming from. If the air around the reaction becomes colder during the reaction, it is obvious that heat is being removed from the air. However, other reactions are not so obvious. Photosynthesis is an endothermic reaction, but the temperature around a leaf does not go down as it performs photosynthesis. Instead its energy comes from the light of the sun.

As an endothermic reaction takes place, where does the energy go? We know that according to the first law of thermodynamics energy cannot be created or destroyed; so we know the energy is not lost, it has just changed form. What form is the energy in now? The energy is stored in the bonds of the products.

All chemical bonds between atoms contain energy. Some bonds are strong and store a greater amount of energy than weaker bonds. Recall that it takes energy to break bonds. It takes more energy to break strong bonds than it does to break weak bonds.

Also recall that energy is released when bonds are formed. When weak bonds are formed a small amount of energy is released. When strong bonds are formed more energy is released. This energy is usually released as either heat or light.

So, when a chemical reaction takes place some energy is required to break the bonds of the reactants. The amount of energy required depends on how many bonds must be broken and how strong those bonds are. Then as the products are formed the new bonds release energy. Again the amount of energy released depends on how many bonds are formed and how strong those bonds are. If the amount of energy required to break the bonds in the reactants is greater than the amount of energy released by the bonds formed in the products, then the reaction is endothermic. If the amount of energy

Fireworks are the result of an exothermic reaction that produces light, heat, and sound.

released is greater than the amount required to break the bonds then the reaction is exothermic.

The amount of energy stored in a substance's bonds is called its **enthalpy** (EN-thal-pee). Many scientific experiments have been conducted to determine the enthalpy contained in various compounds. These values can be found in chemistry charts. To determine if a chemical reaction is endothermic or exothermic, scientists do not have to test the temperature of the surroundings. Instead, they can calculate the enthalpy of all of the reactants and the enthalpy of all of the products and decide if the enthalpy is higher before or after the reaction. If the enthalpy of the reactants is higher than the enthalpy of the products, then energy was released and the reaction was exothermic. If the enthalpy of the reactants is less than the enthalpy of the products, then energy was absorbed during the reaction and it was endothermic.

Now you get to do an experiment to determine if a reaction is endothermic or exothermic. Perform the experiment described on the "Endothermic or Exothermic?" worksheet.

UNIT 5

Acids & Bases

21 Chemical Analysis
22 Acids
23 Bases
24 Salts

◊ **Distinguish** between the properties of acids and bases.
◊ **Describe** the result of mixing acids and bases.
◊ **Demonstrate** how to test for acids and bases.

Properties of Atoms & Molecules • 347

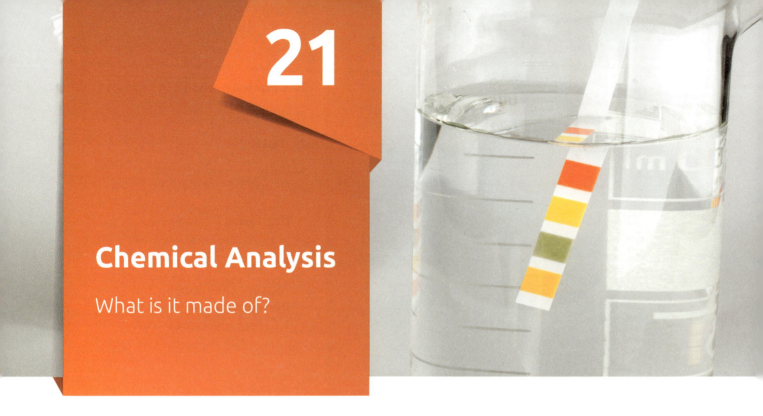

21

Chemical Analysis

What is it made of?

What kinds of chemical tests are there?

Words to know:
chemical analysis
pH scale
indicator

How can you know what chemicals a substance contains? A good scientist always starts with observation. You can tell many things about a sample by observing its color, texture, state, mass, boiling and freezing points, and other physical characteristics. However, tasting and smelling chemicals can be very dangerous. Touching can also be dangerous as some chemicals are corrosive and can burn your skin. So direct observation of physical characteristics is limited. Therefore, often the most useful way to determine what kind of matter a sample is made of is to test it with chemical reactions. This type of testing is called **chemical analysis**.

Some chemical analysis techniques can be very involved and require a greater understanding of chemistry than we will cover in this book. Others are dangerous and should only be done in a laboratory. For example, one type of chemical analysis is called a flame test. A small sample of the substance is heated until it burns. The color of the flame can indicate what elements were in the sample. Other tests require expensive equipment. A spectroscope is a piece of equipment that passes a spectrum of light through a sample. The color of light that passes through indicates what the sample is made of. Although these tests are interesting, they are not easy to do at home. Fortunately, there are some chemical analysis tests that are fun and easy to do at home.

One common type of chemical test that is easy to do is to use indicators. An **indicator** is a chemical

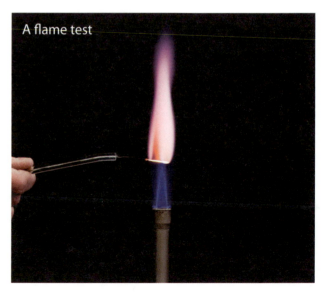

A flame test

compound that changes color when it reacts with certain other chemicals. Iodine is a liquid that is normally red or orange-brown. But in the presence of starch, iodine reacts to form a liquid that is blue or green. This is a simple way to test if your sample contains starch. Other chemicals change color when they are in the presence of protein or sugar. Indicators are an important part of chemical analysis.

One of the most common uses of indicators is to test for acids and bases. There are several chemicals that change color in the presence of an acid or a base. One of the most famous is litmus. Litmus is a chemical produced from certain lichens that are native to California. Litmus is naturally blue but it turns red in the presence of an acid. Blue litmus paper is thus used to test for acids in liquids. Once the paper turns red it stays red until it reacts with a base. Then it becomes blue again. So red litmus paper can be used to test for a base.

pH indicator paper

Other chemicals turn different colors depending on the strength or weakness of the acid or base. These chemicals can be dried together on one sheet of paper to form what is called a universal pH indicator paper, or mixed together to form a universal indicator liquid. The **pH scale**, which stands for per hydronium, or power of hydrogen, indicates the strength of the acid or base. The pH scale goes from 0 to 14 with 0 being a very strong acid, 7 being neutral (neither acid nor base), and 14 being a very strong base. The universal indicator will change many different colors depending on the strength of the acid or base. One color would indicate a pH of 1, a different color would be a pH of 2, and so on.

If it is not critical to know the strength of an acid or base, other chemicals can be used just to

Fun Fact

Stomach acid is hydrochloric acid (HCl) and can be a 1 on the pH scale—a very strong acid. Lye, which is used in some cleaners and is sometimes used in soap, is a strong base and can be a 14 on the pH scale. Pure water, blood, and eggs are a 7 on the pH scale; they are neutral.

🧪 Making an acid/base indicator

Purpose: To make your own acid/base indicator solution

Materials: purple cabbage, water, microwave oven or sauce pan and stove

Procedure:

1. Combine 1 cup of chopped purple cabbage with 1 cup of water.
2. Heat in the microwave for 2–3 minutes or bring water to a boil and boil for 5 minutes on the stove.
3. Drain and save the water. The water should have a definite purple color. If the water is very light, add it back in with the cabbage and boil for a few more minutes.
4. Store your indicator in a sealed container and keep it in the refrigerator for future use.

Conclusion: The purple pigment from the cabbage chemically reacts with acids to form a pink liquid, and it chemically reacts with bases to form a blue or blue/green liquid. If the substance is neutral, the indicator will remain purple. Use your new indicator to test several substances around the house to see if they are acids, bases, or neutral by combining a few drops of indicator with the substance to be tested. The substance to be tested must be a liquid or a solid that is dissolved in water.

LESSON 21 **Properties of Atoms & Molecules • 349**

Fun Fact

Hydrangeas are flowering plants. If the soil that the plant is growing in is acidic, the flowers will be blue. If the soil is basic, the flowers will be pink. Soil formed from chalk or limestone tends to be basic and soil formed from sandstone tends to be acidic.

indicate the presence of acids and bases. Bromothymol blue is a substance that is blue in basic solutions, green in neutral solutions, and yellow in acidic solutions. Several plants can also be used to make acid/base indicators. One of the easiest to use is red cabbage. The liquid that is drained after boiling red or purple cabbage is purple. In the presence of an acid, it will turn pink and in the presence of a base it will turn blue or green.

Chemical analysis is very useful and can be fun. In the next few lessons, we will use an acid/base indicator to learn more about the chemicals around us.

What did we learn?

- What is chemical analysis?
- List three different types of chemical analysis.
- What is a chemical indicator?
- What is the pH scale?
- What does a pH of 7 tell you about a substance?

Taking it further

- Why is it important to periodically test the pH of swimming pool water?
- Name at least one other use for testing pH of a liquid.

Chemical analysis methods

There are a variety of chemical analysis methods that are used today. Choose one of the methods listed here. Do some research and see what you can find out about that analysis method. How does it work? What does it tell you? What equipment do you need to do it? Make a presentation of what you have learned.

- Spectroscopy
- Mass Spectrometry
- Colorimetry
- Chromatography
- Electrophoresis
- Crystallography
- Microscopy
- Electrochemistry
- Gravimetry

22

Acids

Does it burn?

What makes a substance an acid?

Words to know:

acid

neutralize

hydronium ion

Challenge words:

electroplating

What do you think of when you hear the word acid? Do you picture a liquid eating through metal and destroying everything in its path? Some acids are very caustic and can eat through metals; however, many acids are weak and are used every day. Let's take a look at what an acid actually is.

Acids and bases are substances that form ions when dissolved in water. Recall that ions are molecules or atoms that have either a positive or negative charge. When an **acid** is dissolved in water, one or more hydrogen atoms break off of the molecule. The original molecule holds tightly to the electrons so the hydrogen atom leaves behind its electron, causing the hydrogen to have a positive charge and the remaining molecule to have a negative charge.

The symbol for a positive hydrogen atom is H^+. The H^+ ion is really just a proton. It is very reactive and won't stay alone for very long. It quickly combines with a water molecule to form H_3O^+, which is called a **hydronium ion**. The formation of hydronium ions in water is what classifies a substance as an acid. Some acids are stronger than others. A strong acid easily gives up its hydrogen atoms to form hydronium ions and a weak acid holds onto most of its hydrogen atoms so it forms fewer hydronium ions.

Acids have specific characteristics. The word *acid* comes from the Latin word *acer* meaning sour, and foods that contain acids have a distinct sour taste. Citrus fruits have citric acid in them, which is why lemons and limes have such a sour flavor.

Fun Fact

A bee sting contains formic acid. When the acid combines with the water in your skin cells it forms hydronium ions, which irritate or hurt your cells. You can help neutralize the acid from the bee sting by covering the area with a paste made from baking soda and water or with toothpaste, both of which are bases.

LESSON 22 **Properties of Atoms & Molecules • 351**

Lemons taste sour because of the citric acid in them.

Soft drinks containing carbon dioxide form carbonic acid, which gives the soda a sour/tangy flavor. Foods that contain vinegar, such as pickles, are sour because of the acetic acid from the vinegar. And rhubarb contains oxalic acid, giving it a sour taste as well.

Acids also have other distinctive characteristics. Because acids easily form ions, they are good conductors of electricity. Most acids will react with metals and many are corrosive and can burn your skin. Acids **neutralize** bases, meaning they react with them to form substances that are neither acid nor base. And finally, acids react with indicators. As discussed in the previous lesson, acids change the color of many different compounds when they are chemically combined.

Acids are found in many places other than food. Hydrochloric acid is found in your stomach and helps to digest your food. Sulfuric acid is used in car batteries. And decaying plants produce humic acid. So, the next time you hear the word *acid*, you don't have to fear a liquid that melts through everything it touches. Just think about your favorite soft drink.

Fun Fact

Sulfuric acid is the most produced chemical in the United States. The biggest use of sulfuric acid is in the making of fertilizers. It is also used to make car batteries, paints, plastics, and many other manufactured items. In fact, sulfuric acid is so important to manufacturing that some economists use a country's use of sulfuric acid as an indicator of how well that country's economy is doing.

What did we learn?

- What defines a substance as an acid?
- What is a hydronium ion?
- How is a weak acid different from a strong acid?
- What are some common characteristics of an acid?
- How can you tell if a substance is an acid?

Taking it further

- Why is saliva slightly acidic?
- Would you expect water taken from a puddle on the forest floor to be acidic, neutral, or basic? Why?
- What would you expect to be a key ingredient in sour candy?

Testing for acids

Purpose: To test for acids

Materials: lemon juice, vinegar, clear soda, milk, saliva, cabbage indicator solution from lesson 21

- Lemon juice
- Vinegar
- Clear soda pop—lemon lime soda works well
- Milk
- Saliva

Procedure:

1. Add a few drops of the indicator you made in the previous lesson to a sample of each of the following items to determine if they are acids.

Questions:

- What color did the indicator become when mixed with each of these items?
- Which items are acidic?

Displacement reaction

Acids and bases have many uses in industry. They can be used for many chemical reactions. Today you will use an acid to move copper atoms from a penny to a steel paperclip. The acid reacts with the copper in the penny and with the iron in the paperclip. This is a displacement reaction so the iron is displaced by the copper.

Purpose: To create a copper-coated paperclip

Materials: 15 pennies, jar with lid, salt, vinegar, steel paperclip

Procedure:

1. Put 15 pennies into a jar.
2. Sprinkle 2 tablespoons of salt over the pennies.
3. Add enough vinegar to cover the pennies.
4. Put a lid on the jar and swirl the solution for 15 seconds. Try to have the pennies end up on top of the salt.
5. Open the jar and drop in a steel paper clip. Close the lid.
6. Observe the pennies and the paperclip after 15 minutes. How do they look? What do you observe happening in the jar? Observe them again after 30 minutes and again after 60 minutes.

7. Allow the solution to sit undisturbed for several days, then remove the pennies and the paperclip. How do they look compared to when they went into the jar?

Conclusion: This process is similar to the process of electroplating. **Electroplating** is the process of passing electrical current through a metallic salt solution to deposit a thin layer of metal onto a conductive object. This can be done for many reasons. Often electroplating is performed to deposit a layer of metal that will not rust onto a steel object to keep it from rusting. You could speed up this process in your experiment if you connected a penny to one terminal of a battery and the paperclip to the other terminal, while the objects are in the acid solution.

23

Bases

The opposite of acids

What makes a substance a base?

Words to know:

base   hydroxide ion

Challenge words:

quantitative measurement   acid/base titration

A base is often described as the opposite of an acid, but what does that really mean? Just as an acid produces ions in a water solution, so also a base produces ions in a water solution. But an acid produces positive hydronium ions and a **base** produces negative hydroxide ions. When a substance that is a base is dissolved in water, it releases OH⁻ ions from its molecule, leaving the rest of the original molecule short one electron, so it has a positive charge and consists of one oxygen atom and one hydrogen atom. It pulls an extra electron with it when it leaves the original base molecule. This gives it a negative charge. Therefore, it is called an ion instead of a molecule. The OH⁻ ion is called a **hydroxide ion**. The hydroxide ion is very reactive.

Just as the ability to produce hydronium ions determines the strength of an acid, so also the ability to produce hydroxide ions determines the strength of a base. The more hydroxide ions a base produces in water, the stronger the base. Weak bases hold onto their OH⁻ ions more tightly than strong bases do.

Another name for a base is an alkali. This is because some of the strongest bases are formed from the alkali and alkaline earth metal elements. These are the elements in columns IA and IIA on the periodic table. These elements include sodium, potassium, and calcium.

Because of their common molecular structures, bases have common characteristics. One characteristic of a base is that bases have a bitter taste. Soap is a base and anyone who has ever gotten soap in his mouth can attest to its bitter aftertaste. Bases also have a slippery feeling. And because bases produce ions in water, they are good conductors of electricity.

Fun Fact

Sodium hydroxide is a base that dissolves wood resin. It is added to wood pulp because it eliminates the resin, leaving the cellulose behind. The cellulose strands are then used to make paper.

Testing for bases

Purpose: To test for bases

Materials: soap, ammonia, baking soda, anti-acid, toothpaste, cabbage indicator

Procedure:

1. Use the cabbage indicator from the previous lessons to test for bases. Add a few drops of indicator to a sample of each of the following items to determine if they are bases.

- Soap
- Ammonia
- Anti-acid (liquid or tablets that are crushed and dissolved in water)
- Baking Soda
- Toothpaste

Questions:

- What color did the indicator become when mixed with each of these items?
- Which items were bases?

Bases, such as soap, are slippery.

Many bases are also caustic. A strong base can burn your skin as easily as a strong acid can.

Some common bases you may encounter include ammonium hydroxide, which is found in many household cleaners; sodium hydroxide, which is lye; and magnesium hydroxide, which is found in anti-acid medications. The reason bases are used in anti-acid medications is that an acid and a base will neutralize each other. The H⁺ from the acid will quickly combine with the OH⁻ from the base to form water. Another place you are likely to find a base is in your toothpaste. Your saliva naturally has acids in it that help digest your food, so toothpastes usually have a base in them to help neutralize the acid in your mouth to help prevent tooth decay.

What did we learn?

- What defines a substance as a base?
- What is a hydroxide ion?
- How is a weak base different from a strong base?
- What are some common characteristics of a base?
- How can you tell if a substance is a base?

Taking it further

- If you spill a base, what should you do before trying to clean it up?
- Do you think that strontium (Sr) is likely to form a strong base? Why or why not?

Fun Fact

A wasp sting contains a base. When the hydroxide ions in the base irritate or hurt your cells, you can help neutralize the alkali from the wasp sting by covering the area with vinegar. But be sure that it is a wasp that has stung you and not a bee. Bee stings contain an acid and vinegar will only increase the problem, not neutralize it.

Acid/base titration

Although you have tested for the presence of acids and bases using your indicator, you have not done any **quantitative measurements**, tests that use actual numbers to measure how much of something there is. **Acid/base titration** is a method for determining the unknown quantity of a base (or acid) by carefully measuring the exact amount of acid (or base) needed to completely neutralize it. This may sound complicated, but an example may help show how easy this is.

If you know how many molecules of acid are in a drop of acid, and you know how many drops of acid it takes to neutralize a base solution, then you know how many molecules of base were in the solution. Learning how to calculate the number of molecules in a sample is beyond the scope of this book; you will learn how to do this in a high school chemistry course. For now, just think of titration as a way to figure out how many drops of base are in a solution by counting how many drops of acid are needed to neutralize it.

Purpose: To better understand how acid/base titration works

Materials: distilled water, clear glass, ammonia, acid/base indicator, vinegar, eyedropper

Procedure:

1. Pour ½ cup of distilled water into a clear glass.
2. Add 2 teaspoons of clear ammonia.
3. Next add one tablespoon of the acid/base indicator. This should result in a blue/green solution.
4. Now use an eyedropper to add vinegar to this solution one drop at a time until you see the solution just start to turn a pale pink. Count the drops as you add them to the solution. This will take a while so don't get impatient.

Conclusion: When the solution turns pink, it indicates that there are no more base molecules for the acid to react with so it is now reacting with the indicator. The number of drops of vinegar gives you an indication of the number of drops of ammonia in the solution. Actual titration involves chemical formulas and calculation of the number of molecules, but you get the basic idea of titration here.

24

Salts

Pass the salt, please.

What are salts, and how are they formed?

Words to know:

salt
normal salt
acid salt
basic salt

Challenge words:

proton donor
proton acceptor

What do you think of when someone says, "Pass the salt, please"? You would probably think of table salt. But sodium chloride, table salt, is not the only salt around. There are many other common salts. A few of these salts are used in cooking. For example MSG, monosodium glutamate, is a salt that is used in many oriental dishes. Other salts are used to make fertilizer, medical supplies, and a number of other chemical products.

A **salt** is formed when an acid and a base mix. Recall that when an acidic material is dissolved in water the molecule breaks up into a positive hydrogen ion (H^+) and a negative acid ion. Similarly, when a base is dissolved the molecule breaks up into a negative hydroxide (OH^-) ion and a positive base ion. We already discussed how the hydrogen ion and the hydroxide ion combine to form water. But the positive base ion and the negative acid ion also combine to form a salt. For example, table salt is formed when a positive sodium ion, Na^+, combines with a negative chlorine ion, Cl^-, to form NaCl. Monosodium glutamate is formed when Na^+ combines with the glutamate ion, $C_5H_8O_4^-$, to form $NaC_5H_8O_4$.

Salt crystals as seen through a microscope

Fun Fact

Many paints receive their color from salts that are added. Vermilion is a salt that is added to make a red paint; cadmium sulfide makes a yellow paint; and malachite is a salt used in some green paints.

LESSON 24 **Properties of Atoms & Molecules** • 357

When an acid and base completely neutralize each other, the resulting salt is called a **normal salt**. However, not all acids and bases completely neutralize each other. If some of the acid remains after all the base is used up, the result is called an **acid salt**. If all of the acid is used up in the reaction but some of the base remains, the result is called a **basic salt**.

Because of their ionic structures, salts generally form crystals. If you examine table salt, you will see that it forms distinct crystal shapes. This is true of most salts. Another characteristic of most salts is the distinctive salty flavor.

Families of salts are named by the acid from which they originate. Sulfates are salts that are made from sulfuric acid. Chlorides come from hydrochloric acid. Nitrates are formed from nitric acid and carbonates come from carbonic acid. You may have heard of some of these salts and not realized what they were. Many fertilizers contain nitrates, phosphates, and potash, which are all salts.

The next time you salt your corn on the cob, remember that there are many different kinds of salt.

What did we learn?

- How is a salt formed?
- What are two common characteristics of salts?
- How are salt families named?
- Name three salt families.

Taking it further

- What do you expect to be the results of combining vinegar and lye?
- Why are some salts still acidic or basic?

Fun Fact

Plaster of Paris, that white powder that is used for many art projects, is actually a salt—calcium sulfate.

🧪 Acid + base = salt + water

When an acid reacts with a base it produces a salt and water. It can also produce other substances such as carbon dioxide. You are going to demonstrate this with an acid/base reaction that you are very familiar with—vinegar and baking soda. Vinegar contains acetic acid which combines with baking soda, which is a base, to produce sodium acetate, which is a salt, as well as water and carbon dioxide. The chemical equation for this reaction is:

$$CH_3COOH + NaHCO_3 \rightarrow CH_3COONa + H_2O + CO_2$$

Acetic acid + baking soda → sodium acetate + water + carbon dioxide

Purpose: To demonstrate that an acid and base reaction produces salt and water

Materials: water, measuring cups and spoons, vinegar, baking soda, cup, cotton swabs

Procedure:

1. Place ½ cup of water in a measuring cup. Add 1 teaspoon of baking soda and stir until the solution becomes clear.
2. Pour 1 tablespoon of the solution in a separate cup.
3. Pour 1 tablespoon of vinegar in the measuring cup with the soda water. Stir the solution until the production of bubbles slows down. Stop stirring and wait for the bubbles to stop completely. You now have two solutions. In one cup you have a solution of baking soda and water. This is a basic solution. In the measuring cup you have a solution that was created when the baking soda reacted with the acetic acid in the vinegar. Which solution do you think will have a saltier taste?
4. Use a swab to place a small amount of the soda water solution on your tongue. Think about how salty this solution is. Use a second swab to place a small amount of the vinegar/soda water solution on your tongue. How salty is this solution compared to the first solution?

Conclusion: You should find that the first solution with baking soda and water is slightly salty. However, after adding the vinegar the new solution should taste even saltier since the chemical reaction produced a salt in addition to water and carbon dioxide.

🏅 Acid/base reactions

You have learned that an acid is defined by how easily it gives up a hydrogen ion and a base is defined by how easily it gives up a hydronium ion, but there is an alternative way to define acids and bases. Recall that a hydrogen ion (H^+) is really just a proton. So an alternative definition for an acid is a **proton donor** and an alternative definition for a base is a **proton acceptor**. Using this definition, a base does not necessarily have to donate a hydronium ion but has to be able to bond with a hydrogen atom.

Now that you have learned about acid/base reactions and chemical equations it is time for you to put what you have learned to work. Show how well you understand these reactions by completing the "Acid/Base Reactions" worksheet.

Batteries

SPECIAL FEATURE

What do an eel, a ray, an African catfish, and the Energizer bunny's battery have in common? They all create electricity by chemical means. It is known that chemicals have been used to generate electricity since Alessandro Volta produced the first modern battery in 1800. Volta discovered that stacks of copper and zinc that were separated by a saltwater solution would create electricity. This early electrochemical cell, or battery, was called a voltaic cell in honor of Volta, and is the basis for all the batteries to follow.

However, it is unclear if Alessandro Volta's batteries were the first batteries ever used in history. In 1938 a German archaeologist, Wilhelm Konig, was working in Khujut Rabu, just outside Baghdad, Iraq. In the artifacts he found a clay jar that measured about five inches (13 cm) long. Inside the jar was a copper cylinder that encased an iron rod. The copper tube and iron rod were held in place with asphalt, but only the iron rod was exposed on the top. It has been shown that if the jar were filled with an electrolyte, like vinegar or wine, a small voltage would be generated between the copper and iron.

This clay jar, and other similar jars, have been dated as early as 200 BC. It is unclear what their purpose might have been. Not all scientists agree that they were used as batteries or to generate electricity. If they were used to generate electricity, they may have been used for medical purposes, or by hooking several of them together they might have generated enough power to electroplate a very thin layer of gold onto silver. So far, the purpose or uses of these *Baghdad batteries*, as they have been called, have not been found in any writings from the past and remain a mystery. So it is commonly accepted that Volta invented the first real battery.

At the heart of every battery is the electrolyte solution. An electrolyte is any solution that conducts electricity. Electricity is conducted by ions or charged particles, so solutions containing acids, bases, or salts make good electrolytes.

In order to conduct electricity, a chemical reaction must take place inside the battery. A simple battery has a center core made out of graphite, which has a positive charge and is attached to the top of the battery. This core is surrounded by an electrolyte paste. The bottom of the battery is attached to a plate of zinc with a negative charge. When the positive and negative terminals are connected, electricity flows from the negative to the positive terminal through the electrolyte paste as a chemical reaction takes place.

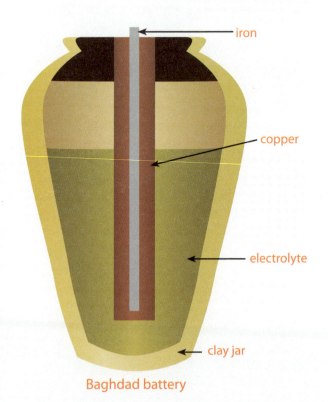

Baghdad battery

Different types of batteries use different metals and different electrolytes, but the idea is the same. A very common example is the car battery, often called a lead-acid battery. In a car battery, negative plates made of lead (Pb) are connected to the anode, or negative terminal, of the battery. The positive plates made of lead dioxide (PbO_2), are connected to the cathode, or positive terminal. These plates are submerged in a sulfuric acid solution (H_2SO_4). When the two terminals are connected, the lead loses two electrons and becomes Pb^{2+}. This atom combines with an SO_4^{2-} ion in the solution to produce $PbSO_4$ (lead sulfate). At the cathode, PbO_2 atoms combine with the H_2 atoms in the electrolyte solution to form Pb (lead) and H_2O (water). This chemical reaction aids in the flow of electrons through the electrolyte solution, which produces electricity that helps start your car.

In recent years, many new designs of rechargeable batteries have been developed. The batteries are recharged by applying a higher voltage in the opposite direction. This causes the chemical reaction to reverse and the battery can again produce electricity.

The next time you replace the batteries in your flashlight, remember that you are holding a chemical reaction in your hand.

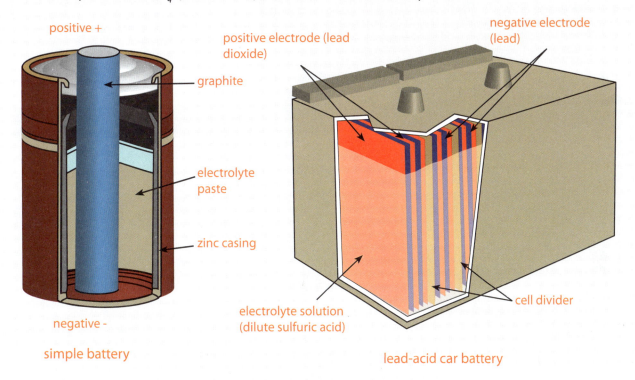

simple battery

lead-acid car battery

UNIT 6

Biochemistry

25 Biochemistry
26 Decomposers
27 Chemicals in Farming
28 Medicines

◊ **Describe** the importance of water, proteins, fats, and carbohydrates to living things.

◊ **Explain** the connection between natural decomposers and fertilizers.

◊ **Describe** how medicines have impacted mankind.

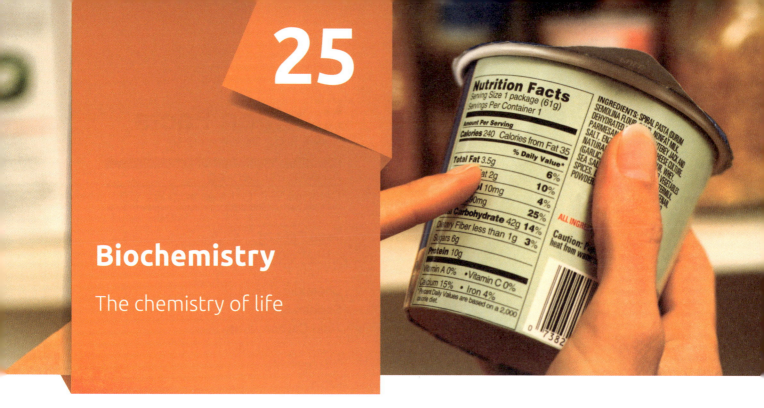

25

Biochemistry

The chemistry of life

How do our bodies use chemistry?

Words to know:

solvent

protein

fat

carbohydrate

Challenge words:

protease

Plants and animals depend on chemical reactions for nearly every function of life. As you have already learned, plants carry on the chemical reaction of photosynthesis, converting water and carbon dioxide into sugar and oxygen, thus providing food for nearly every food chain. Cellular respiration is the chemical reaction that breaks down the sugar into carbon dioxide, water, and energy for the body. Another important chemical reaction that takes place in nearly every animal occurs during breathing. The hemoglobin in the red blood cells reacts with the oxygen in the lungs. This new compound is carried by the blood stream to all parts of the body where it reacts with the muscles to release the oxygen and make it available for other uses, such as cellular respiration.

Chemical reactions are taking place in your body all the time. Most of these chemical reactions require water as a **solvent**. God designed the plasma in your blood to be mostly water, which is used to dissolve the many chemical compounds that your body needs so they can be easily transported throughout your body. About two-thirds

Our bodies need about ½ gallon of water each day to function properly.

LESSON 25 **Properties of Atoms & Molecules • 363**

🧪 A balanced diet

Because your body needs particular chemicals to function, it is necessary that you eat a well-balanced diet. You need to be sure to include carbohydrates, proteins, and fats each day. To help you in deciding what to eat, food manufacturers include information on food labels that tell you how much of each of these chemicals you will get in a serving. The labels also tell how many calories are in a serving so you know how much energy the food contains. And because some people must watch their sodium intake, most food labels include the amount of sodium as well. Examine the label shown here to see what information is included.

According to the USDA (United States Department of Agriculture), each day everyone should eat foods from each of the following food groups: grains, fruits, vegetables, dairy, and meat. This is because the various food groups are composed of the different chemicals your body needs. Grains, fruits, and vegetables have the different carbohydrates your body needs. Dairy products and meats primarily have proteins and fats.

Americans often eat foods high in fat and don't eat enough fruits and vegetables. Note that many pre-packaged foods contain a large amount of fat and salt—often more than your body needs. Fresh foods that you prepare yourself are generally healthier for you. Design a well-balanced meal that includes all the necessary chemicals to help your body stay healthy.

Nutrition Facts
Serving Size 1 Package (42.5g)

Amount Per Serving
Calories 220   Calories from Fat 110

	% Daily Value*
Total Fat 12g	18%
Saturated Fat 11g	55%
Trans Fat 0g	
Cholesterol 0mg	0%
Sodium 50mg	2%
Total Carbohydrate 29g	10%
Dietary Fiber less than 1g	4%
Sugars 24g	
Protein 1g	

Vitamin A 0%	•	Vitamin C 0%
Calcium 0%	•	Iron 2%

* Percent Daily Values are based on a 2,000 calorie diet. Your daily values may be higher or lower depending on your calorie needs:

	Calories	2,000	2,500
Total Fat	Less than	65g	80g
Sat. Fat	Less than	20g	25g
Cholesterol	Less than	300mg	300mg
Sodium	Less than	2,400mg	2,400mg
Total Carbohydrate		300g	375g
Dietary Fiber		25g	30g

of your body weight is water. This is about 70 pints (33 l) of fluid in an average person.

Because water is so vital to most chemical functions in your body, it is necessary to make sure you drink enough water to replace the water used each day. Most people need about 4.4 pints (2 l) of water each day. On average, 1.2 pints (0.6 l) of water come from the food you eat, 2.6 pints (1.2 l) come from the liquids you drink, and the chemical processes that take place in your body produce about 0.5 pints (0.2 l) of water. This should equal the amount of water you lose each day through urine, feces, sweat, and breathing.

The major chemicals that are used by your body are **proteins**, **fats**, and **carbohydrates**. In order to be able to use these chemicals, your body produces over 50 different chemicals that are used in digestion. Most of these chemicals are enzymes, which are catalysts that are used to speed up the digestion process. Most of these chemicals are supplied by the gall bladder, liver, and pancreas. The chemical processes of digestion start in the mouth where the enzyme salivary amylase reacts with starch to break the starch molecules into sugar molecules. Once the food reaches the stomach, chemicals are added that break down proteins and fats into smaller molecules. In the small intestine, more chemicals are added to break the smaller protein molecules down into amino acids, and also break complex sugars into simple sugars. These simple sugars are then transported to all the cells in your body where cellular respiration occurs, breaking the sugars apart to release the stored energy and produce CO_2 and water.

There are many other chemical reactions taking place in your body besides the ones required for breathing and digestion. Your body releases chemicals called hormones which tell your body how to grow and how to change as you grow up. Your body also releases chemicals that make you sleepy at bedtime and chemicals that help you feel awake in the morning. Your body is an amazing chemical factory.

What did we learn?

- List at least two chemical functions performed inside living creatures.
- What is the chemical reaction that takes place during photosynthesis?
- What is the main chemical reaction that takes place during digestion?
- What substance is necessary for nearly every chemical reaction in living things?
- Name the three major chemicals your body needs that are found in the foods we eat.

Taking it further

- Why did God design your body to have enzymes?
- With what you know about chemical processes, why do you think it is important to brush your teeth after you eat?
- Can you think of other chemical processes in your body besides the ones mentioned in this lesson?

Fun Fact

Each day your body produces about 12 pints (5.6 l) of digestion fluids. This includes 2.6 pints (1.2 l) of saliva, 1.7 pints (0.8 l) of bile, 2.6 pints (1.2 l) of pancreatic juice, and 5.1 pints (2.4 l) of intestinal juices. Most of the fluid in these liquids is recycled throughout your body.

Enzymes

Enzymes are critical to the chemical functions that are happening in living organisms so it is important that they be able to do their jobs. Several factors affect the rate at which enzymes work. What factors have you learned about that affect reaction rates of chemical reactions in general? You have learned that temperature, density, and surface area of the reactants, and the presence of catalysts and inhibitors can all affect the rate of a chemical reaction. There are similar things that affect how effective enzymes are at speeding up a chemical reaction. Two of these factors are temperature and pH (how acidic or basic the surrounding tissues are).

Many of the enzymes that are in your body can be found in plants as well. One enzyme that works to break down protein is called **protease**. This enzyme is found in several fruits including pineapple, kiwi, and papaya. Today you are going to do an experiment using fresh pineapple juice to demonstrate the effects of temperature and pH on protease.

Gelatin is mostly made of protein. This is what causes the gelatin to set or become thick. If you add protease to gelatin it will break down the protein and the gelatin will not set. If protease becomes too hot or is in a very acidic environment it will be destroyed. The following experiment will show how temperature and acid affect protease.

Purpose: To demonstrate the effects of temperature and pH on protease

Materials: box of gelatin mix, fresh pineapple (do not use canned or frozen), vinegar, four cups, measuring spoon, stove, sauce pan, refrigerator, "Enzyme Reaction" worksheet

Procedure:

1. Peel and cut up a fresh pineapple. Place some of the fruit in a blender and blend until you have a pulpy liquid.
2. Label four cups with the numbers 1–4.

LESSON 25 **Properties of Atoms & Molecules**

3. Place 1 tablespoon of pineapple juice/pulp in cup number 1.
4. Place 1 tablespoon of pineapple juice/pulp and 1 tablespoon of vinegar in cup 2. Stir to mix the two liquids.
5. Using a clean spoon, place the rest of the pineapple juice in a sauce pan. Heat on medium heat on the stove until the liquid has boiled for 5 minutes. Using a clean measuring spoon, place 1 tablespoon of the heated juice/pulp in cup 3. It is very important that none of the raw pineapple juice or vinegar are mixed in with this heated juice.
6. In a bowl, prepare the gelatin according to the package directions. Pour ½ cup of the gelatin into each of the four cups. Place all four cups in the refrigerator.
7. Complete the hypothesis section of the "Enzyme Reaction" worksheet.
8. Observe each cup of gelatin after 30 minutes. Write your observations of how the gelatin looks on the worksheet.
9. Repeat your observations after 60 minutes, 90 minutes, 2 hours, and 3 hours.
10. At the end of 3 hours answer the questions on the worksheet.

Conclusion: You should have found that the liquid in cup 1 did not gel. This is because the protease in the pineapple broke down the protein in the gelatin preventing it from gelling. The gelatin in cups 2 and 3 should have gelled or at least gotten very thick compared to cup 1. This is because the acid in cup 2 and the heat applied to the pineapple juice in cup 3 damaged the protease

Pineapples contain the enzyme protease.

preventing it from breaking down the protein in the gelatin. Why do you think we needed cup 4? Whenever you do an experiment, you need a control. This is the item that you compare all of the other tests against. This shows what happens if you do not do anything. Cup 4 is a regular cup of gelatin. You can then compare each of the cups to cup 4 to see what happens when you make one change at a time.

26

Decomposers

Ultimate recycling

What is the nitrogen cycle?

Words to know:

decomposition   decomposer

scavenger

Have you ever thought about what happens to a plant after it stops growing and withers? Have you ever seen a dead animal beside the road and wondered what would become of it? God has provided a way for the chemicals in dead plants and animals to be recycled. This process of recycling is called **decomposition**.

One of the most important elements that is recycled is nitrogen. Nitrogen is necessary for plant growth, and plants absorb nitrogen from the soil. Some of this nitrogen is then absorbed into an animal's body when the animal eats the plant. If another animal later eats that animal, the nitrogen passes on to the larger animal's body. Once the animal dies, it then decays and the nitrogen is returned to the soil.

Consider what would happen if plants and animals did not decay. The nitrogen in their bodies would be "lost" because it could not be reused, and eventually the earth would run out of nitrogen and new plants could no longer grow. Instead, there are special organisms to help break down dead plants and animals so that the nitrogen and other chemicals can be reused.

Scavengers are the first animals that help in the decomposition process. **Scavengers** are animals that eat dead animals. One common scavenger is the vulture. Lions, bears, jackals, hyenas, and komodo dragons are also scavengers. Sea sponges and insects are scavengers as well. Once the scavenger has eaten the dead animal, some of the chemicals become part of its body, but many of the chemicals are eliminated through its dung.

Organisms called **decomposers** then break down the nitrogen compounds and other complex

Scavengers are the first animals that help in the decomposition process.

LESSON 26 **Properties of Atoms & Molecules • 367**

🧪 The nitrogen cycle

Purpose: To draw a picture of the nitrogen cycle

Materials: drawing paper, colored pencils

Procedure:

1. Draw a picture of a plant in the center of the page, showing that nitrogen is being absorbed from the ground by the plant's roots.
2. Draw an arrow to a small animal, such as a mouse, that eats the plant.
3. Draw an arrow to a larger animal, perhaps an eagle or an owl, that eats the smaller animal.
4. The next arrow should point from the large animal to a scavenger, such as a vulture.
5. From the vulture there should be an arrow to dung on the ground. This picture needs to be labeled as having bacteria.
6. The final arrow should point to the plant, showing that the nitrogen has completed the cycle and is ready to be absorbed by a plant again.
7. Label the picture with the words "Nitrogen Cycle."

Conclusion: This picture demonstrates the path that nitrogen takes as it is used to support life on earth. Be sure to note that people absorb nitrogen when they eat plants as well. The nitrogen cycle is God's plan for recycling nitrogen. In addition to nitrogen, other chemicals are recycled in a similar manner. Recall the carbon cycle you learned about in lesson 9.

The nitrogen cycle is actually much more complex than we have explained in the lesson. Nitrogen is recycled in several different ways. Interestingly, lightning also plays a role in the nitrogen cycle by moving nitrogen from the air into the soil.

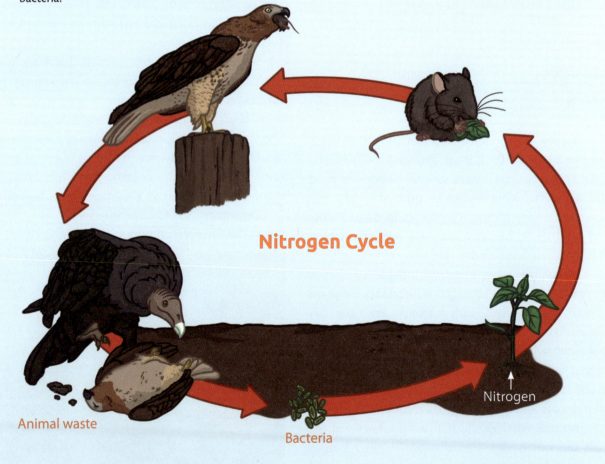

Mold is a fungus that can grow quickly on bread.

chemicals in the animal waste into simple compounds. These decomposers are simple organisms like bacteria and fungi. Decomposers also work directly on dead animals and plants to break them down into elements that can be reused by plants growing nearby.

God created the first man, Adam, from the dust of the ground to live forever (Genesis 1:27). Adam's sin brought death into the world (Romans 5:12) as part of God's curse on the world. Now when animals or people die, their bodies return to the dust of the ground (Genesis 3:19) and are recycled, like plants. God has even designed the effects of man's sin to be a blessing to the earth through the nitrogen cycle.

What did we learn?

- What is a scavenger?
- What is a decomposer?
- What is this way of recycling nitrogen called?

Taking it further

- Why are decomposers necessary?
- Were there animal scavengers in God's perfect creation, before the Fall of man?
- Explain how a compost pile allows you to participate in the nitrogen cycle.

Rate of decomposition

The most common decomposers are microscopic bacteria and fungi. These organisms are found on nearly every surface in the world. You can't get away from them even if you wanted to, but you can control and affect the rate at which decomposers work. Think about mold growing on your food. What conditions seem to encourage mold growth? What conditions seem to discourage it? You can test the effects of various conditions on decomposers by completing the following experiment in which yeast, which is a fungus, will be used to aid in the decomposition of a banana.

Purpose: To test the effects of various conditions on the rate of decomposition

Materials: three plastic zipper bags, banana, baking yeast, "Rate of Decomposition" worksheet

Procedure:

1. Label three plastic zipper bags with the conditions in which you will be testing decomposition:
    - Cold and dark
    - Warm and dark
    - Warm and light
2. Peel a banana and divide it into three equal pieces. Place one piece in each bag.
3. Sprinkle ½ teaspoon of baking yeast over each banana piece then seal each bag.
4. Place the bag testing cold and dark conditions in a location in a refrigerator where it will not be disturbed.
5. Place the bag testing warm and dark conditions in a cupboard where it will not be disturbed.
6. Place the bag testing warm and light conditions in a sunny location where it will not be disturbed.
7. Predict which bag you expect to see the fastest decomposition in. Write your hypothesis on the "Rate of Decomposition" worksheet.
8. Observe each bag daily for one week. Write your observations on your worksheet. Be sure to include how the banana looks, how the bag looks, how the banana feels, and any smells that you observe.
9. At the end of one week, compare your observations with your hypothesis and see if you were right.

27

Chemicals in Farming

Helping plants grow

How do farmers use chemicals?

Words to know:

crop rotation
fallow
fertilizer
hydroponics
pesticide
fungicide
herbicide
organic farming
genetic engineering
genetic modification

As you learned in the last lesson, decom- posers help nitrogen and other elements return to the soil to be reused by plants. This system works very well in natural areas. However, continuous farming of a piece of land uses up the nutrients, particularly nitrogen, faster than decomposers can replace them. In order to maintain healthy usable soil, the nutrients must be replaced.

There are several methods for replacing the nutrients in the soil. One of the oldest methods of keeping the soil viable is to use **crop rotation**. Different crops use nutrients in different amounts, and some crops, such as beans and alfalfa, actually return nitrogen to the soil, so planting different crops from year to year helps to keep the soil useful.

Also, allowing a portion of the land to lie **fallow**, without crops, allows the bacteria in the soil to have enough time to return the needed nitrogen.

Crop rotation has been used for centuries. In the nineteenth century, parts of Europe, and England in particular, followed the Norfolk 4-course crop rotation plan. In the first year, the farmer would plant a root crop such as turnips. Year two, he would plant barley. Year three, he would plant grass or clover and allow animals to graze on it. Then, in the fourth year, he would plant wheat. Farmers repeated this cycle, and it helped to keep their land productive.

A second method for adding nutrients back to

Nitrogen, phosphorous, and other chemicals can be added to the soil with man-made fertilizers.

the soil is through fertilization. Nitrogen, phosphorous, and other chemicals can be added to the soil in one of two ways. One way is through the application of animal dung. Much of the nitrogen that is in the plants that animals eat comes out through the animal's waste. So, applying animal waste to the land is one way to improve the soil's productivity. A second way to add nutrients back to the soil is to add artificial or man-made **fertilizers**. These are chemicals that are prepared in a laboratory or factory and then applied to the soil.

The third method of improving the soil is through burning. This method is used primarily in tropical areas where plant growth is rapid. Much of the nitrogen and other chemicals become tied up in unwanted plants that grow in the fields. Farmers burn these plants, and the ash returns the chemicals to the soil. Then the fields can be planted with desired crops.

Chemistry is important to farming in many ways other than just the nutrients in the soil. In fact, one type of farming does not use soil at all. **Hydroponics** is a type of farming in which plants are attached to some sort of supporting framework and the roots are bathed in a water solution containing boron, calcium, nitrogen, phosphorous, potassium, and other chemicals. Hydroponics was first used on a large scale by U.S. troops in the Pacific Islands during World War II. Canada began selling hydroponic tomatoes

A crop of organic hydroponic tomatoes

to consumers in 1988. Today, it is often more economical to grow many flowers and vegetables using hydroponics than in the traditional manner.

In addition to the chemicals needed by the plants, farmers also use chemicals as pesticides, fungicides, and herbicides. **Pesticides** are chemicals that are used to control insects and other pests that might damage crops. **Fungicides** are chemicals that kill fungi, which often cause diseases in plants. **Herbicides** are chemicals that kill unwanted plants, like weeds, without damaging the desired crops. Researchers are always trying to find better chemicals to help crops grow without leaving behind chemicals that will hurt the consumer.

🧪 The effects of fertilizer

Purpose: To demonstrate the effect of fertilizer on plants

Materials: two identical plants, plant food

Procedure:

1. Obtain two identical plants. Label one plant as plant A and the other as plant B.
2. Prepare a solution of water and plant food according to the manufacturer's directions. Save this solution to use each day to water one of the plants.
3. Pour ¼ cup of water without plant food on the soil of plant A each day. Pour ¼ cup of the water and plant food solution on the soil of plant B each day. If ¼ cup is too much or too little water to keep the soil moist but not soggy, adjust the amount as needed. However, be sure that both plants are receiving the same amount of liquid each time.

Questions: Based on what you have learned, which plant would you expect to grow faster? Why?

Conclusion: After a few days it should become obvious that the plant receiving the plant food is growing better than the plant without it. This is because the additional chemicals in the plant food provide the necessary nutrients for plant growth. If you did not see a significant difference, why do you think the plant growth was the same? (If the soil already had as much nutrients as the plants could use, adding more nutrients would not increase its growth. You can transplant both plants into less productive soil and try the experiment again.)

Because many people have concerns about eating crops that are treated with chemicals, some farmers have gone organic. Those who follow **organic farming** methods grow their crops without the use of any artificial chemicals. These crops are often more expensive because the organic farmer has to deal with insects and weeds more than other farmers do. But many people feel that organic produce is a healthier choice. You and your family will have to decide for yourselves.

Genetic engineering is another method that scientists are developing to help avoid the use of so many chemicals. Many crops have been modified with genes that make them resistant to certain diseases or undesirable to certain pests. **Genetic modification** can also make some crop plants more resistant to herbicides so only the weeds will be killed, but the plant will not be harmed. A large percentage of produce in our supermarkets is genetically modified in some way.

What did we learn?

- What are three ways that farmers ensure their soil will have enough nutrients for their crops?
- What is hydroponics?
- How are chemicals used in farming other than for nutrients for the plants?
- How is an organic farm different from other farms?

Taking it further

- Why did the farmers let cattle graze on their land once every fourth year in the Norfolk 4-course plant rotation method?
- How does hydroponics replace the role of soil in plant growth?

Organic farming

Organic farming has become more popular in recent decades. Organic farmers reject the use of artificial chemicals as well as genetically modified organisms. They claim to use more natural ways to control weeds and insects. There is much controversy and little agreement about the advantages and disadvantages of organic farming. See what you can find out about the pros and cons on each of the following aspects of organic farming:

1. Controlling pests—are organic methods effective?
2. Productivity—which way produces more crops?
3. Labor required—which way requires more labor?
4. Genetically modified organisms—are they good or bad?
5. The environment—which farming method is friendlier to the environment?
6. Food quality—which is better?
7. Food health—which is healthier?

There are many other issues and controversies surrounding organic farming and the use of chemicals, but starting with these questions will give you a good introduction to organic farming. Try to find out the arguments on both sides of each issue. Share what you learn with your class or family.

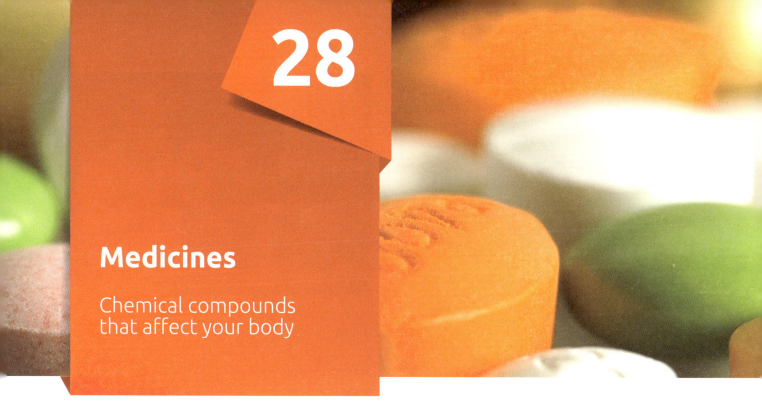

28

Medicines

Chemical compounds that affect your body

How do chemical compounds help heal your body?

Words to know:

pharmaceutical

medicine

herb

anesthetic

synthetic

antibiotic

vaccine

ethnobotanist

Challenge words:

chemotherapy

Your body is a living chemistry lab. We have already discussed some of the chemical processes involved in your body, such as breathing and digestion. Because your body works with chemicals, adding different chemical compounds to your body can greatly affect how you feel and how your body responds. For this reason, you need to be very careful what chemicals you allow into your body. Some chemicals can be very harmful and others can be very helpful. Some helpful chemicals are **pharmaceuticals** or **medicines**.

Herbs were the first plants to be used for healing purposes. From after the Fall until today, people have used various plants as cures for different illnesses. Some have been more effective than others, but many have been very effective even if the user did not know why. Most herbal remedies were discovered by trial and error.

The bark of certain trees has been used to cure headaches. This remedy has been found in the records of the Egyptians, the Chinese, and the Sumerians, as much as 3,000 years ago. But it wasn't until the nineteenth century that scientists discovered that the tree bark contained salicylic acid, which today is the main ingredient in aspirin. The Bayer Company began selling aspirin as a medication in 1899.

The different effects of chemical compounds began to be studied in earnest in the nineteenth century, and since then many helpful discoveries have been made. In 1799 Sir Humphry Davy discovered that nitrous oxide could be used as an **anesthetic** to stop pain. Today we know this chemical as laughing gas, and it is frequently used in dental offices

Fun Fact

Americans spend $5 billion each year on medicines that are derived from plants.

LESSON 28 **Properties of Atoms & Molecules • 373**

to eliminate the pain of dental procedures. In 1831 chloroform was discovered, and in 1842 ether was used during an operation for the first time. These and other anesthetic chemicals make it possible for doctors to perform operations without their patients feeling pain during the procedure.

When a patient takes a painkiller such as aspirin, or receives an anesthetic, the chemical in the medication locks onto the nerve cells in the brain. This prevents the pain signals from getting through, so the patient does not feel the pain. This process is like a chemical lock and key. The chemical is a key to locking up your pain receptors.

Shortly after these discoveries, many other drugs were discovered. The first **synthetic**, or man-made, drug was developed in 1910. Then, in 1930, a group of drugs called sulpha drugs was developed to kill some bacteria. But the really big break in medicine came in 1928 with the discovery of penicillin by Sir Alexander Fleming. The process for mass producing penicillin was discovered by Andrew Moyer. Penicillin was the first antibiotic. An **antibiotic** is a substance produced by living organisms that is used to kill bacteria. Since the 1940s, dozens of other

Andrew Moyer discovered the process for mass producing penicillin.

antibiotics have been discovered, such as amoxicillin and tetracycline.

In addition to antibiotics, another important medical use of chemicals is vaccinations. **Vaccines** are chemical solutions that have been developed to help stimulate your body's natural defenses against certain diseases. This helps to keep you from developing those diseases. Edward Jenner tested the first vaccine in 1796 when he gave an injection containing cowpox to people to help them develop immunity to smallpox.

The search for new medicines continues. A person who studies plants in order to develop new medicines is called an **ethnobotanist**. These scientists try to find plants with chemical compounds that are beneficial to the human body. Only about 0.5% of all plants in the world today have been tested for

Fun Fact

Genetic engineers are working on developing new organisms that can quickly produce needed chemicals for medications.

Common herbs

Man has used herbal remedies for centuries. Ginger is commonly used to settle an upset stomach. Certain oils have been used to improve digestion. Garlic is a natural antibiotic and has been shown to help lower cholesterol. Ginseng is an herb that increases your energy level. The list goes on and on. You can enjoy the benefits of some of these natural herbal substances by making a fun snack of garlic bread and ginger ale.

Spread butter or margarine on a piece of bread. Sprinkle a small amount of garlic powder on the bread. Toast under the broiler of your oven until golden. Enjoy this natural antibiotic with a cool glass of ginger ale.

Ginger root   Garlic

medicinal purposes. There may be many helpful substances that God has placed in the world just waiting to be discovered and used to benefit mankind.

What did we learn?

- Why are chemicals used as medicines?
- What were the earliest recorded medicines?
- What was Sir Alexander Fleming's important discovery?

Taking it further

- If plants have the potential of supplying new medicines, where might a person look to find different plants?
- What other sources might there be for discovering new medicines?

Chemotherapy

Chemicals have been used to treat various kinds of cancer since the 1950s. This type of treatment is called **chemotherapy**. Chemotherapy originally meant any kind of chemical used to treat a disease, but today it specifically applies to chemicals used to treat cancer.

There are many different kinds of cancer, but all cancers are diseases that involve cells that are growing and reproducing out of control. Drugs that treat cancer have been developed to target cells that are reproducing very quickly. Cells in your body can be in one of several states with respect to reproduction. The cells can be resting, meaning they are performing their designed functions, they can be preparing to reproduce, or they can be in the process of reproducing. Noncancerous cells spend much of their time in the resting phase; however, cancer cells spend a large part of their time preparing to reproduce or actually reproducing. Therefore, chemotherapy drugs have been designed to target cells that are either preparing to reproduce or actually reproducing.

These drugs chemically interact with reproducing cells to prevent them from completing the reproduction cycle, and destroy the cells. This can be very effective in eliminating cancer cells. However, the drugs cannot distinguish cancer cells from noncancer cells; they only react with reproducing cells and leave resting cells alone. Thus, other cells in the body that reproduce quickly are more affected by the chemotherapy than slowly reproducing cells.

Hair cells reproduce very quickly and are often destroyed by the chemicals used in chemotherapy. This is why many cancer patients lose their hair

Hair cells reproduce very quickly and are often destroyed by the chemicals used in chemotherapy.

during treatment. Also, the cells in the lining of the stomach and intestines are replaced very quickly, so these cells are also harmed by the chemotherapy, often resulting in upset stomach and vomiting.

Overall, doctors must balance the use of the chemicals to destroy the cancer cells with the amount of side effects the patient can withstand. Decades of testing and treatment have resulted in very effective uses of chemicals to treat many forms of cancer.

Chemicals administered by doctors are not the only treatments that people use for cancer. Some people have had success in using herbs and special diets to change the chemical balance in their bodies and thus slow down or eliminate some cancers. Our bodies are very complex, and chemicals have drastic effects on many functions. Therefore, we must be careful what chemicals we put into our bodies. On the other hand, God has miraculously designed our bodies to heal themselves in many ways, by fighting disease and rebuilding damaged tissues. So we can see the hand of God in our bodies every day.

LESSON 28 Properties of Atoms & Molecules • 375

Alexander Fleming
1881–1955

SPECIAL FEATURE

Alexander Fleming is a name you may not know, but you can be thankful for what he did. He discovered penicillin, a medicine made from a mold that kills harmful bacteria. Who was this man that would try such a thing? Alexander Fleming was born in Scotland in 1881, the seventh of eight children. The family worked an 800-acre farm where the children spent much of their time roaming the countryside. Later in life Alexander said, "We unconsciously learned a great deal from nature."

His father passed away when Alexander was 14. His oldest brother took over the farm, and Alec, as he was called, along with four of his siblings left Scotland and moved to London. After completing his schooling, Alec went to work for a shipping company, but he did not like it very much. So when the Boer War in South Africa broke out in 1900, Alexander and two of his brothers joined a Scottish regiment. This unit was never sent to war but spent most of its time playing different sports, such as shooting, swimming, and water polo.

It was around this time that Alec's uncle died and left an inheritance of about 250 British pounds to each of the Fleming children. Tom, Alec's brother, encouraged him to use his inheritance to study medicine. Alexander made top scores on the entrance exam and won a scholarship to St. Mary's, the school he preferred because he had played water polo against them. After completing his training, he could have left St. Mary's and taken a position as a surgeon. However, the captain of St. Mary's rifle club wanted Alec to stay and be part of his rifle team, so he encouraged Alec to switch from surgery to bacteriology, which he did.

In 1915 Alexander married an Irish woman named Sarah Marion, and they had a son who became a physician. When World War I started, Alec, and most of the people who worked in the lab

he was in, went to France and set up a battlefield hospital lab. There on the battlefield he saw firsthand how small wounds could become infected and often lead to death. He felt that there must be some chemical solution that could help prevent this. He made many innovations in the treatment of the wounded during the war that helped decrease the mortality rate, but he had yet to discover something that would prevent infection.

After the war, Fleming spent most of his time researching different chemicals and made some important discoveries, but none as important as what he found in 1928. Fleming had been growing mold and bacteria in several different petri dishes. He had stacked several of them in the sink and did not get around to cleaning them up right

Mold growing in a petri dish

away. When he finally did get around to cleaning up his experiments, he looked at each one before putting it into the cleaning solution. One dish made him stop and say, "That's funny." In this dish, some mold had grown, and around the mold the staph bacteria had died. He sampled the mold and found it to belong to the penicillium family. He labeled his discovery penicillin.

He published his findings in 1929, but it raised little interest. Fleming found it difficult to process the penicillin and a group of chemists took over the work for him. Eventually, even this work stopped when several of the chemists either died or moved away. Fleming's work on penicillin did not advance much more until World War II started. At that time, Howard Florey and Ernst Chain started up the work again.

They were able to demonstrate how effective penicillin is against several infectious diseases and described the mechanisms for how it works. For this work Florey, Chain, and Fleming shared the 1945 Nobel Prize for Chemistry.

However, they were still unable to produce enough penicillin for commercial use. And with World War II still raging, it was important to find a way to produce it so that soldiers could be treated in the battlefield and not die from infections. So in 1941, Florey went to the United States and worked with Andrew Moyer. Together they developed a method for mass producing penicillin. In 1940 penicillin was so rare it was nearly priceless. In 1943 it was $20 per dose, and by 1946 it was only $0.55 per dose. All of these men were responsible for discovering a way to save millions of lives.

In 1944 Fleming was knighted for his discovery of penicillin. In 1947 Dr. Fleming became the director of the Wright-Fleming Institute of St. Mary's Hospital. He died in 1955 at the age of 73, and his body now rests in St. Paul's Cathedral in London.

Properties of Atoms & Molecules • 377

UNIT 7
Applications of Chemistry

29 Perfumes
30 Rubber
31 Plastics
32 Fireworks
33 Rocket Fuel
34 Fun With Chemistry—Final Project
35 Conclusion

◊ **Describe** ways in which chemistry can be used for the benefit of mankind.
◊ **Demonstrate** uses of chemistry in industry.

29

Perfumes

What's that smell?

How are the scents in perfumes made?

Words to know:
perfume
aromatic
solvent extraction
steam distillation

Have you ever walked through a depart-ment store and smelled the perfumes near the cosmetic area? You may have liked some scents and disliked others. Perfumes and colognes are very popular with many people. But you probably never realized that chemistry plays an important role in making these scents. In the next several lessons, we will be learning about many different ways that chemistry is used in the world around us, and we will begin with a look at perfume.

A **perfume** is a liquid with a pleasant smell. It is **aromatic**; this means that its smell is easily detected by the human nose. In order for the liquid to be aromatic, it must contain scent molecules that are light enough to float in the air so they will reach the nose. Molecules that are water soluble or fat/oil soluble are the most aromatic.

Most perfumes are made by extracting scent molecules from plants that are considered to have a pleasant smell. Flowers are the most common source of scent molecules used in the making of perfumes. And this is where an understanding of chemistry becomes important.

The scent of a flower petal comes from the oil in the petals and this oil must be removed in order to make the perfume. There are two main methods for extracting these scent molecules from plants. The first is called **solvent extraction**. In this method, the flower petals are soaked in a solvent. The solvent is a chemical in which the oil containing the scent will dissolve. The flower parts are then removed. Next, the solvent is allowed to evaporate. This leaves the fragrant oil behind.

LESSON 29 **Properties of Atoms & Molecules • 379**

Steam distillation aparatus

The second method of scent extraction is called **steam distillation**. In this method, steam is passed through the petals causing the oil to vaporize. The oil moves with the steam through a tube to another container where both the oil and water condense and become liquids. The water sinks to the bottom of the container and the oil floats on the top of the water. The oil is then skimmed off the water and removed.

Once the fragrant oil is obtained, the perfume is made by combining the oil with an alcohol. This is done so that the oil is easily sprayed onto your skin. The alcohol quickly evaporates, leaving behind the desired fragrance.

What did we learn?

- What is a perfume?
- What must be removed from flower petals to make perfume?
- Describe the two main methods for removing oil from flower petals.

Taking it further

- Why should you test a new perfume on your skin before you buy it?
- Why wasn't it necessary to use one of the methods described in the lesson to make your homemade perfume?

Making your own perfume

You can easily make your own perfume. Many of you performing this experiment are probably thinking, "I'm a guy, and guys don't wear perfume." Well this may be true, but many men wear cologne, which is a scent designed for a man. So you can make cologne instead of perfume. You will be using cloves, which is a scent that is not particularly feminine or masculine.

Purpose: To make your own perfume/cologne

Materials: cloves, jar with lid, rubbing alcohol

Procedure:

1. Place 15 whole cloves in the bottom of a jar.
2. Add ¼ cup of rubbing alcohol to the jar and close the lid.
3. Allow this solution to sit for seven days.
4. At the end of one week, remove the cloves from the alcohol and you have your own perfume.

Conclusion: The perfume may not smell like cloves when you sniff the jar, but take a small amount of the liquid and place it on your skin. The alcohol quickly evaporates, leaving behind a pleasant clove scent.

 Scents

If you have ever been to a department store, you will find that there are hundreds of different scents available for perfumes and colognes. This is because the same scent may not smell the same on two different people. Not only do scent molecules evaporate from your skin, but they can chemically combine with the oils on your skin to produce a different scent. So the scent you like so much on your best friend might smell terrible on you.

Also, each person's nose is sensitive to different scents. Although you may love the smell of coffee, another person might hate it. So a whole industry has developed producing hundreds of different scents for people to wear.

The scents in perfume are actually much more complex than just a single scent like the clove perfume you made. Most perfumes contain several scents combined together. Often a perfume will be made of at least three scents. The first scent that you smell is called the head. This is usually a strong scent that is easily detected but easily evaporates and lasts for only a short period of time. Orange is a common head scent. The second scent is called the heart. This is usually a floral scent and is the smell you most associate with the perfume. The final scent is called the base. This is the smell that lasts the longest and is usually more subtle than the other scents. Common bases include sandalwood and vanilla. The combination of smells results in a pleasant experience. You immediately smell the head scent, but then the smell melts into a combination of the heart and base scents, slowly changing over time.

You can try to produce your own personalized perfume or cologne. Try soaking different items in alcohol for several days to make a scent you really like. You can choose from the following list, or come up with your own ideas. Try several scents and see which ones you like best.

- ginger root
- mint leaves
- cinnamon sticks
- dried fruit
- different flowers
- allspice

Some cooking extracts may make pleasant perfumes or colognes. You might consider:

- peppermint oil
- almond extract
- vanilla extract

Today many households have essential oils on hand. Essential oils are oils that have been extracted from plants, usually by distillation just like we described in the lesson. Many of these oils are safe to use on the skin. If you have essential oils that are safe for the skin, you could combine small amounts of two or three oils to make a perfume as well.

Applications of Chemistry

30

Rubber

Do you have a rubber band?

How is rubber made, and what is it used for?

Words to know:

rubber
latex
vulcanization
polymer

Challenge words:

rosin
lac
silk
casein
keratin

You place it on the end of your finger and slowly stretch it back; then suddenly you release it and it goes flying across the room! Who hasn't experienced the thrill of launching a rubber band? It's great fun to try to improve your aim and hit a target. (Just don't hit your brother or sister!) But did you know that there is a lot of chemistry involved in making a rubber band?

Rubber is made from a naturally occurring substance called latex. **Latex** is a sticky, milky-colored material that is found in rubber trees. Rubber products became very popular in the early 1800s.

However, two major problems occurred with the original rubber products. First, when the rubber got cold, like during a winter storm, it became brittle and cracked. Second, when the rubber got very warm, like in the middle of summer, it became sticky. Rubber products were not very useful if they could only be used in mild temperatures. A car with rubber tires wasn't very useful if you could only drive it in the spring and the fall.

So scientists worked very hard to find a way to make rubber more useful. In 1839 a chemist named Charles Goodyear discovered how to

Tapping latex from a rubber tree

382 • Properties of Atoms & Molecules   LESSON 30

improve rubber's performance. He found that adding sulfur to the latex caused it to become elastic and eliminated the bad reactions to extreme temperatures. This process is called **vulcanization**.

Rubber is a chemical that forms long chains of carbon and hydrogen molecules into what are called **polymers**. These linked molecules can pivot around the carbon atoms, allowing rubber to change it shape. Naturally, these long chains of molecules are a random mess. But when an outside force is added, the chains can be stretched out and become parallel to each other. The chemical attraction of the covalent bonds pulls the molecules back into their original shape when the outside force is gone. Vulcanized rubber adds sulfur atoms to the polymer chains, connecting them together so that they act like one big molecule. This helps the rubber retain its elasticity, even at extreme temperatures.

Once vulcanization was discovered, rubber became very popular. People began trying to make nearly everything out of rubber. Of course, one of the most important uses for rubber was automobile tires, but clothing, shoes, balls, grommets, erasers, and many other items were made of rubber as well.

This dependence on rubber became a real problem during World War II. After attacking Asia, Japan controlled most of the rubber tree plantations there,

Fun Fact

Synthetic rubber became widespread after World War II. In 1960 the use of synthetic rubber surpassed the use of natural rubber and this trend continues today.

and the Axis powers together controlled nearly 95% of all natural rubber supplies. This created a crisis for America. To make a Sherman tank required ½ ton of rubber, not to mention all of the other rubber products required to keep an army functioning. So America did two things. First, the whole country did a major recycling campaign to provide enough rubber to keep the army going for at least a year or two. And second, the president asked scientists to develop an economical synthetic rubber.

Synthetic rubber had been discovered in 1875, but it was too expensive to make so it had not been developed. During World War II it became crucial for the United States to find an inexpensive synthetic rubber, and that is exactly what the scientists did. The process they discovered is still used today. First, scientists extract a chemical called naphtha from petroleum. Naphtha is then sent through a chemical process to change it into polymers that

Many everyday objects are made from rubber.

LESSON 30 **Properties of Atoms & Molecules • 383**

are very similar to natural rubber. This discovery helped America win the war and led to a boom in the number of uses for both natural and synthetic rubber.

Today, a large percentage of the rubber used around the world is produced from petroleum instead of latex from rubber trees. Synthetic and natural rubber are used in everything from cars to bikes, and from running shoes to clothing. Whether the rubber is natural or man-made, it is still fun to shoot a rubber band.

What did we learn?

- What is natural rubber made from?
- What is synthetic rubber made from?
- What is vulcanization?
- What is a polymer?

Taking it further

- Why is it difficult to recycle automobile tires?
- What advantages and disadvantages are there to using synthetic rubber instead of natural rubber?

Fun Fact

In 1943 James Wright was trying to produce a synthetic rubber, but only succeeded in producing a thick putty-like material. He put it on the shelf and forgot about it. A few years later a salesman picked it up and used it to entertain some customers. The value of the putty as a toy was soon recognized and in 1957 it was introduced to the world as Silly Putty.

Playing with rubber

Purpose: To understand how rubber polymers act

Materials: latex balloon, marker, wide rubber band

Activity 1—Procedure:

1. Inflate a balloon, but do not tie it closed.
2. While you hold the inflated balloon, draw a picture or write a message on the balloon (you may need someone to help with this step).
3. Let the air out of the balloon and look at the picture. Has it changed?

Conclusion: Most balloons are made from latex. This latex may be natural or synthetic, but it has the same effect. The polymer fibers of the balloon are all coiled up when it is deflated. As you inflate the balloon, the fibers are forced to stretch out and become straight. After you released the air the stretched molecules returned to their original shape, and the picture has shrunk and changed shape along with it.

Activity 2—Procedure:

1. Stretch a wide rubber band and quickly place it against your forehead. How does it feel? (It should feel warm.)
2. Remove the rubber band from your forehead and release the pressure on the rubber band, then quickly place it against your forehead again. How does it feel now? (It should feel cool.)

Conclusion: The polymers release energy when they are stretched out so the rubber band feels warm for a few seconds after it is stretched out. The polymers gain energy when they recoil, so the rubber band feels cool for a few seconds after it returns to its normal shape.

Polymers

Latex is the natural polymer from which rubber is made, but latex is not the only natural polymer. Many other products are made from natural polymers as well. Dead wood and wood pulp contain a polymer called **rosin**. Rosin is used to make varnish and soap. It is also used by violinists to keep their bows smooth and by gymnasts who want to keep their grip on the equipment.

Animals also produce polymers that are useful. An insect called the lac (*Kerria lacca*) produces a polymer that is also called **lac**. Lac is used to make lacquer, which is used as a shiny coating on furniture and other surfaces. Silkworms produce **silk** which is also a polymer. The strands of a silkworm cocoon are unwound and woven into soft beautiful cloth. This natural polymer was one of the most important trade items between Europe and Asia for many centuries.

Cattle also produce a couple of polymers that are very important. Cow's milk contains a protein called **casein**. This natural polymer is used in making cheese as well as in making artificial gems. Its long molecules work as a glue to hold cheese together and to hold together the ingredients for artificial gems.

Finally, horns of several animals contain

Cow's milk contains casein.

keratin. Your body also contains keratin. You find it in your hair and fingernails as well as in your skin cells. This polymer keeps your body watertight. There are many other natural polymers. If you find this interesting, you can do more exploration and find out more information about natural polymers such as cellulose and amber.

Silkworm cocoon

A rhino's horns contain keratin.

LESSON 30 **Properties of Atoms & Molecules** • 385

Charles Goodyear
1800–1860

SPECIAL FEATURE

The use of rubber is not new. The Indians of Central and South America were using it for centuries before Columbus found it and introduced it to western culture. The Indians called it *caoutchouc* from the word *cahuchu* which means *weeping tree*. But the Europeans called it *rubber* because it could be used to rub out pencil marks, what today we would call an eraser. However, rubber did not have many uses until after 1839 because it had two main problems. In the heat of the summer it turned into a gooey mess, and in the winter it became stiff and brittle. In 1839 this all changed, possibly by accident, but certainly through hard work.

Nine years earlier, Charles Goodyear, a bankrupt hardware merchant, walked into the Roxbury India Rubber Company in New York to sell them a new valve, but this company was about to go bankrupt. The problem was that summer had come and all the products they had been selling were being returned because the rubber was melting. The directors, in an effort to keep things quiet, had even met in the middle of the night to bury $20,000 worth of the melting products in a pit. So Goodyear was unable to sell his valve. But after hearing this, Charles, at the age of 34, decided to take his first good look at rubber to try and understand how it worked.

Since his trip was unsuccessful, he returned to Philadelphia and was put in jail for not paying his debts. While in jail, he put his time to good use. He had his wife bring him some raw rubber and a rolling pin, and he did his first experiments with the gummy mess.

He tried adding drying powders like magnesia to the rubber with some promising results. Once out of jail, Charles, his wife, and their daughters made up hundreds of pairs of magnesia-dried

rubber overshoes in their kitchen. Unfortunately, the heat of summer came before he could sell them and the overshoes turned into a shapeless paste. And the Goodyears were again penniless.

Because his neighbors complained about Goodyear's smelly work, he felt compelled to move. So he moved to New York. There, Goodyear tried adding two drying agents, magnesia and quicklime, to the rubber. He was getting much better results, well enough that he received a medal at a New York trade show.

A further advancement came by accident. Charles put designs on the products he sold. And one time he used nitric acid to remove his design and found that it turned the rubber black. He threw this piece away only to retrieve it later because he remembered it also made the rubber smooth and dry. It was a better rubber than anyone had ever made. A businessman advanced him several thousand dollars to start producing this rubber. However, the 1837 financial panic wiped out the business and Charles and his family again lost everything. They ended up camping in an abandoned rubber factory on Staten Island and eating the fish they caught. But Goodyear did not give up.

Eventually, Goodyear moved to Boston and got financial backing to make mailbags for the government using his nitric acid rubber. He made 150 bags and left them in a warm storeroom while he took his family on a month-long vacation. When he returned he found that the mailbags had melted and were useless.

Goodyear again hit rock bottom and was dependent on the kindness of farmers to give his children milk and half-grown potatoes to survive. Next, he started using sulfur in his work with rubber. The details of how he made his famous discovery are not clear or consistent, but the most reliable story says that one cold day in February he went into the general store in Woburn, Massachusetts to show off his latest gum-and-sulfur rubber when the customers sitting around the cracker-barrel started snickering at him. He must have been on very hard times because he was normally a mild-mannered man, but this day he got excited and started waving a fistful of his gummy substance in the air. It flew out of his hand and landed on a hot stove and proceeded to cook. He went over to scrape it off the surface, thinking it would melt like molasses. Instead, what he found was a leathery rubber that had elasticity. This rubber was remarkably different from the other rubbers he had tried before; it was a weatherproof rubber. Goodyear denied this incident and said his discovery was not an accident.

That winter took its toll on Charles and his family. Due to failing health, he hobbled around doing his experiments on crutches. But he now knew that heat and sulfur held the answer he was looking for. The questions remained: how much sulfur, what temperature, and how long to heat it? After extensive experimentation, Goodyear at last found the right combination of sulfur, heat, and time. He found that pressurized steam at 270°F (132°C) for four to six hours gave the best results. This process is now called vulcanization.

Charles evidently wrote his wealthy brother-in-law about his discovery and his brother-in-law took an immediate interest in using the rubber as a textile. Two factories were put into production as the new form of rubber became a worldwide success. Unfortunately, Goodyear disposed of his manufacturing interest as soon as he could, and he never saw the millions he might have made.

Charles Goodyear was a good inventor, but not a good businessman. For instance, the people holding rights to his rubber made $3.00 a yard on the rubber they sold, while Mr. Goodyear made only 3 cents per yard. He was also in 32 different patent infringement cases. Some court cases went as high as the U.S. Supreme Court. In one of the cases, Goodyear hired Daniel Webster, at that time the Secretary of State, and paid him $15,000 to temporarily step down from office and work as his lawyer. This was the largest sum paid to a lawyer at that time. Mr. Webster's two-day speech won a permanent injunction against further patent infringements in the U.S., securing some rights for Goodyear. In spite of this victory however, Charles died in 1860, $200,000 in debt.

Goodyear's family did eventually recover some of the royalties due them and eventually made a comfortable living. They finally reaped the reward for their father's undying commitment to the production of rubber. Today, we can't imagine life without rubber tires, rubber boots and shoes, rubber seals, and a host of other rubber products; all because Charles Goodyear did not give up.

31

Plastics
The wonder material

How are plastics made, and what are they used for?

Words to know:
plastic
thermoplastic
thermosetting resin

What would your life be like without plastic? Look around your home and notice how many items are made from plastic. Your mixer or blender, the handle on your refrigerator, buttons, light switch covers, even your clothes are probably made from some kind of plastic. Plastic is a modern wonder of chemistry. Since its discovery in the 1860s, plastic has revolutionized how many items are made.

Plastics are polymers. Polymer comes from the Greek words *poly* meaning many and *mer* meaning parts. A polymer is a giant molecule, which is really a long chain of molecules connected end to end. As you already learned, many polymers occur naturally, including rubber, wool, silk, and DNA. However **plastic** is a synthetic, or man-made, polymer.

The first artificial polymer was made in 1862. It was called *celluloid* because it was made from the cellulose found in cotton fibers. Later, polymers were made from coal tar. Today, plastic polymers are made from petroleum, or oil. The main ingredient is ethylene (a gas) which is combined with a catalyst that causes the molecules to form long chains. The first petroleum-based plastic was made in 1907. It was called Bakelite and was a dark brown moldable plastic that was mainly used for radio cabinets and later for TV cabinets or cases. Other famous plastics include nylon, which was invented by the DuPont Corporation in 1930, polyester, and acrylic.

One very interesting way that plastic is used is in the making of clothing. Polymer chips are melted. Then the liquid plastic is forced through very tiny holes and drawn into a thread. The thread is very thin and very flexible. This plastic thread is then woven into cloth to be used for clothing and other products. Check the labels of your clothes and see how many of them contain polyester or nylon. Other man-made fibers include acetate and rayon, which are made from cellulose, not from petroleum.

Most plastics fall into two categories. One kind is thermoplastic. **Thermoplastic** is made by melting

Fun Fact
Some plastics are ten times harder than steel.

Applications of Chemistry

388 • Properties of Atoms & Molecules   LESSON 31

plastic chips, and then injecting the liquid into a mold. As the plastic cools, it hardens and keeps its form. Thermoplastics will melt again if they are reheated. There are many uses for thermoplastics including trash cans and PVC pipe. However, it is very inconvenient in some uses for plastics to become soft or begin to melt when they become hot. Therefore, scientists have also developed thermosetting resin.

Thermosetting resin melts when it is heated initially. This liquid is placed in a mold under pressure in a process called compression molding. This kind of plastic hardens under pressure. Then once it cools, it will not melt again even if it becomes hot. This is much better for making items like coffee cups and baking dishes. Thermosetting resin is also used for car bodies and boat hulls, as well as a great number of other applications.

Today, millions of tons of plastic are produced each year. And the uses for plastic are innumerable. Plastics are used for counter tops, plastic bowls, and cooking utensils. Plastics are also used in carpet and other floor coverings. Pens, paintbrush handles, and many other art supplies are made from plastic. Plastics are used to make film. Plastics are used for fishing equipment and tricycles. And plastics are even used to make artificial joints for people. These are only a few of the many uses for plastics. If you look around you, you will find that plastics affect every area of your life.

One of the biggest concerns about all of the plastic that is used every day is the fact that plastic does not decompose. Once the plastic is thrown away, it often ends up in a landfill where it will sit for hundreds of years before breaking down. To help reduce this problem, most cities now offer recycling programs. Plastic, glass, metal, paper, and cardboard can all be recycled. While glass, metal, paper, and cardboard have been recycled for decades, recycling of plastics on a large scale is a relatively new

Many everyday items are made from plastic.

idea. Recycling non-plastic materials is relatively easy compared to plastic recycling. Because of the shape of the plastic polymers, it is difficult to break the molecules apart, so different types of plastic cannot easily be recycled together and must be separated from each other.

Most plastic items are stamped with a recycling number which indicates what type of plastic it contains. When items reach a recycling center they must be sorted. In many cases this is done by hand. Workers pull out cardboard, paper, and plastics. The plastics are then sorted by type. The remaining materials are sent on to machines that use magnets to separate steel cans from aluminum cans. Then glass is separated by hand according to the various colors. This is a very labor intensive process, but a load of recyclables is usually processed in about an hour.

To reduce the costs and time required for sorting, companies have been developing better ways to sort recyclables. Many recycling centers now have machines that use spectrophotometry and other scientific principles to automatically separate various materials. These machines shine a halogen light on the materials as they pass by on a conveyor belt. Each type of material reflects back a unique spectrum of infrared light. The computer analyzes this light to determine the size, shape, color, and location of the item as well as what type of plastic, paper, or glass the item is. A jet of air then pushes the item onto the correct conveyor belt. This is a much more efficient way to sort materials.

Most water and soft drink bottles are made from plastic number 1, which is PET (polyethylene terephthalate). This plastic is often recycled into

Chemical word search

Using the "Chemical Word Search," review the meaning of each word then find the words in the puzzle.

Recycling numbers indicate what type of plastic is used.

plastic furniture, carpet, tote bags, and other items. Other bottles and containers are plastic number 2, which is HDPE (high density polyethylene). This type of plastic gets recycled into pens, other bottles, and plastic lumber. For many years these were the only plastics that most recycling programs would accept. However, with improved recycling processes most plastics are now accepted.

Plastic number 4 is found in shopping bags and other plastic bags. These are usually collected separately from other plastics because they tend to get caught in the machinery that separates the other material that has been collected. Plastic 6 is Styrofoam. This type of plastic is very difficult to recycle, and most recycling programs will not accept it.

Although most people would agree that recycling plastic is a good idea, there is still controversy surrounding plastic recycling. Some people argue that the costs involved in collecting, sorting, and recycling outweigh the benefits. Others argue that the methods used to recycle plastics put toxic materials back into the environment. And other people just don't want to be bothered with sorting their trash before getting rid of it. These issues are all being addressed, and plastic recycling has become much more common in recent years. This means we use less petroleum to make plastic items, and we often use less wood and other materials that those plastic items now replace. So, keep your eyes open for recycled plastic, and see where it shows up around you.

What did we learn?

- What is plastic?
- What was celluloid, the first artificial polymer, made from?
- What is the difference between thermoplastic and thermosetting resin?
- Why are people concerned about throwing plastic items away?
- What does the recycling number on a plastic item mean?
- Why are plastic bags usually recycled separately from other plastics?

Taking it further

- Name three ways that plastic is used in sports.
- What advantages do plastic items have over natural materials?

Polymer ball

There are many fun experiments that you can perform that involve polymers. Today you will make a polymer bouncy ball.

Purpose: To make a polymer ball

Materials: borax powder, water, plastic zipper bag, two cups, white glue, cornstarch, markers

Procedure:

1. Mix 2 tablespoons of warm water with ½ teaspoon of borax powder in a cup.
2. In a second cup, combine 1 tablespoon of glue, ½ teaspoon of the borax mixture you just made, and 1 tablespoon of cornstarch. Do not stir. Allow the ingredients to react together for 15 seconds.
3. After 15 seconds begin stirring the ingredients together.
4. Once it becomes impossible to stir the mixture, remove it from the cup and roll it between your hands. Continue rolling the ball until most of the stickiness is gone.
5. Decorate your ball with markers.
6. Enjoy bouncing your ball. Store it in a plastic zipper bag to keep it fresh.

32

Fireworks

Is it the fourth of July?

How are fireworks made?

One of the most exciting parts of any Independence Day celebration is watching the fireworks display after the sun goes down. Fireworks have become an American Fourth of July tradition. But the earliest fireworks were the invention of the Chinese in the tenth century. These early fireworks were more like the later Roman Candles and were used in warfare rather than for entertainment. In 1242 an English monk named Roger Bacon wrote down the first European recipe for black powder, a key ingredient in fireworks. Then, in 1353, black powder was used by the Arabs in the first gun. But the first known use of fireworks for entertainment purposes was in France, when Louis XIV amazed his guests with fireworks sometime in the late seventeenth century. Later, the Italians added color to the fireworks and one of the most enjoyable forms of entertainment was born.

Making fireworks is a very specialized and somewhat dangerous use of chemical reactions. The recipe for each firework is unique. Most fireworks companies are family owned and the recipes for their fireworks are strictly guarded secrets, passed down from generation to generation. Still, there are some common steps in making and firing fireworks.

Each fireworks shell is packed with balls of chemicals. How the balls are packed determines how it will explode and what the explosion will look like. Different chemicals emit different colors when burned, so the type of chemical determines the color that will be given off. Sodium gives off a yellow light, copper salt produces blue, lithium and strontium salts produce red, and barium nitrate produces green.

The chemical balls are layered inside the shell and held in place by rice or corn. The shell is then wrapped in brown paper and paste and allowed to dry. Charges of black powder and fuses are then added to the shell. The charge on the bottom lifts the shell into the air. A bursting charge is added as well. Each charge has its own fuse, which determines when it will blow so that the shell does not burst before it reaches an appropriate altitude.

Fun Fact

Fireworks are usually stored in bunkers that are separated by 20-foot mounds of sand. This way if one bunker somehow explodes, the others would still be safe.

Pipes being set up for a fireworks show

are used as mortars for firing the larger shells and PVC pipes are used to hold the smaller shells. One pipe is used for each shell to be fired. Sand is packed around and in between each pipe to hold it in place. An appropriately sized mortar is set in each specific place according to the plan for when each shell is to be fired.

Next, a wire is run from a firing board to each mortar. The wire is then connected to the firing fuse of the shell. The firing board is then used to ignite each shell in the proper order. The size and order of each shell is determined ahead of time to guarantee a spectacular display.

When the bursting charge explodes, the energy released excites the electrons in the outermost shell of the atoms, moving the electrons to a higher energy level than they would normally be in. This is an unstable situation so the electrons quickly go back to their original energy level. When the electrons return to their normal energy levels, they release energy in the form of light. Different chemical compounds release different wavelengths of light, thus we enjoy a wide variety of different colors of fireworks.

Nearly as much work goes into setting up a fireworks display as goes into making the shells. It can take up to two days to set up everything for a 30–60 minute fireworks show. First, copper pipes

What did we learn?

- What are the key ingredients in a fireworks shell?
- Why does a fireworks shell have two different black powder charges?
- How do fireworks generate flashes of light?
- What determines the color of the firework?

Taking it further

- How can a firework explode with one color and then change to a different color?
- Why would employees at a fireworks plant have to wear only cotton clothing?

Designing a fireworks display

Each firework has a special design to give just the desired effect. You can design your own fireworks display by making a fireworks picture. Decide the shapes and colors of the exploding fireworks and then draw the shapes on a piece of construction paper with glue. Next, sprinkle colored glitter on the glue and allow it to dry. Be creative and design fireworks that you would like to see.

 ## Colored flames

You can experience different colored flames in your own campfire. You can soak pinecones in different chemical solutions. When they are dry you can burn them in a campfire. They will burn with different colors of flame because of the different chemicals they are coated with.

Purpose: To observe the different colors produced by different salts

Materials: pinecones, container, water, as many of the chemicals in the chart as you can obtain

Procedure:

For each color you wish to produce:

1. Dissolve as much of the colorant chemical as you can dissolve in 2 cups of water. It is okay if some chemical settles to the bottom of the container.
2. Place a pinecone in the solution and allow it to soak for several hours.
3. Remove the pinecone and allow it to completely dry. It is then ready to burn.
4. **Adult supervision is required for this step.** Burn your pinecones in a safe area, where there is no danger of catching anything else on fire. Do not cook food over the pinecones. The different colored flames are beautiful to watch but could make your food toxic.

Color	Chemical
Orange	Calcium Chloride (a bleaching powder)
Yellow	Sodium Chloride (table salt)
Yellowish Green	Borax
Green	Copper Sulfate
Blue	Copper Chloride
Purple	Potassium Chloride
White	Magnesium Sulfate (Epsom salts)

33

Rocket Fuel

Do you need a rocket scientist?

What is rocket fuel made from?

Words to know:
Newton's third law of motion

Challenge words:
solid rocket fuel hypergolic fuel
cryogenic fuel

Have you ever watched a spacecraft lift off and fly into space? The engines ignite, and steam and fire billow out the bottom as the spacecraft lifts off the ground, picking up speed every second. This is an exciting thing to watch. And a lot of chemistry went into making the rocket fuel needed for that exciting liftoff.

Combustion is a chemical reaction that produces large amounts of heat. This reaction is what provides the thrust that pushes the rocket off the ground. In most modern rockets, the fuel of choice is the combination of liquid oxygen and liquid hydrogen. These two elements are kept under pressure and then combined in a combustion reaction inside the rocket engine. At very high temperatures the oxygen and hydrogen combust and turn into steam. The resulting gases are forced out the back of the rocket engine at very high speeds. The water molecules travel at about 1,250 miles per hour (560 meters/second). But not all of the hydrogen and oxygen atoms combine to form water; some atoms just evaporate and then leave the engine at high speeds, providing additional thrust. The O_2 molecules travel at up to 950 mph (425 m/s) and the hydrogen atoms move at up to 3,580 mph (1600 m/s).

Newton's third law of motion states that for every action there is an equal and opposite reaction. So, as the gases escape out the back of the

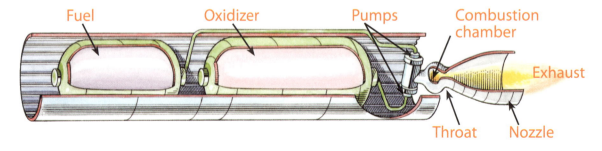

one of the first rocket fuels. The gas produced in the combustion of kerosene is carbon dioxide. Carbon dioxide is heavier than steam so it does not move as quickly, thus it produces less thrust. Therefore, oxygen and hydrogen are more efficient rocket fuels.

The next time you see a video of the space shuttle or other rocket taking off, remember that the white clouds that are billowing around the rocket are not smoke as you might think, but actually clouds of steam from the oxygen/hydrogen reaction.

rocket engine, they are pushing the rocket forward with the same amount of thrust. This is the basic physics behind a rocket.

Not all of the liquid hydrogen is used in the combustion reaction. Because the reaction produces so much heat, some of the liquid hydrogen is piped around the inside of the engine where it absorbs heat to keep the parts from melting.

Earlier rocket engines did not use liquid hydrogen and liquid oxygen as their propellants. Kerosene was

What did we learn?

- What is combustion?
- What two elements are combined in most modern rocket fuel?
- What compound is produced in this reaction?
- How does combining oxygen and hydrogen produce lift?
- What is Newton's third law of motion?

Taking it further

- Why is oxygen and hydrogen a better choice for rocket fuel than kerosene was?

Balloon rocket

Purpose: To learn how rockets work

Materials: balloon, straw, string, tape, two chairs

Procedure:

1. Blow up a balloon and then release it. What happens? Tape a straw to the balloon.
2. Thread a string through the straw.
3. Tape the ends of the string to opposite walls of a room or to two chairs that are several feet apart.
4. Blow up the balloon and pull it to one end of the string.
5. Release the balloon.

Conclusion: When you first released the balloon, it flew around the room. This is because the air molecules inside the balloon were forced out the end. And as we already learned about motion, for every action there is an equal and opposite reaction. So, as the air molecules rushed out the back of the balloon they produced a force on the front of the balloon that made it move.

The string provided a guidance system for the rocket, and the balloon should have flown forward along the string.

A real rocket has a very exact engine that carefully controls how the molecules exit the engine so that the rocket lifts off straight and does not fly all over the place like your balloon did.

Fuel types

Although most modern rockets use the combustion of liquid hydrogen and liquid oxygen as their propellant, there are actually three different categories of rocket fuel that are used for different purposes.

First there are **solid rocket fuel** engines. A solid rocket fuel begins with a combustible material. Originally, gunpowder was used as the fuel for rockets, but modern rockets use fuels with more energy. This fuel is bound together with an oxidizer, which is a chemical that provides the necessary oxygen for combustion. The fuel and oxidizer are formed into grains which are compressed to form the core of the engine. Solid rocket fuel is usually ignited by an electrical charge.

Solid rocket fuel is not used in most space rockets, but is commonly used for model rockets. Also, because solid rocket fuel can be stored for indefinite periods of time, many military missiles use solid rocket fuel.

The most common type of rocket fuel is **cryogenic fuel**. Cryogenic fuel must be stored under very high pressure and at very low temperatures. This is the type of liquid fuel discussed earlier in the lesson. Hydrogen, which is the fuel, and oxygen, which is the oxidizer, are stored under great pressure and then released at a specific rate into the combustion chamber where they are burned. This type of engine is more complicated than a solid rocket engine, but it allows for more control of the rate at which the propellants are burned. A solid rocket engine must continue burning until all of the fuel is used up, but a liquid fuel engine can be turned on and off.

A third type of fuel is called **hypergolic fuel**. Hypergolic refers to two substances that ignite when they come in contact with each other. Hypergolic fuel rockets can be controlled the same way as a hydrogen/oxygen engine; however, they produce more toxic chemicals in their combustion reaction. Therefore, they are used only in a few types of rockets.

34

Fun with Chemistry: Final Project

Understanding chemical reactions

Have fun with chemical reactions.

Now that you have learned about atoms and molecules, you should have a better understanding of what happens during chemical reactions. Take a few minutes to review what you have learned. Especially review lessons 17–20 on chemical reactions. Now, let's have some fun!

What did we learn?

- What was your favorite chemical reaction?
- Why did you like that reaction?

Taking it further

- What do you think will happen if you use skim milk in the first activity?
- What colors would you expect to see separate out of orange ink? Brown ink?
- Why is it important not to inhale the sodium polyacrylate from the diaper?

LESSON 34 **Properties of Atoms & Molecules • 397**

🧪 Fun chemical reactions

For each activity, complete the appropriate section of the "Fun With Chemistry" worksheet.

Older students should make a hypothesis about what they expect to see in each experiment before performing the experiment.

Purpose: To experiment with various chemical reactions

Materials: "Fun with Chemistry" worksheet, milk, dish, food coloring, water, sink, liquid dish soap, paper towel, markers, unused disposable diaper, scissors, large glass, eyedropper, white glue, liquid starch, plastic zipper bag

Activity 1—Moving molecules

Procedure:

1. Pour one cup of milk into a dish.
2. Place a drop of blue food coloring, red food coloring, and green food coloring equally spaced around the edge of the dish.
3. Observe the movement of the colors for several seconds.
4. Now add a drop of liquid dish soap to the center of the milk and watch the movement of the colors. Write your observations and explanation on the worksheet.

Activity 2—More moving molecules

Procedure:

1. Fold a paper towel in half.
2. Using a black water-soluble marker, draw a 1-inch wide bar about one inch from the narrow edge of the paper.
3. Draw a green bar and other colors of bars also one inch from the edge. Make sure none of your bars touch each other.
4. Put about ½ inch of water in a sink.
5. Carefully place the edge of the paper towel in the water so the bars are just above the water level.
6. Set the paper towel against the edge of the sink and place a heavy object on the top edge to hold it in place.
7. Allow the water to slowly wick up the paper towel for one hour.
8. Observe the ink after one hour.
9. Remove the paper from the water and allow it to dry. What did you observe about the ink? Write your observations on the worksheet.

Activity 3—Super absorbent molecules

You will be removing the powder that is found in an unused disposable baby diaper. **During this activity be very careful not to inhale or swallow any of the powder, and be careful not to get it in your eyes.**

Procedure:

1. Cut away the plastic outer covering of a disposable diaper.
2. Carefully place a large section of the inner padding in a plastic zipper bag.
3. Wash your hands and seal the bag.
4. Hold the padding toward the top of the bag and shake until you have about ½–1 teaspoon of powder in the bottom of the bag.

5. Carefully remove the padding and discard it. Again, wash your hands.
6. Carefully pour the powder into in a large glass or mixing bowl.
7. In a separate cup, combine one cup of water and a few drops of food coloring.
8. Using an eyedropper, add the colored water one drop at a time to the powder. What do you see happening?
9. After adding several droppers of water, begin adding water 1 teaspoon at a time. How much water do you predict the powder/gel will be able to absorb? See if you can add the whole cup of water to the gel. Write your observations on the worksheet.

Activity 4—Making your own goop
Procedure:

1. Pour ½ cup of white glue into a mixing bowl.
2. Add a few drops of food coloring and stir in the food coloring until it is well mixed.
3. Slowly add ¼ to ½ cup of liquid starch to the glue, mixing well until you have a gooey mixture that is easy to handle. What do you see happening? You can play with your goop.
4. When you are done, place the goop in a plastic zipper bag for later playing.
5. Wash your hands and record your observations on the worksheet.

Explanation for each activity
Activity 1—Moving molecules

Milk is a colloid, a liquid with solids suspended in it. Most of the solids in milk are fat molecules. These fat molecules keep the food coloring molecules separated from each other for the most part. When you add the soap, the soap molecules attract the fat molecules, allowing the food coloring to dissolve in the milk.

Activity 2—More moving molecules

This is an activity called chromatography. Black, green, and many other colors of ink are a combination of different colored ink molecules. These different molecules have different weights so they move up the paper towel at different rates. This allows you to see the different colors of molecules in the ink.

Activity 3—Super absorbent molecules

The powder in disposable diapers is a polymer called sodium polyacrylate. These molecules have the ability to absorb over 100 times their weight in water. This makes it very useful for diapers. This chemical is also used as a soil additive for some potted plants. These molecules absorb the water and then slowly release it, allowing you to water your plants less often, and to use less water.

Activity 4—Making your own goop

Starch and glue molecules combine to form a polymer. Recall that a polymer is a long flexible chain of molecules. This flexibility is what makes the goop so much fun.

More experiments

There are many more fun chemistry experiments on the Internet. Choose one that sounds interesting and perform it. Always obtain permission from your parent or teacher before conducting your own experiment.

35

Conclusion

Appreciating our orderly universe

Thank God for His amazing universe!

We live in a world that operates according to specific natural laws that were set in motion by God. We have learned how God designed our world to recycle all of the elements so that the matter in the universe is not used up. God designed our bodies to perform chemical reactions that are complementary to the reactions performed by plants; plants provide food and oxygen for us, and we provide carbon dioxide for photosynthesis.

All that you have learned about chemistry should point out that God is the master designer of our world. Read some Scripture verses that describe God's work in our world. Read Psalm 148, Isaiah 42:5–7, Colossians 1:16–17, and Job 42:1–2.

Now take a few moments and thank God for creating a world that obeys His rules, even at the molecular and atomic levels.

Properties of Atoms & Molecules — Glossary

Acid   Substance the easily forms hydronium ions by producing H+ ions in water

Acid salt   Salt formed when there is more acid than base

Activation energy   The energy required for a chemical reaction to take place

Alloy   Resultant metal when a small amount of one metal is added to another

Anesthetic   Chemical that stops or blocks pain

Antibiotic   Substance produced by living organisms to kill bacteria

Aromatic   Smell is easily detected

Atomic mass unit/amu   Mass of a proton or neutron

Atomic mass/Mass number   Number of protons plus the number of neutrons in an atom

Atomic number   Number of protons in an atom

Atom/Element   Smallest part of matter that cannot be broken down by ordinary chemical means

Base   Substance that easily forms hydroxide ions

Basic salt   Salt formed when there is more base than acid

Carbon cycle   Recycling of carbon for reuse by plants and animals

Catalyst   Substance that increases the reaction rate by lowering activation energy

Ceramic   Inorganic nonmetallic material formed by the action of heat

Chemical analysis   Use of chemical reactions to determine what a substance is made of

Chemical equations   Symbolic representation of a chemical reaction

Chemical formula   Symbolic representation of the atoms in a molecule

Chemical reaction   Occurs when atomic bonds are formed or broken

Chemistry   Study of the basic building blocks of matter

Chemists   Scientists who study how matter reacts to other matter

Composition reaction   Two or more atoms combine to form a molecule

Compound   Bonding of two or more different kinds of atoms

Covalent bonding   Molecules formed by sharing of electrons

Crop rotation   System of growing different crops each season to replace nutrients in soil

Crystal   Substance whose atoms form in an orderly pattern

Decomposers   Organisms that break down dead plant and animal cells into simple chemical compounds

Decomposition reaction   A molecule breaks apart into two or more atoms

Decomposition   Process that breaks down dead cells to recycle chemicals

Dehydrogenation   Removal of hydrogen atoms from a molecule

Diatomic molecule   Two of the same type of atom bonded together

Ductile   Able to be drawn into a wire

Edge   Where two faces of a crystal meet

Electrolysis   Separating of atoms by passing electrical current through them

Electron energy levels   The various distances at which an electron orbits the nucleus of an atom

Electronegativity   A measure of how strongly an element attracts electrons to itself

Electrons   Negatively charged particles orbiting the nucleus of an atom

Endothermic reaction   Reaction in which energy is stored

Enzyme   Catalyst involved in biological functions

Ethnobotanist   Person who studies plants to develop new medicines

Exothermic reaction   Reaction in which energy is released

Face   Flat surface of a crystal

Fallow   Not planting any crops for one or more years

Family   Column of the periodic table

Fertilizer   Chemicals added to make soil more productive

First law of thermodynamics   Matter and energy cannot be created or destroyed; they can only change forms

Free electron model   Metals share electrons on a large scale

Fungicide   Chemical that kills fungus

Genetic engineering/Genetic modification   Manipulation of a plant's or animal's genes to produce offspring with desired characteristics

Halogens   Top four elements in column VIIA on the periodic table
Healing herbs   Plants used as cures for illnesses
Herbicide   Chemical that kills unwanted plants
High electronegativity   An element does not easily give up its electrons
Hydrogenation   Addition of hydrogen atoms to a molecule
Hydronium ion   Water molecule that has reacted with an H^+ ion
Hydroponics   Growth of plants without soil
Hydroxide ion OH^- ion

Indicator   Substance that changes color when it reacts with a particular chemical
Inert   Non-reactive
Inhibitor   Substance the slows down the reaction rate
Ionic bond/Ionic bonding   A bond formed by transferring of electrons
Ionic compound   Molecules formed by ionic bonding
Ion   An element with an electrical charge due to missing or extra electrons

Latex   Sticky, milky substance found in rubber trees
Low electronegativity   An element easily gives up its electrons

Malleable   Able to be hammered into a shape
Matter   Anything that has mass and takes up space
Metallic bonding   Substance formed when thousands of atoms freely share their electrons
Metalloids/Semi-conductors   Elements on the periodic table between the metals and nonmetals
Molecule   Group of chemically connected atoms

Neutrons   Electrically neutral particles in the nucleus of an atom
Noble gases   Elements in column VIIIA, non-reactive elements
Normal salt   Salt formed when an acid and base completely neutralize each other
Nucleus   Tight mass of protons and neutrons in the center of an atom

Organic farming   Growing plants without the use of artificial chemicals
Oxidation   Adding of oxygen to a molecule

Period   Row of the periodic table
Perfume   Liquid with a pleasant smell
Pesticide   Chemical that kills unwanted insects
Pharmaceuticals/Medicines   Chemicals used to cure illnesses
pH scale   Indicates the strength of an acid or base
Plastic   Substance made from polymers formed from petroleum
Polymers   Long chains of molecules connected end to end
Products   Atoms or molecules formed by a chemical reaction
Proteins, fats, and carbohydrates   Main chemicals used by the body
Proton   Positively charged particle in the nucleus of an atom

Reactants   Atoms or molecules present at the beginning of a chemical reaction
Reaction rate   The rate or speed at which a chemical reaction takes place
Reactive   Easily combines chemically with other elements
Reduction   Removal of oxygen from a molecule
Rubber   Flexible substance made from latex

Salt   Substance formed when an acid and a base combine
Scavenger   Animal that eats dead animals
Smelting   Refining ore through high temperatures
Solvent   Substance used to dissolve other substances
Solvent extraction   Removing of scent molecules by a solvent
Steam distillation   Use of steam to vaporize scent molecules
Synthetic   Man-made, not naturally occurring

Thermoplastic   Plastic that can be melted over and over again
Thermosetting resin   Plastic that will not melt after it initially cools and hardens
Third law of motion   For every action there is an equal and opposite reaction
Transition elements   Elements in the center columns, labeled with the letter B

Vaccination/Vaccine Chemical solution that stimulates your body's immune system against a particular disease

Valence electrons Electrons in outermost energy level of an atom

Vulcanization Use of sulfur to make rubber elastic in all temperatures

Properties of Atoms & Molecules — Challenge Glossary

Acid/base titration Method for determining the unknown quantity of a base (or acid) by carefully measuring the exact amount of acid (or base) needed to completely neutralize it

Active ceramic Bioceramic that chemically reacts with tissues in the body

Allotrope Different forms or molecules made from the same element

Anhydrous Substance that has had the water removed

Anion Negatively charged ion

Bioceramic Ceramic that is used in the body

Buckminsterfullerene/Buckyball Carbon atom in the shape of a soccer ball

Carbon nanotube Extremely tiny thread-like tube created from matrix of carbon atoms

Casein Polymer found in cow's milk

Cation Positively charged ion

Chemotherapy Use of chemicals to treat cancer

Cryogenic rocket fuel Liquid fuel stored under pressure at very low temperatures

Dehydration The process of removing water

Double displacement reaction Reaction in which two substance trade places

Electroplating Depositing a thin layer of metal on a conductor using electricity

Enthalpy Energy stored in a molecule's bonds

Heterogeneous catalyst Catalyst in a different phase from the reactants

Homogeneous catalyst Catalyst in the same phase as the reactants

Hydrated Bonded with water

Hydrate Crystalline substance that is bonded with water

Hydrogen fuel cell Technology that combines hydrogen and oxygen to produce electricity

Hypergolic rocket fuel Liquid fuel that combusts when two substances come in contact with each other

Inert ceramic Ceramic that does not react with the body

Isotope Same type of atom with different numbers of neutrons

Keratin Polymer found in animal horns and human hair and nails

Lac Polymer produced in the insect called a lac

Nanotechnology The manipulation of matter on the atomic or molecular level for technological uses

Neutralize To make an acid of base chemically neutral

Noble metals Those that do not easily react with other metals

Poor metals Those to the right of the transition metals, between transition metals and metalloids

Protease Enzyme that breaks down proteins

Proton acceptor Alternate definition of a base when the molecule accepts a proton ion

Proton donor Alternate definition of an acid when the molecule donates a proton ion

Quantitative measurement One in which actual numbers are used

Reactivity series List of metals from most to least reactive

Resorbable ceramic Bioceramic that breaks down inside the body

Rosin Polymer found in dead wood and wood pulp

Silk A polymer produced by silk worms

Single displacement reaction Reaction in which one substance take the place of another

Solid rocket fuel Rocket fuel bound with an oxidizer in solid form

Super alloy Alloy that is strong even at high temperatures

Valence Number of electrons an element needs to lose or gain to be stable

Water of crystallization Water that is bonded in a hydrate

Index

abiotic   142-144, 270
acid   47, 72, 88, 98, 100, 113, 124, 132, 189, 279, 300, 340, 342, 345, 349-359, 361, 365-366, 373, 386-387, 401-403
acid rain   244, 246-247, 258-260, 263, 270
activation energy   341, 401
adaptation   228, 236-238, 270
adaptive radiation   236, 238, 272
algae bloom   196-198, 270
alkali-earth metal   294, 298, 315
alkali metal   294, 298, 302-303, 312
allotrope   305, 307, 403
alloy   321-323, 401, 404
alpine   204-205, 217-219, 270
alternative energy   260, 263
amorphous solid   53, 55, 137
anion   312, 315, 403
Antarctic   144, 163-164, 183, 204-205, 270, 272
antibiotic   373-374, 401
antifreeze   101, 103
antioxidant   124, 126, 136
artificial flavor   120
aspartame   98-100
atmospheric pressure   59-60, 136
atoll   187, 243, 270
atom   25, 53-54, 59, 67-68, 71, 73-74, 76-77, 79, 83-84, 134, 136, 277, 280-284, 287-289, 292, 302, 305, 307-308, 312-317, 319, 334, 338-339, 351, 354, 359, 361, 401-403
atomic mass unit/AMU   283-284, 401
atomic number   70, 283-284, 291, 294, 302-303, 305-306, 401
banyan   180-181
barrier reef   187-188, 270
base   23, 113, 155, 209, 219, 279, 349-350, 352, 354-359, 381, 401-403
bauxite   321-322, 324
beach   182, 190-192, 196, 270, 273
Beaufort scale   26, 29, 137
bioceramic   403
biogeographic realms   144, 272
bioluminescent   183, 186, 272
biome   142-143, 150, 162-164, 169, 171, 177, 210, 270

biosphere   142, 144, 270
biotic   142-144, 270
boiling point   49-50, 57, 79, 101, 103, 105, 132, 136, 302
boreal forest   176-177, 270
Boyle, Robert   62, 65
Boyle's law   62, 136
buckminsterfullerene   305, 307, 403
buffering capacity   258-260, 270
buoyancy   30, 42-44, 47, 100, 136
calorie   99-100, 116, 118, 137-138
camouflage   206, 233-234, 270
canopy   173-174, 270
captive breeding   248, 250, 270
carbohydrate   112-113, 116-118, 127, 136, 317, 362-364, 402
carbon   16, 36, 47-48, 50-51, 53-54, 61, 67, 69-70, 72-75, 77, 79, 82-84, 98-99, 104-105, 112-113, 127, 132, 134, 136-137, 153-154, 159, 185, 216, 239, 254-255, 258, 261-264, 271, 279, 284, 288, 290, 294, 300, 305-307, 309-310, 312, 321-322, 326, 332, 334-335, 338, 340-341, 343-344, 352, 359, 363, 368, 383, 395, 400-401, 403
carbon cycle   305-306, 368, 401
carnivore   150-152, 170, 226, 241, 270
carrying capacity   150, 152, 155, 170, 272
catalyst   304, 310, 336, 339, 341-343, 388, 401, 403
cation   312, 315, 403
cellulose   302, 354, 385, 388
centrifuge   79, 81, 137
ceramic   306, 330-332, 401, 403
Charles, Jacque   62
Charles's law   62-64, 136
chemical analysis   111, 116-117, 132-133, 347-350, 401
chemical formula   73-74, 120, 123, 136, 312, 318, 401
chemical property   47, 136
chemistry   14-15, 25, 47, 65, 71, 74, 90, 94, 111-112, 121, 277-279, 286, 307, 324, 330-331, 346, 348, 356, 363, 371, 373, 377-379, 382, 388, 394, 397-401
chemotherapy   373, 375, 403
chlorofluorocarbons/CFCs   254-255, 257, 272-273

• 405

chocolate   28, 41, 119, 122-123, 125
chromatography   79, 81, 138, 350, 399
classification   67-68, 136
climate   142-144, 146, 162-165, 168-169, 171, 173, 211, 213, 216-217, 220-221, 238, 248, 261, 263, 270, 272
colloid   92-93, 136, 399
combustion   82-83, 309, 343-345, 394-396
commensalism   156-158, 270
community   26, 147-148, 238, 270
competition   146, 148, 156-158, 231, 239, 246, 270
composition reaction   334-335, 339-340, 401
compost   266, 369
compound   67-69, 73-76, 79-80, 83-84, 87, 125, 136, 287-288, 302, 306, 312, 314-318, 320, 324, 339-340, 349, 363, 395, 401-402
concentration   61, 89, 101, 103-106, 136, 138, 185, 263, 336, 341, 343
condense   49, 78, 81, 84, 107, 121, 136, 215, 380
conductivity   104-105, 138, 296
conservation of mass   30, 34-36, 137, 153, 159, 271
consumer   150-152, 154, 216, 239-241, 270, 272, 371
coral   143, 182, 184, 187-189, 191, 194-195, 211, 243, 270-272
courtship   229, 231-232, 272
covalent bond   317
crystal   54, 90, 196, 305, 307, 314, 326-331, 357-358, 401
crystalline solid   53, 138
Curie, Marie   285
DDT   242
deciduous   162, 173, 176-178, 181, 217-219, 270
decomposer   141, 153-155, 159, 161, 241, 270, 367, 369
decomposition   153-154, 270, 334-335, 339-340, 342, 367, 369, 401
deforestation   261-262, 270
dehydration   326, 329, 403
density   29-30, 40-45, 47, 51-53, 79, 100, 105, 132, 134, 136, 194, 196-197, 365, 390
desalination   107
desert   144, 149, 163-164, 169, 203, 210-216, 235, 254, 256, 270-272
diatomic molecule   67-68, 73, 83, 136, 287-289, 308, 316-317, 339, 401

diffusion   59, 61, 138
displacement method   37-39, 41, 136
distillation   79, 81-82, 84, 107, 121, 138, 379-381, 402
dominion   247
dormant   169, 233, 235, 272
dynamic equilibrium   190-191, 270
echolocation   202, 224, 226, 272
ecology   142, 144, 149, 167-168, 243-245, 247, 253, 270
ecosystem   141-147, 155-156, 161, 163, 165, 169, 171, 173-174, 183, 185, 188, 190-193, 195, 198-201, 205, 213-215, 217-219, 221-222, 224-225, 227, 239, 241-243, 245, 254, 258, 268, 270-273
ecotone   163, 165, 180, 272
electrolysis   74, 321, 325, 401
electrolyte   360-361
electron   69-70, 280-282, 284, 289, 292, 294, 299, 302, 313-317, 319-320, 334, 351, 354, 401
electronegativity   312-314, 401-402
electroplating   351, 353, 403
element   26, 51, 54, 65-71, 73, 75, 79-80, 82-83, 87, 90, 104, 110, 117, 136, 138, 161, 282-284, 286-287, 289, 291-293, 295, 299, 302-303, 307-309, 314-315, 322-323, 326, 335, 340, 401-404
emulsifier   92-94, 136
endangered   244, 248-251, 270
endothermic   333, 344-346, 401
enthalpy   344, 346, 403
enzyme   85, 88, 98, 128, 138, 341-343, 364-366, 401, 403
ephemeral   204, 270
epiphyte   156-157, 180-181, 270
estivation   228-229, 231, 270
estuary   193-195, 270-272
evaporate   34, 49-50, 57-58, 81, 107, 128, 136, 138, 193, 301, 328-329, 379, 381, 394
evergreen   173, 176-178, 217-218, 270
exothermic   333, 344-346, 401
extinct   196, 242, 248, 250, 262, 270-271
fat   229-230, 363-364, 379, 399
fats   85-87, 91, 93-94, 112-113, 116-118, 128, 136, 317, 362, 364, 402
fauna   142-144, 167, 271
fermentation   122-123, 127, 136
fire cues   221, 223

first law of thermodynamics   34, 136, 333-335, 340, 346, 401
flavor enhancers   112, 124-125
Fleming, Alexander   374-376
flora   142-144, 167, 271
foam   44, 85, 87, 136, 279
Food and Drug Administration/FDA   115, 124-125, 136
food chain   150-155, 186, 196, 240, 242, 246, 249, 258, 271, 363
food web   150-152, 154, 240, 271
forest   82, 157, 165-166, 173-178, 180-181, 190, 192-194, 201, 217-219, 245, 253, 258, 262, 270-273, 352
fractional distillation   82, 84, 121, 138
free electron model   319-320, 401
fringing reef   187-188, 271
fruit bat   226
Fujita scale   26, 29, 138
gas pressure   59, 136
Genetically modified organism/GMO   239, 242, 273
genetic engineering   370, 372, 402
global warming   207, 244, 261-263, 271
gluten   127-128, 130, 136
Goodyear, Charles   382, 386-387
Great Barrier Reef   188
greenhouse effect   261, 263-264, 271
habitat   142-144, 148, 198-199, 214, 245-246, 248-249, 251, 262, 271
Hall, Charles Martin   324
halogens   73, 75, 138, 299, 301-303, 313, 389, 402
heartwood   173, 175, 273
Henry, William   96
herbivore   150, 152, 271
heterogeneous   79-80, 91, 136, 341, 343, 403
hibernation   228-229, 231, 271
Himalayas   219-220, 262
homogeneous   79-80, 90-91, 136-137, 341, 343, 403
homogenization   66, 85-87, 136
hydrate   326, 329, 403-404
hydrogenation   93, 302-303, 342, 402
hydrogen fuel cell   302, 304, 403
hydronium   349, 351-352, 354, 359, 401-402
hydroponics   370-372, 402

hydroxide   354-355, 357, 401-402
hypothesis   17-20, 44, 96-97, 132, 136, 144, 264, 318, 342, 366, 369, 398
immiscible   92-93, 136-137
indicator   116-117, 132, 136, 348-350, 352-353, 355-356, 402
inert   299-301, 330, 332, 402-403
inhibitor   341-343, 402
ion   283, 312-315, 351-352, 354-355, 357, 359, 361, 402-403
ionic bond   312-314, 402
ionic compound   312, 315, 320, 402
isotope   283-284, 293, 403
kinetic energy   49, 51, 53, 56-57, 59, 136
lake   42-44, 57-58, 105, 159-160, 163, 178, 182, 195-200, 213, 218, 227, 249, 253-254, 258, 271
litmus   349
Lord Kelvin   24-25, 286
mangrove forest   193-194, 271
maritime forest   190, 192, 273
mass   14, 22, 26-28, 30-37, 40-43, 47, 51-53, 69-70, 132, 136-138, 153-155, 159, 271, 277-278, 281, 283-284, 286, 291, 293-294, 348, 350, 374, 377, 401-402
Maxwell, James Clerk   24, 45
melting point   49-50, 55, 137
Mendeleev, Dmitri   295
meniscus   37-38, 137
metallic bonding   311, 319-320, 402
metalloid   294, 296-297, 299, 402-403
metric system   13, 26-29, 32-33, 118, 136-137
microscope   21, 23, 138, 241, 280, 357
migration   144, 229-231, 271
mineral   29, 73, 75, 79-80, 85, 98, 104-107, 110, 112-113, 116, 124-125, 128, 137-138, 212
mixture   67, 69, 79-85, 87, 90-93, 102-103, 105, 136-137, 322, 325, 390, 399
Mohs scale   26, 29, 138
molecule   67-68, 73-77, 83-84, 91, 136, 257, 287-289, 301, 308, 310, 312-314, 316, 334-335, 338-340, 351, 354, 357, 383, 388, 401-403
Mount St. Helens   254
mutualism   156-158, 189, 271
nanotechnology   305, 307, 403
native   73, 75, 138, 198, 211, 245-246, 249, 349
naturalism   14, 16, 138

• 407

natural selection   236-238, 271
neutralism   156-157, 271
neutron   280-284, 291, 302, 305, 308, 313, 317, 320, 401
Newton, Sir Isaac   32
niche   147-149, 237, 268, 271
nitrogen   34, 51, 68, 70, 73, 75, 82-83, 104-105, 112, 153-154, 159, 161, 185, 239, 246, 255, 258-260, 273, 287, 294, 302, 343, 367-371
nitrogen cycle   159, 161, 273, 367-369
noble gas   292, 300-301
nocturnal   210-211, 227, 271
nucleus   280-283, 291-293, 302, 305, 308, 313, 317, 401-402
omnivore   150, 152, 271
operational/observational science   14-16, 35, 138
organic   71, 115, 124, 225, 266, 305-306, 370-372, 402
origins science   14-16, 138
osmium   70, 294
overturn   196-198, 271
oxidation   223, 308, 310, 322-323, 335, 343, 345, 402
oxide   73, 303, 308, 310, 322, 325, 338, 373
oxygen cycle   159-160, 239, 241, 271
ozone   83, 256-257, 261, 272, 308-309
pampas   169-171, 271
parasitism   156, 271
Pasteur, Louis   85-86
pasteurization   66, 85-87, 137
penicillin   374, 376-377
percolate   76, 138
periodic table   69-70, 75, 138, 277, 282-284, 289-296, 298-299, 302, 308, 313-315, 318, 339, 354, 401-402
permafrost   204-205, 218, 271
pesticide   242, 245, 370, 402
pH   241, 348-350, 365, 402
pharmaceutical   373
phase change   49-51, 137
phloem   173, 175, 273
phosphates   73, 75, 138, 332, 358
photosynthesis   48, 77-78, 113, 150, 157, 159-160, 175, 183-185, 187, 189, 199, 211, 214-215, 226, 255, 262, 271, 305-306, 334-335, 339-341, 344-346, 363, 365, 400

physical property   47, 49, 137
plankton   183-185, 197, 271
plastic   15, 44, 64, 67, 102, 109, 129, 143, 191, 212, 215, 255-257, 261, 264, 267, 273, 297, 369, 388-390, 398-399, 402
polar bear   205-207, 218, 236, 262
polymer   265, 273, 331, 382-385, 388, 390, 399, 403-404
population   147-148, 152, 155, 186, 198, 237, 240-242, 247, 249-250, 260, 271-272
prairie   152, 169-170, 233-234, 240-241, 271
precipitate   95-96, 108, 110, 137
predator   146, 151-152, 156, 180, 184, 198, 206-207, 225, 233-234, 240-242, 246, 249, 271
preservative   112, 124-125, 137, 327
prey   146, 156, 174, 187, 207, 233-234, 240, 271
producer   148, 150-151, 155, 183, 239, 241, 271
protein   85, 93, 112-113, 116-118, 127-128, 137, 302, 317, 349, 363-366, 385, 402-403
proton   280-281, 283-284, 291, 293, 302, 304, 308, 313-215, 317, 320, 351, 357, 359, 401-403
Prout, William   71
qualitative observation   103
quantitative observation   103
reactant   334-337, 339-341, 343-346, 365, 402-403
reaction   22, 34, 36, 48, 74, 83, 88, 113, 127, 133, 136, 161, 189, 256, 264, 271, 278-279, 293, 300, 310, 321-322, 329, 331, 334-346, 353, 358-361, 363, 365-366, 394-397, 401-404
reactivity series   296, 298, 403
recycle   154, 159, 194, 255, 265-267, 384, 390, 400-401
reduction reaction   321, 335
resources   148, 157, 216, 246, 250, 253, 263, 270, 273
respiration   72, 159-160, 271, 305, 310, 363-364
reverse osmosis   107
riparian zone   199-200, 271
rocket fuel   303, 378, 394-396, 403-404
Saffir-Simpson scale   26, 29, 138
Sahara Desert   144, 210, 212-214
salinity   104-106, 138

salt   23, 48, 50, 67, 69, 73, 79-81, 86, 90-91, 93, 95, 97, 100, 102-107, 110, 116, 124-125, 127, 129, 138, 185, 192-195, 197, 212-213, 272, 288-289, 313-315, 318, 326-329, 353, 357-359, 364, 391, 393, 401-402
saturated   90, 96-97, 101, 137, 328
scavenger   153-154, 184, 272, 367-369, 402
scientific method   13, 15, 17-20, 132, 137
sedimentation   108, 110, 137
semiconductor   296-297, 300, 307, 320, 328
silicates   73, 75, 110, 138
smelting   321, 402
solubility   81, 89-91, 95-97, 136-137
solute   90-91, 95-96, 101, 136-137
solution   17, 64, 89-93, 95-97, 101-105, 117, 123, 136-137, 318, 321, 328-329, 349, 353-354, 356, 359-361, 371, 376-377, 380, 393, 403
solvent   76-77, 89-91, 95-97, 105, 108, 134, 137, 363, 379, 402
stabilizer   124
steam   34, 48-50, 71, 77, 81, 107, 121, 329, 379-380, 387, 394-395, 402
steppe   169-170, 272
structural integrity   85-86, 137
sublimation   49-50, 137
succession   163, 165-166, 192, 273
succulent   210, 272
sulfides   73, 75, 138
survival of the fittest   236-237, 271
suspension   90, 92-94, 136-137
symbiosis   156-158, 189, 243, 272
synthetic   88, 293, 373-374, 383-384, 388, 402
temperate   162-165, 176-178, 181, 217-219, 271-272
territoriality   239-240, 272
third law of motion   394-395, 402
timberline   217-218, 272
titration   354, 356, 403
transition element   291-292, 402
transpiration   76, 78, 138, 210-211, 214-215, 272
tropical   122, 146, 162-165, 170, 173, 176, 179-181, 187, 218, 272, 371
tundra   149, 163-164, 203-207, 211, 217-219, 270, 272
universal solvent   76-77, 97, 108, 137
U.S. Fish and Wildlife Service   251
vaccine   373-374, 403
valence electron   280-282, 287, 289, 293, 296-297, 299, 308, 313-317, 319-320, 404
vanilla   86-87, 99, 102, 119-123, 125, 381
viscosity   45, 56-57, 132, 137
volume   21, 26, 28-30, 37-43, 46-47, 51-53, 55-56, 59, 62-64, 71, 118, 136-137
vulcanization   382-384, 387, 403
water cycle   34, 57, 159-160, 183, 272
watershed   193, 195, 198, 201, 273
weight   21, 26, 30-33, 44, 71, 137-138, 175, 207, 255, 364, 399
xylem   173, 175, 273
yeast   18, 127-130, 136, 344, 369
zooplankton   183-185, 196, 249, 272

Photo Credits

- 1 Shutterstock.com
- 3 Getty Images /Jupiterimages
- 5 Getty Images/iStockphoto
- 7 © Galyna Andrushko | Dreamstime.com
- 9 Getty Images/iStockphoto
- 11 Getty Images /Jupiterimages
- 13 Getty Images/iStockphoto
- 14 ©2008 Jupiterimages Corporation
- 17 ©2008 Jupiterimages Corporation
- 19 ©2008 Jupiterimages Corporation
- 21T ©2008 Jupiterimages Corporation
- 21T Getty Images/Istock
- 22 ©Razvanjp | Dreamstime.com
- 23T ©2008 Jupiterimages Corporation
- 23B Answers in Genesis
- 24 Public domain
- 26 ©2008 Jupiterimages Corporation
- 30 ©2008 Jupiterimages Corporation
- 31T ©2008 Jupiterimages Corporation
- 31B Getty Images/iStockphoto
- 32 ©2008 Answers in Genesis
- 33 ©2008 Answers in Genesis
- 34 ©Fromac | Dreamstime.com
- 37T ©2008 Jupiterimages Corporation
- 37B Getty Images/iStockphoto
- 38 ©2008 Jupiterimages Corporation
- 40 ©2008 Jupiterimages Corporation
- 42T ©2008 Jupiterimages Corporation
- 42B ©Mg7 | Dreamstime.com
- 43 ©Hemera Technologies, Inc.
- 45 Public domain
- 46 ©2008 Jupiterimages Corporation
- 47 ©2008 Jupiterimages Corporation
- 47 ©Mk74 | Dreamstime.com
- 49 ©2008 Jupiterimages Corporation
- 50 ©2008 Jupiterimages Corporation
- 51T Getty Images/iStockphoto
- 51B ©Richard Lawrence
- 53 ©2008 Jupiterimages Corporation
- 54L Getty Images/Zoonar RF
- 54M Getty Images/iStockphoto
- 54R Getty Images/iStockphoto
- 55 Getty Images/iStockphoto
- 56 Getty Images/iStockphoto
- 57 Getty Images/iStockphoto
- 59 ©2008 Jupiterimages Corporation
- 62 Getty Images/iStockphoto
- 63 Courtesy NOAA/John Bortniak
- 65 Creative Commons | Wellcome Images
- 66 ©2008 Jupiterimages Corporation
- 67 ©2008 Jupiterimages Corporation
- 68 ©Richard Lawrence
- 71 Public domain
- 73 ©2008 Jupiterimages Corporation
- 74 ©2008 Answers in Genesis
- 76T ©2008 Jupiterimages Corporation
- 76B Getty Images/Dorling Kindersley RF
- 77 ©2008 Jupiterimages Corporation
- 78 ©2015 Answers in Genesis
- 79 Getty Images/iStockphoto
- 81 Public domain
- 82 ©2008 Jupiterimages Corporation
- 84 ©2008 Answers in Genesis
- 85T Getty Images/iStockphoto
- 85B Getty Images/Hemera
- 86 Getty Images/iStockphoto
- 88 Getty Images/Photodisc
- 89 ©2008 Jupiterimages Corporation
- 90 ©2008 Jupiterimages Corporation
- 92 ©2008 Jupiterimages Corporation
- 94 ©2008 Jupiterimages Corporation
- 95 ©2008 Jupiterimages Corporation
- 96 Getty Images/iStockphoto
- 98 Getty Images/Stockbyte
- 101 Getty Images/iStockphoto
- 102T ©2008 Jupiterimages Corporation
- 102B Getty Images/iStockphoto
- 104 ©2008 Jupiterimages Corporation
- 106 Getty Images/iStockphoto
- 107 Creative Commons | Octal
- 108 ©2008 Jupiterimages Corporation
- 109 ©2015 Answers in Genesis
- 111 ©2008 Jupiterimages Corporation
- 112 ©2008 Jupiterimages Corporation
- 114 © Lightpoet | Dreamstime.com
- 115 ©istockphoto.com/SimplyCreativePhotography
- 116T ©2008 Jupiterimages Corporation
- 116B Getty Images/iStockphoto
- 118 ©Richard Lawrence
- 119 ©2008 Jupiterimages Corporation
- 122 Getty Images/Zoonar RF
- 123 Getty Images/iStockphoto
- 124 Getty Images/Photodisc
- 127T Getty Images/Ingram Publishing
- 127B Getty Images/iStockphoto
- 128 Getty Images/iStockphoto
- 130 ©2008 Jupiterimages Corporation
- 132 Getty Images/Blend Images
- 134 ©2008 Jupiterimages Corporation
- 139 Getty Images/iStockphoto
- 141 ©Desertsolitaire | Dreamstime.com
- 142 ©2008 Jupiterimages Corporation
- 143 ©Kopstal | Dreamstime.com
- 145 ©2008 Answers in Genesis
- 147T ©Goosey | Dreamstime.com
- 147B ©Nantela | Dreamstime.com
- 149 ©2008 Jupiterimages Corporation
- 150 ©2008 Jupiterimages Corporation
- 151 Getty Images/iStockphoto
- 152 ©2008 Jupiterimages Corporation
- 153 ©Bcollet | Dreamstime.com
- 153 ©2008 Jupiterimages Corporation
- 154 Getty Images/iStockphoto
- 155 Getty Images/iStockphoto
- 156 ©2008 Jupiterimages Corporation
- 157 ©2008 Jupiterimages Corporation
- 159 ©2008 Jupiterimages Corporation
- 160 ©2008 Answers in Genesis
- 162 Getty Images/Fuse
- 163 ©Sherwoodimagery | Dreamstime.com
- 164 Josh BrockMap Resources
- 166 ©2008 Jupiterimages Corporation
- 167 Public domain
- 169 ©2008 Jupiterimages Corporation
- 170 ©Dmytro | Dreamstime.com
- 172 ©2008 Jupiterimages Corporation
- 173T ©2008 Jupiterimages Corporation
- 173B ©Phillip Lawrence
- 174 ©Phillip Lawrence
- 175 ©2008 Answers in Genesis
- 176 ©2008 Jupiterimages Corporation
- 179 Getty Images/iStockphoto
- 180 ©Kcphotos | Dreamstime.com
- 182 ©2008 Jupiterimages Corporation
- 183 ©2008 Jupiterimages Corporation
- 185 Getty Images/iStockphoto
- 186 ©2008 Jupiterimages Corporation
- 187 ©Ptoone | Dreamstime.com
- 188 Credit NASA/GSFC/LaRC/JPL, MISR Team
- 190 ©2008 Jupiterimages Corporation
- 191 ©Iofoto | Dreamstime.com
- 192 ©Nem0fazer | Dreamstime.com
- 193T ©2008 Jupiterimages Corporation
- 193B ©2008 Jupiterimages Corporation
- 195 Image courtesy NASA/GSFC/LaRC/JPL, MISR Team
- 196 ©Uolir | Dreamstime.com
- 197 ©Vaida | Dreamstime.com
- 198 Getty Images/Stocktrek Images
- 199 ©Phillip Lawrence
- 201 ©Tony1 | Dreamstime.com
- 203 Getty Images/iStockphoto
- 204 ©Outdoorsman | Dreamstime.com
- 205 ©Phillip Lawrence
- 207 Getty Images/Fuse
- 208 Credit NOAA
- 210 Getty Images/iStockphoto
- 211T Getty Images/iStockphoto
- 211B Public domain
- 212 ©2008 Jupiterimages Corporation
- 214 ©2008 Jupiterimages Corporation
- 215 ©Pancaketom | Dreamstime.com
- 217 ©Roadbully | Dreamstime.com

218	©Phillip Lawrence	
220	Getty Images/Hemera	
221	©Stockshooter | Dreamstime.com	
222	©Slateriverproductions | Dreamstime.com	
223	©Carve1 | Dreamstime.com	
224	©2008 Jupiterimages Corporation	
225	©Phillip Lawrence	
227	©Aje | Dreamstime.com	
228	Getty Images/liquidlibrary	
229	©2008 Jupiterimages Corporation	
230	©2008 Jupiterimages Corporation	
231	©2008 Jupiterimages Corporation	
232T	©2008 Jupiterimages Corporation	
232M	Getty Images/iStockphoto	
232B	Getty Images/iStockphoto	
233	©2008 Jupiterimages Corporation	
234	©2008 Jupiterimages Corporation	
235	©2008 Jupiterimages Corporation	
236	©2008 Jupiterimages Corporation	
238	Getty Images/iStockphoto	
239	©2008 Jupiterimages Corporation	
240	©2008 Jupiterimages Corporation	
242	©2008 Jupiterimages Corporation	
244	Getty Images/Stockbyte	
245	©2008 Jupiterimages Corporation	
246	Getty Images/iStockphoto	
248T	Creative Commons | Charles W. Hardin	
248B	Getty Images/iStockphoto	
249	©2008 Jupiterimages Corporation	
250	©Jgroup | Dreamstime.com	
252	Public domain	
254	©2008 Jupiterimages Corporation	
255	©2008 Jupiterimages Corporation	
258T	©2008 Jupiterimages Corporation	
258B	©Catcha | Dreamstime.com	
259	©2008 Answers in Genesis	
260	©2008 Jupiterimages Corporation	
261	©2008 Jupiterimages Corporation	
265	©2008 Jupiterimages Corporation	
266	©2008 Jupiterimages Corporation	
268	©2008 Jupiterimages Corporationd	
269	©2008 Jupiterimages Corporation	
275	© Galyna Andrushko | Dreamstime.com	
277	©2008 Jupiterimages Corporation	
278	©2008 Jupiterimages Corporation	
279	©Richard Lawrence	
280	Getty Images/iStockphoto	
281	Public domain	
283	Getty Images/iStockphoto	
285	Public domain	
287T	©istockphoto.com/dra_schwartz	
287B	Public domain	
288T	©2008 Jupiterimages Corporation	
288B	Getty Images/Creatas RF	
290	©2008 Jupiterimages Corporation	
291	Getty Images/iStockphoto	
295	Public domain	
296	©2008 Jupiterimages Corporation	
297	Getty Images/iStockphoto	
299	©2008 Jupiterimages Corporation	
300T	Getty Images/iStockphoto	
300B	©Phillip Lawrence	
301	Getty Images/iStockphoto	
302	©istockphoto.com/Elke Dennis	
304L	© Richard Lawrence	
304R	Creative Commons | Sludge G	
305	Getty Images/iStockphoto	
307LT	Getty Images/Zoonar RF	
307LB	Getty Images/iStockphoto	
307R	Getty Images/iStockphoto	
308T	©istockphoto.com/Matthew Hull	
308B	Public Domain	
310	©Phillip Lawrence	
311	Getty Images/Stockbyte	
312T	Getty Images/Hemera	
312B	Public Domain	
314	©Richard Lawrence	
315	©istockphoto.com/Nancy Nehring	
316	©2008 Jupiterimages Corporation	
319	©2008 Jupiterimages Corporation	
321	©2008 Jupiterimages Corporation	
322L	©2008 Answers in Genesis	
322R	©2008 Jupiterimages Corporation	
323	© Richard Lawrence	
324	Public domain	
326T	Getty Images/iStockphoto	
326B	Getty Images/Stockbyte	
327	©2008 Jupiterimages Corporation	
330T	Getty Images/iStockphoto	
330B	©2008 Jupiterimages Corporation	
331	©2008 Jupiterimages Corporation	
333	Getty Images/iStockphoto	
334T	Getty Images/iStockphoto	
334B	©Richard Lawrence	
338T	Getty Images/iStockphoto	
338B	©Richard Lawrence	
341	Getty Images/iStockphoto	
344	©2008 Jupiterimages Corporation	
346	©2008 Jupiterimages Corporation	
347	Getty Images	
348T	Getty Images/iStockphoto	
348B	©Tommounsey | Dreamstime.com	
349	©Sudo | Dreamstime.com	
351	©2008 Jupiterimages Corporation	
352	Getty Images/iStockphoto	
353	©Phillip Lawrence	
354	©2008 Jupiterimages Corporation	
355	Getty Images/iStockphoto	
357T	©2008 Jupiterimages Corporation	
357B	©Kivig | Dreamstime.com	
358	©2008 Jupiterimages Corporation	
360	©Richard Lawrence	
361	©2008 Answers in Genesis	
362	©2008 Jupiterimages Corporation	
363T	Getty Images/Photodisc	
363B	Getty Images/Fuse	
364	Getty Images/Hemera	
366	©2008 Jupiterimages Corporation	
367T	Getty Images/iStockphoto	
367B	Getty Images/iStockphoto	
368	©2015 Answers in Genesis	
369	©Lockstockbob | Dreamstime.com	
370T	Getty Images/iStockphoto	
370B	Getty Images/Zoonar RF	
373	©2008 Jupiterimages Corporation	
374T	Courtesy USDA	
374B	©2008 Jupiterimages Corporation	
375	Getty Images/iStockphoto	
376	Public domain	
377	© Tine Grebenc | Dreamstime.com	
378	Getty Images/iStockphoto	
379T	Getty Images/iStockphoto	
379B	©istockphoto.com/kemie	
380	Creative Commons | Izmaelt	
381	Getty Images/iStockphoto	
382T	©2008 Jupiterimages Corporation	
382B	©Highlanderimages | Dreamstime.com	
383	©2008 Jupiterimages Corporation	
385L	©Seesea | Dreamstime.com	
385RT	Getty Images/Goodshoot RF	
385RB	Getty Images/iStockphoto	
386	Public domain	
388	©2008 Jupiterimages Corporation	
389	Public domain	
390	Creative Commons | Woakamkurhram	
391	©2008 Jupiterimages Corporation	
392	©Mhryciw | Dreamstime.com	
394T	©istockphoto.com/Dennis Sabo	
394B	Getty Images/Dorling Kindersley RF	
395	Getty Images/iStockphoto	
397	Getty Images/iStockphoto	
400	Credit NASA/JPL-Caltech, D. Figer	

CHARLOTTE MASON INSPIRED
ELEMENTARY CURRICULUM THAT CONNECTS CHILDREN TO AMERICA'S PAST... AND THEIR FUTURE!

Through this unique educational style, children develop comprehension through oral and written narration, and create memories through notebooking and hands-on crafts. This is not just facts and figures; this is living history for grades 3 through 6.

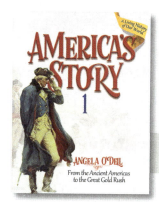

FROM THE ANCIENT AMERICAS TO THE GREAT GOLD RUSH

Part 1: Begins at the infancy of our country and travels through the founding of our great nation, catching glimpses of the men who would become known as the Founding Fathers.

America's Story Vol 1 *Teacher Guide*
978-0-89051-979-0   978-0-89051-980-6

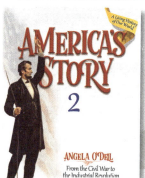

FROM THE CIVIL WAR TO THE INDUSTRIAL REVOLUTION

Part 2: Teaches students about the Civil War, the wild West, and the Industrial Revolution.

America's Story Vol 2 *Teacher Guide*
978-0-89051-981-3   978-0-89051-982-0

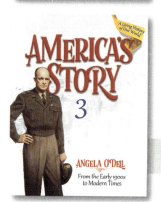

FROM THE EARLY 1900s TO OUR MODERN TIMES

Part 3: Carries the student from the turn of the 20th century through the early 2000s.

America's Story Vol 3 *Teacher Guide*
978-0-89051-983-7   978-0-89051-984-4

VISIT MASTERBOOKS.COM — Where Faith Grows! — TO SEE OUR FULL LINE OF FAITH-BUILDING CURRICULUM OR CALL 800-999-3777

GOD'S DESIGN FOR SCIENCE SERIES

EXPLORE GOD'S WORLD OF SCIENCE WITH THESE FUN CREATION-BASED SCIENCE COURSES

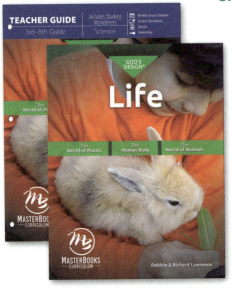

GOD'S DESIGN FOR LIFE
GRADES 3-8

Learn all about biology as students study the intricacies of life science through human anatomy, botany, and zoology.

GOD'S DESIGN FOR HEAVEN & EARTH
GRADES 3-8

Explore God's creation of the land and skies with geology, astronomy, and meteorology.

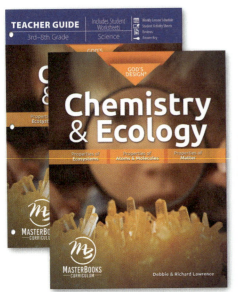

GOD'S DESIGN FOR CHEMISTRY & ECOLOGY
GRADES 3-8

Discover the exciting subjects of chemistry and ecology through studies of atoms, molecules, matter, and ecosystems.

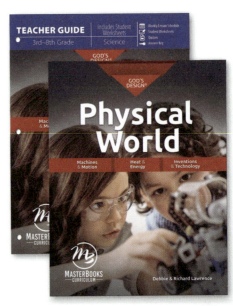

GOD'S DESIGN FOR THE PHYSICAL WORLD
GRADES 3-8

Study introductory physics and the mechanisms of heat, machines, and technology with this accessible course.

AVAILABLE AT MASTERBOOKS.COM 800.999.3777
& OTHER PLACES WHERE FINE BOOKS ARE SOLD

CHARLOTTE MASON INSPIRED
ELEMENTARY CURRICULUM THAT CONNECTS CHILDREN TO AMERICA'S PAST... AND THEIR FUTURE!

Through this unique educational style, children develop comprehension through oral and written narration, and create memories through notebooking and hands-on crafts. This is not just facts and figures; this is living history for grades 3 through 6.

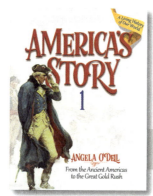

FROM THE ANCIENT AMERICAS TO THE GREAT GOLD RUSH

Part 1: Begins at the infancy of our country and travels through the founding of our great nation, catching glimpses of the men who would become known as the Founding Fathers.

America's Story Vol 1 *Teacher Guide*
978-0-89051-979-0   978-0-89051-980-6

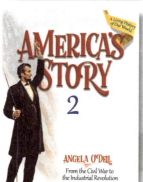

FROM THE CIVIL WAR TO THE INDUSTRIAL REVOLUTION

Part 2: Teaches students about the Civil War, the wild West, and the Industrial Revolution.

America's Story Vol 2 *Teacher Guide*
978-0-89051-981-3   978 0 89051-982-0

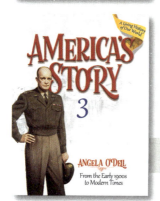

FROM THE EARLY 1900S TO OUR MODERN TIMES

Part 3: Carries the student from the turn of the 20th century through the early 2000s.

America's Story Vol 3 *Teacher Guide*
978-0-89051-983-7   978-0-89051-984-4

VISIT **MasterBooks.com** *Where Faith Grows!* TO SEE OUR FULL LINE OF FAITH-BUILDING CURRICULUM OR CALL 800-999-3777

GOD'S DESIGN FOR SCIENCE SERIES

EXPLORE GOD'S WORLD OF SCIENCE WITH THESE FUN CREATION-BASED SCIENCE COURSES

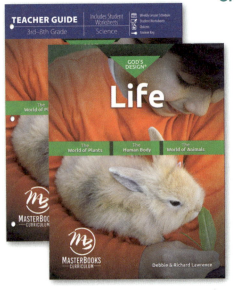

GOD'S DESIGN FOR LIFE
GRADES 3-8

Learn all about biology as students study the intricacies of life science through human anatomy, botany, and zoology.

GOD'S DESIGN FOR HEAVEN & EARTH
GRADES 3-8

Explore God's creation of the land and skies with geology, astronomy, and meteorology.

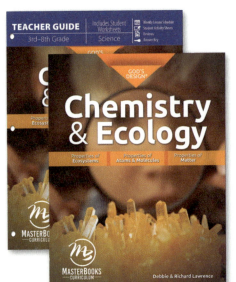

GOD'S DESIGN FOR CHEMISTRY & ECOLOGY
GRADES 3-8

Discover the exciting subjects of chemistry and ecology through studies of atoms, molecules, matter, and ecosystems.

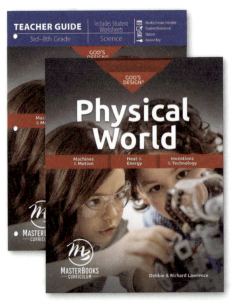

GOD'S DESIGN FOR THE PHYSICAL WORLD
GRADES 3-8

Study introductory physics and the mechanisms of heat, machines, and technology with this accessible course.

AVAILABLE AT MASTERBOOKS.COM 800.999.3777
& OTHER PLACES WHERE FINE BOOKS ARE SOLD